T0180812

Lecture Notes in Computer Science 13480

More information about this series at https://link.springer.com/bookseries/558

Andreas Holzinger · Peter Kieseberg ·
A Min Tjoa · Edgar Weippl (Eds.)

Machine Learning and Knowledge Extraction

6th IFIP TC 5, TC 12, WG 8.4, WG 8.9, WG 12.9
International Cross-Domain Conference, CD-MAKE 2022
Vienna, Austria, August 23–26, 2022
Proceedings

 Springer

Editors
Andreas Holzinger [iD]
University of Natural Resources and Life
Sciences Vienna
Vienna, Austria

Medical University Graz and Graz University
of Technology
Graz, Austria

University of Alberta
Edmonton, Canada

A Min Tjoa [iD]
TU Wien
Vienna, Austria

Peter Kieseberg [iD]
St. Pölten University of Applied Sciences
St. Pölten, Austria

Edgar Weippl
SBA Research
Vienna, Austria

University of Vienna
Vienna, Austria

ISSN 0302-9743   ISSN 1611-3349 (electronic)
Lecture Notes in Computer Science
ISBN 978-3-031-14462-2   ISBN 978-3-031-14463-9 (eBook)
https://doi.org/10.1007/978-3-031-14463-9

This Springer imprint is published by the registered company Springer Nature Switzerland AG
The registered company address is: Gewerbestrasse 11, 6330 Cham, Switzerland

Preface

The International Cross Domain Conference for Machine Learning & Knowledge Extraction CD-MAKE is a joint effort of IFIP TC 5, TC 12, IFIP WG 8.4, IFIP WG 8.9 and IFIP WG 12.9, and is held in conjunction with the International Conference on Availability, Reliability and Security (ARES) – this time in beautiful Vienna, Austria, during August 23–26, 2022. Thanks to the current improved situation of the COVID-19 pandemic, which affected us all heavily, we were happy that we could meet all our international colleagues and friends in person again.

For those who are new to our conference, the letters CD in CD-MAKE stand for "Cross-Domain" and describe the integration and appraisal of different fields and application domains to provide an atmosphere to foster different perspectives and opinions. We are strongly convinced that this cross-domain approach is very fruitful for new developments and novel discoveries. The conference fosters an integrative machine learning approach, considering the importance of data science and visualization for the algorithmic pipeline with a strong emphasis on privacy, data protection, safety, and security. It is dedicated to offering an international platform for novel ideas and a fresh look on methodologies to put crazy ideas into business for the benefit of humans. Serendipity is a desired effect for the cross-fertilization of methodologies and transfer of algorithmic developments.

The acronym MAKE stands for "MAchine Learning and Knowledge Extraction", a field of artificial intelligence (AI) that, while quite old in its fundamentals, has just recently begun to thrive based on both novel developments in the algorithmic area and the availability of vast computing resources at a comparatively low cost.

Machine learning (ML) studies algorithms that can learn from data to gain knowledge from experience and to generate decisions and predictions. A grand goal is in understanding intelligence for the design and development of algorithms that work autonomously (ideally without a human-in-the-loop) and can improve their learning behavior over time. The challenge is to discover relevant structural and/or temporal patterns ("knowledge") in data, which is often hidden in arbitrarily high dimensional spaces, and thus simply not accessible to humans. Knowledge extraction is one of the oldest fields in AI and is seeing a renaissance, particularly in the combination of statistical methods with classical ontological approaches.

AI is currently undergoing a kind of Cambrian explosion and is the fastest growing field in computer science today thanks to the successes in machine learning to help to solve real-world problems. There are many application domains, e.g., in agriculture, climate research, forestry, etc. with many use cases from our daily lives, which can be useful to help to solve various problems Examples include recommender systems, speech recognition, autonomous driving, cyber-physical systems, robotics, etc.

However, in our opinion the grand challenges are in sensemaking, in context understanding, and in decision making under uncertainty, as well as solving the problems of human interpretability, explainability, and verification.

Our real world is full of uncertainties and probabilistic inference enormously influences AI generally and ML specifically. The inverse probability allows us to infer unknowns, to learn from data, and to make predictions to support decision-making. Whether in social networks, recommender systems, health applications, or industrial applications, the increasingly complex data sets require a joint interdisciplinary effort bringing the human-in-control and to manage ethical and social issues, accountability, retractability, explainability, causability and privacy, safety, and security!

The International Federation for Information Processing (IFIP) is the leading multinational, non-governmental, apolitical organization in information and communications technologies and computer sciences. IFIP is recognized by the United Nations (UN) and was established in 1960 under the auspices of UNESCO as an outcome of the first World Computer Congress held in Paris in 1959.

IFIP is incorporated in Austria by decree of the Austrian Foreign Ministry (September 20, 1996, GZ 1055.170/120-I.2/96) granting IFIP the legal status of a non-governmental international organization under the Austrian Law on the Granting of Privileges to Non-Governmental International Organizations (Federal Law Gazette 1992/174). IFIP brings together more than 3500 scientists without boundaries from both academia and industry, organized in more than 100 Working Groups (WGs) and 13 Technical Committees (TCs).

To acknowledge all those who contributed to the efforts and stimulating discussions at CD-MAKE 2022 would be impossible in a preface like this. Many people contributed to the development of this volume, either directly or indirectly, so it is impossible to list all of them. We herewith thank all local, national, and international colleagues and friends for their positive and supportive encouragement. Finally, yet importantly, we thank the Springer management team and the Springer production team for their professional support. This year CD-MAKE received 45 submissions, which all have been carefully reviewed by our program committee in a double-blind review. Finally 23 papers have been accepted and were presented at the conference in Vienna.

Thank you to all! Let's MAKE it cross-domain!

August 2022

Andreas Holzinger
Peter Kieseberg
Edgar Weippl
A Min Tjoa

Organization

Conference Organizers

Andreas Holzinger	University of Natural Resources and Life Sciences Vienna, Medical University of Graz, and Graz University of Technology, Austria/University of Alberta, Canada
Peter Kieseberg	FH St. Pölten, Austria
Edgar Weippl (IFIP WG 8.4 Chair)	SBA Research and University of Vienna, Austria
A Min Tjoa (IFIP WG 8.9. Chair, Honorary Secretary IFIP)	TU Vienna, Austria

Conference Office Team

Bettina Jaber	SBA Research, Austria
Daniela Freitag David	SBA Research, Austria
Barbara Friedl	St. Pölten University of Applied Sciences, Austria

Proceedings Manager

Bettina Jaber	SBA Research, Austria

Program Committee

Amin Anjomshoaa	National University of Ireland Galway, Ireland
John A. Atkinson-Abutridy	Universidad de Concepcion, Chile
Frantisek Babic	Technical University of Košice, Slovakia
Christian Bauckhage	University of Bonn, Germany
Smaranda Belciug	University of Craiova, Romania
Mounir Ben Ayed	Ecole Nationale d'Ingenieurs de Sfax, Tunesia
Chris Biemann	Universität Hamburg, Germany
Guido Bologna	Université de Genève, Switzerland
Ivan Bratko	University of Ljubljana, Slovenia
Federico Cabitza	Università degli Studi di Milano-Bicocca, Italy
Andre Calero-Valdez	RWTH Aachen University, Germany
Andrea Campagner	University of Milano-Bicocca, Italy
Angelo Cangelosi	University of Manchester, UK
Mirko Cesarini	Università degli Studi di Milano-Bicocca, Italy
Nitesh V. Chawla	University of Notre Dame, USA

Roberto Confalonieri	Free University of Bozen-Bolzano, Italy
Tim Conrad	Zuse Institute Berlin, Germany
Gloria Cerasela Crisan	Vasile Alecsandri University of Bacau, Romania
Beatriz De La Iglesia	University of East Anglia, UK
Javier Del Ser	Universidad del País Vasco/Euskal Herriko Unibertsitatea, Spain
Natalia Dias Rodriguez	University of Granada, Spain
Josep Domingo-Ferrer	Universitat Rovira i Virgili, Catalonia
Max J. Egenhofer	University of Maine, USA
Massimo Ferri	University of Bologna, Italy
Hugo Gamboa	PLUX Wireless Biosignals and Universidade Nova de Lisboa, Portugal
Panagiotis Germanakos	SAP, Germany
Siegfried Handschuh	University of Passau, Germany
Pitoyo Hartono	Chukyo University, Japan
Barna Laszlo Iantovics	George Emil Palade University of Medicine, Pharmacy, Science and Technology of Targu Mures, Romania
Igor Jurisica	IBM Life Sciences Discovery Centre and Princess Margaret Cancer Centre, Canada
Epaminodas Kapetanios	University of Westminster, UK
Andreas Kerren	Linköping University and Linnaeus University, Sweden
Freddy Lecue	Inria, France
Lenka Lhotska	Czech Technical University in Prague, Czech Republic
Shujun Li	University of Kent, UK
Brian Lim	National University of Singapore, Singapore
Ljiljana Majnaric-Trtica	University of Osijek, Croatia
Bradley Malin	Vanderbilt University, USA
Ernestina Menasalvas	Polytechnic University of Madrid, Spain
Fabio Mercorio	University of Milano-Bicocca, Italy
Yoan Miche	Nokia Bell Labs, Finland
Vasile Palade	Coventry University, UK
Jan Paralic	Technical University of Kosice, Slovakia
Francesco Piccialli	University of Naples Federico II, Italy
Camelia-M. Pintea	Technical University of Cluj-Napoca, Romania
Fabrizio Riguzzi	Università di Ferrara, Italy
Luca Romeo	Università Politecnica delle Marche and Istituto Italiano di Tecnologia, Italy
Irena Spasic	Cardiff University, UK
Ivan Štajduhar	University of Rijeka, Croatia
Bharath Sudharsan	National University of Ireland Galway, Ireland

Contents

Explain to Not Forget: Defending Against Catastrophic Forgetting with XAI

Sami Ede[1], Serop Baghdadlian[1], Leander Weber[1], An Nguyen[2], Dario Zanca[2], Wojciech Samek[1,3,4(✉)], and Sebastian Lapuschkin[1(✉)]

[1] Fraunhofer Heinrich Hertz Institute, 10587 Berlin, Germany
{wojciech.samek,sebastian.lapuschkin}@hhi.fraunhofer.de
[2] Friedrich Alexander-Universität Erlangen-Nürnberg, 91052 Erlangen, Germany
[3] Technische Universität Berlin, 10587 Berlin, Germany
[4] BIFOLD – Berlin Institute for the Foundations of Learning and Data, 10587 Berlin, Germany

Abstract. The ability to continuously process and retain new information like we do naturally as humans is a feat that is highly sought after when training neural networks. Unfortunately, the traditional optimization algorithms often require large amounts of data available during training time and updates w.r.t. new data are difficult after the training process has been completed. In fact, when new data or tasks arise, previous progress may be lost as neural networks are prone to catastrophic forgetting. Catastrophic forgetting describes the phenomenon when a neural network completely forgets previous knowledge when given new information. We propose a novel training algorithm called *Relevance-based Neural Freezing* in which we leverage Layer-wise Relevance Propagation in order to retain the information a neural network has already learned in previous tasks when training on new data. The method is evaluated on a range of benchmark datasets as well as more complex data. Our method not only successfully retains the knowledge of old tasks within the neural networks but does so more resource-efficiently than other state-of-the-art solutions.

Keywords: Explainable AI · Layer-wise Relevance Propagation (LRP) · Neural network pruning · Catastrophic forgetting

1 Introduction

While neural networks achieve extraordinary results in a wide range of applications, from the medical field to computer vision or successfully beating humans on a variety of games [37,41], the established training process typically relies on a large amount of data that is present at training time to learn a specific task. For example, the famous ImageNet dataset [7] consists of more than 14

S. Ede and S. Baghdadlian—Contributed equally.

A. Holzinger et al. (Eds.): CD-MAKE 2022, LNCS 13480, pp. 1–18, 2022.
https://doi.org/10.1007/978-3-031-14463-9_1

million images which results in a size of more than 150 GB, while the authors of [30] collected a dataset of 400 million images that make up more than 10 TB of data [35]. Large amounts of samples can help models generalize better by avoiding overfitting in single examples, but in turn make model training extremely expensive. If more data is later added, and the model should be able to correctly predict on both new and old data, usually it has to be finetuned or trained from scratch with the expanded dataset as opposed to only the new data. Otherwise, *catastrophic forgetting* [14] can occur when learning multiple consecutive tasks or from non-stationary data. One prominent example of this is reinforcement learning, in which an agent continuously interacts with its environment, using a stream of observations as training data. As the observations change with the agent's actions in the environment, the data distribution becomes non-i.i.d., leading to catastrophic forgetting in the agent that is usually countered with an "experience buffer", in which earlier observations are saved. These saved observations are then randomly repeated during training. Other applications would also benefit from solutions to continuous or lifelong learning, e.g., medical applications such as skin cancer detection, where more targets could be added after additional samples have been obtained. The issue of catastrophic forgetting is especially pronounced when previous data is not accessible anymore, e.g., due to being proprietary, making retraining impossible.

Recently, techniques of Explainable Artificial Intelligence (XAI) [33] have been proposed which are able to identify the elements of a neural network model crucial for solving the problem a model has been optimized for. One such method is Layer-wise Relevance Propagation (LRP) [3], which assigns relevance scores to latent network structures through modified backpropagation. In the recent past, this information has been used with great success to efficiently and effectively reduce neural network complexity without sacrificing performance [5, 44].

In this paper, we are proposing *Relevance-based Neural Freezing (RNF),* a novel approach to alleviate catastrophic forgetting that builds upon the aforementioned pruning technique. Instead of compressing the network, the information about unit importance is used to freeze the knowledge represented by the learned network parameters by inhibiting or completely stopping the training process for those parts of the model that are relevant for a specific task, while the remaining units are free to learn further tasks. We evaluate our method on several commonly used datasets, i.e., MNIST [8], CIFAR10 and CIFAR100 [22], ImageNet [7], and the challenging Adience dataset for facial categorization [9], which is a dataset of photos shot under real-world conditions, meaning different variations in pose, lighting conditions and image quality.

2   Related Work

In this section we briefly review the theoretical background of explainable AI and catastrophic forgetting.

2.1   Explainable Artificial Intelligence

In recent years, XAI has gotten more and more attention as the urgency to understand how "black box" neural networks arrive at their predictions has become more apparent. Especially in applications that have far-reaching consequences for humans, like the prediction of cancer (e.g. [6,11,18]), it is not only important to know what the network predicted, but also why a certain decision has been made. Generally, methods from XAI can be roughly divided into two categories:

Global explanations provide general knowledge about the model, its feature sensitivities and concept encodings. Some approaches aim to identify the importance of specific features, concepts, or data transformations (e.g. [15,16,19]) by analyzing the model's reaction to real or synthetic data, while others try to assess their model by finding important neurons and their interactions [17], or by finding the concepts encoded by hidden filters through synthesizing the input that maximizes their activation [10,26,27].

Instead of providing insight into the model's general understanding of the data, *local* explanation methods aim at making individual model predictions interpretable, i.e., by ranking the importance of features w.r.t. specific samples. By attributing importance scores to the input variables, these explanations can be illustrated as heatmaps with the same dimensions as the input space. Among the local explanation methods, there are again multiple approaches, some still treating the model as a "black box", approximating the local explanations via separately trained proxy models [31,45] or otherwise applying perturbation or occlusion techniques [13,47]. Other methods use (augmented) backpropagation in order to compute the importance ranking of the input or latent features, such as [3,4,25,39]. Our proposed method leverages the advantages of Layer-wise Relevance Propagation [3], as this method's ability to measure the per-prediction usefulness and involvement of (also latent) network elements has recently shown great success [5,44] in applications for model improvement.

2.2   Catastrophic Forgetting

Unlike humans or animals, neural networks do not have the inherent ability to retain previously attained knowledge when they are presented with new information while being optimized. This effect is characterized by a drastic performance decrease on tasks trained earlier when progressing on a new task or dataset. This phenomenon is described by the term catastrophic forgetting [14]. As neural networks are generally assumed to train on i.i.d. data, adding a new task to be learned can violate this assumption, causing the gradient updates to override the weights that have been learned for the previous tasks and causing the aforementioned loss of old knowledge. One way to combat catastrophic forgetting is experience replay [23], in which the old data is interspersed with the new data, simulating i.i.d. data such that the network retains the old knowledge. However, this approach is inefficient, does not allow online-learning, and may even be impossible if access to the old data is not available. Therefore, numerous approaches have been proposed to tackle this problem more efficiently. The approach of [42] learns masks defining subnetworks in untrained models that are responsible for a given task, while [36] concurrently learn binary attention vectors to retain the knowledge obtained in previous tasks. Other approaches (e.g., [20,46]) propose constraints on weight updates for neurons that have been identified as being pertinent for a previous task. Dynamically Expandable Networks [24] increase the network capacity when training a new task.

In this paper, we propose a training algorithm that—motivated by the successful XAI-based pruning method described in [44]—uses LRP in order to identify those neurons that are relevant for a given task. After finding the important neurons, they are given a lower elasticity for learning subsequent tasks, such that the network efficiently retains the knowledge from previous tasks while still being able to learn additional tasks.

3 Relevance-Based Neural Freezing

As a local attribution method, LRP has shown (see [32,44]) to not only deliver accurate and interpretable explanations about the input variables, the conservatory property of the local distribution rules also allows to gain insights on the importance of individual latent neurons and their associated filters. Additionally, LRP is scalable w.r.t. network depth, easy to implement through existing frameworks (e.g., [1,2]), and efficient with a linear computational cost w.r.t. a backpropagation pass. It works by treating the prediction of the model $f(\mathbf{x})$ w.r.t. a network output of interest as the total sum of importance, or *relevance* R, that is then redistributed towards the input variables: After the forward pass, the layers of a classifier are reversely iterated, redistributing the relevance among its neurons proportionally to their contributions in the preceding forward pass. This redistribution process follows a conservatory constraint, meaning that the sum of relevance in each layer is equal to the total amount of relevance at the model head:

$$f(\mathbf{x}) = \cdots = \sum_{d_i \in (l+1)} R_{d_i}^{(l+1)} = \sum_{d_j \in (l)} R_{d_j}^{(l)} = \cdots = \sum_{d_k \in (l_0)} R_{d_k}^{(l_0)}, \qquad (1)$$

where $f(\mathbf{x})$ is the model output and R_d^l is the relevance score of unit d in layer l. Depending on the application and layer type, various propagation rules with specific purposes have been proposed. For example, the LRP-ε rule [3,34] is defined as

$$R_{j \leftarrow k} = \sum_k \frac{a_j w_{jk}}{\sum_{0,j} a_j w_{jk} + \varepsilon} R_k. \qquad (2)$$

with a_j being the input activation at neuron j, w_{jk} being the learned weights between neurons j in the lower- and k in the upper layer and ε being a small term of the same sign as the sum in the demoniator. This rule is typically used in upper layers to filter out weaker or contradictory values via the added ε in the denominator, which results in smoother, less noisy heatmaps and prevents numerical instabilities. A discussion of other rules and their applications can be found in [33]. In this paper, we use the LRP-ε rule for fully-connected layers and the LRP-z^+ rule for convolutional layers, as recommended in [21].

The proposed method aims to prevent catastrophic forgetting by decreasing the plasticity of neurons rated as important for a given, already optimized task. The general procedure is as follows: After training the model on the first task, the relevant units are identified by using LRP on a small, randomly sampled subset of the (test-) data, from here on called the **reference dataset**, similarly as in [44]. Note, however, that in our application this reference data is selected specific to each task, containing only samples from classes that are learned in the current task. LRP-attributions are computed w.r.t. the respective ground truth label of each sample. Until the model performance on the test set decreases by a predetermined threshold, the units with the lowest relevance (computed on the reference dataset) are repeatedly selected and then pruned by setting their outgoing connections to zero. Once the threshold is reached, the remaining units are assigned a lower learning rate for any subsequent tasks, as they were the most important ones for the current task. To completely freeze the units, the learning rate is set to zero, but it is also possible to just lower the elasticity to a fraction of the original learning rate. To continue training, the connections to the less relevant units are restored to their state before the pruning. This procedure is outlined in Algorithm 1, and an intuitive illustration can be found in Fig. 1.

Algorithm 1. Relevance-based Neural Freezing

Require: untrained model **net**, reference data x_r, task specific training data x_t, pruning threshold t, pruning ratio r, task number N_t, learning rate lr, learning rate for relevant units lr_{frozen}, learning rate for irrelevant units $lr_{irrelevant}$, epochs N_e, network **units**, with **unit** $\in$ {**neurons, filters**}.

  for task in N_t **do**
    for epoch in N_e **do**
      ▷ train **net** on x_t using lr
    end for
    for all layer in **net do**
      for all units in layer **do**
        ▷ compute importance of **units** using *LRP*
      end for
    end for
    ▷ sort **units** in descending order w.r.t. their global importance to the task
    while t not reached **do**
      ▷ zero out r **units** from **net where importance** is minimal
    end while
    ▷ mark remaining **units** in **net** as **relevant units**
    ▷ lower elasticity of **relevant units** for current task
    $lr_{relevant\ units} \leftarrow lr_{frozen}$
    $lr_{irrelevant\ units} \leftarrow lr_{irrelevant}$
    ▷ restore zeroed out connections to continue training.
  end for

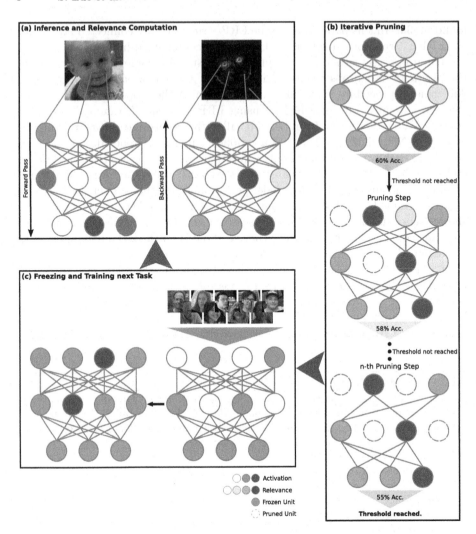

Fig. 1. Illustration of our method to identify relevant neurons and freeze their training progress after training on a task. (a) Forward pass through the network trained on all previous tasks with a reference image of the current task (*left*) and computation of network unit importance using LRP (*right*). (b) Iteratively set connections of irrelevant neurons/units to zero until performance threshold is reached. (c) Freezing the knowledge about the current task by setting the learning rate of the remaining units to zero and training on the next task. The process is repeated after training each task.

The stopping criterion for the unit selection can be determined freely and is not limited to network performance. For instance, it could also be the number of remaining free units in the network or even the energy consumption of the network when running inference. In the following experiments, the learning rate for all frozen units is set to zero.

4   Experiments

We show the effectiveness of the proposed technique on a number of increasingly difficult datasets and tasks. We start by using well-known toy and benchmark datasets, namely MNIST [8], as well as a combination of CIFAR10 and CIFAR100 [22], to illustrate the method and its conceptual functionality before also showing effectiveness in larger benchmark and real-world datasets. We set the pruning threshold to 2% accuracy lost relative to the accuracy after finetuning on the task, which we found to be optimal during our experiments. The optimal value for this parameter can be determined using grid search. Using 2% results in enough free network capacity to learn the additional tasks while keeping the accuracy of the classifier as high as possible. Results on MNIST, CIFAR10, and CIFAR100 are averaged over 20 random seeds, while results on the ImageNet and the Adience dataset are averaged over five random seeds. In addition to the accuracy, we also evaluate the free **capacity** of the model, which we define as the percentage of unfrozen network units after each task. Details on each experimental setup can be found in the Appendix A. In all experiments, new tasks are introduced with a *task-incremental (Task-IL)* setup [28]. In a task-incremental setup, the neural network is additionally informed about the task identity during training as well as during inference. Each task has its own (separate) head, while the rest of the neurons are shared among all tasks.

4.1   MNIST-Split

The first series of experiments is performed on the popular MNIST dataset. The dataset is split up into five tasks, each task being the classification of two digits, e.g., the first task consists of the digits 0 and 1, task two contains the digits 2 and 3, and so on. The model is trained on the first task and then finetuned sequentially using a task-incremental setup, which we refer to as naive finetuning.

Figure 2b shows the effect of RNF on the MNIST-Split dataset. The mean test accuracy over all tasks is increased by about 4%, compared to naive finetuning (Fig. 2a), which is the most evident in the accuracy for both task one and task two. Instead of a drop of 30% in accuracy, the model can still classify task one with an accuracy of almost 90% and retains an accuracy for task two of over 90%. The increase in baseline (naive fine-tuning) accuracy after task five may be attributed to a similarity between the shapes of digits in tasks one and five: both the digits 8 and 9 have rounded forms which makes it easier to distinguish between a 0 and a 1. In the following experiments, all changes to the accuracy are reported in comparison to the naive finetuning baseline.

4.2   MNIST-Permuted

The MNIST-Permuted setup increases the complexity of the MNIST dataset by introducing random pixel permutations to the digits. It is commonly used (see [12]) in a ten-task configuration such that the unpermuted dataset poses as the first task while the remaining nine tasks consist of different permutations of the

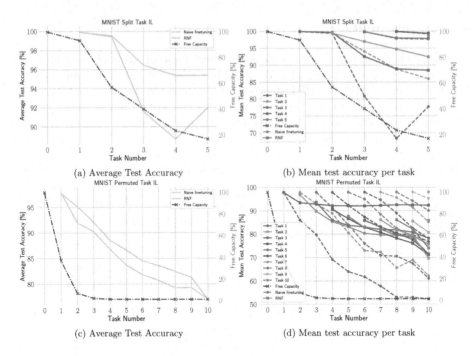

Fig. 2. Results on the MNIST dataset. The left vertical axis shows the mean test accuracy over all already seen tasks. The horizontal axis shows task progression. The right vertical axis shows the model's free capacity. (a) shows the average test accuracy progression for both the naive finetuning and RNF on the MNIST Split dataset. (b) shows the mean test accuracy progression for each task on the MNIST Split dataset. (c) shows the average test accuracy progression for both the naive finetuning and RNF on the MNIST Permuted dataset. (d) shows the mean test accuracy progression for each task on the MNIST Permuted dataset.

original digits. The permutations are the same for all classes but change between tasks.

Even though applying RNF slightly lowers the mean accuracy from 76.99% to 75% compared to the naive finetuning baseline, Fig. 2c shows that the average accuracy over all *seen* tasks is above the naive finetuning baseline during training. For a closer inspection of task performance, Fig. 2d shows the individual task accuracy over the training. It can be seen that especially task one, two and three benefited the most from RNF, whereby the small capacity model did not suffice to successfully learn more tasks. Nevertheless, it suggests that parameters can be re-used for new tasks: even though the free capacity drops to below ten percent after the first two tasks, the model can still learn the remaining tasks with reasonable accuracy by an apparent re-utilization of the frozen filters that have been deemed relevant for the previous tasks.

4.3   CIFAR10 and CIFAR100

For this dataset, the networks' first task is to solve the prediction problem defined by the entire CIFAR10 dataset, after which five further tasks are trained sequentially, each containing 10 randomly selected classes from the CIFAR100 dataset. We expand on this approach and instead split CIFAR100 into ten tasks, each containing ten random classes. Using RNF, we achieve an increase of more than 40% of mean test accuracy compared to the naive finetuning baseline as shown in Fig. 3a. This effect is also displayed in Fig. 3b. Even though the network capacity limit seems to be reached at task 5, as in previous experiments, the model's ability to still learn various tasks suggests that the knowledge attained in the previous tasks is enough to facilitate the learning of the remaining classes due to the random assignment of classes to the individual tasks which prevents a semantic bias towards the underlying concepts of the classes within the tasks. Again, the ability of the model to learn new tasks despite a limited residual free capacity signals a high amount of filter re-use of the already frozen filters.

Another, more complex experiment is performed by manually ordering the classes in semantic groups with seven superclasses, containing several subclasses each: Flowers and trees, (land) animals, aquatic mammals and fish, random objects, small insects, nature scenes and vehicles. Like in the random grouping experiment, applying RNF gains almost 30% in accuracy compared to the baseline, as demonstrated in Fig. 3c. Again, it can be seen that the model's free capacity is getting close to 0% after training the first four tasks, but the model is able to re-use previous abstractions from previous tasks to still learn the remaining tasks despite the difficulty of the semantic grouping. While the network is not able to achieve an initial accuracy for new tasks that is as high as the naive finetuning baseline (as shown in Fig. 3d), RNF can strongly mitigate the loss of accuracy that default finetuning displays when even more tasks are learned. In fact, the discrepancy between random grouping (Fig. 3b) and semantic grouping (Fig. 3d) strongly supports the hypothesis of frozen filters being reused for later tasks, since the random grouping has less semantic bias, making filters obtained while learning the first tasks far more useful in subsequent tasks than when grouping semantically.

4.4   ImageNet Split

For the ImageNet [7] split the naive finetuning baseline displays significant catastrophic forgetting even after training each task for only 10 epochs, as is evident in Fig. 4a. On this dataset, our method not only preserves the performance of the model on previous tasks but even leads to an increase in accuracy, which in turn results in an overall gain of 35.31% of accuracy over all tasks compared to the naive finetuning.

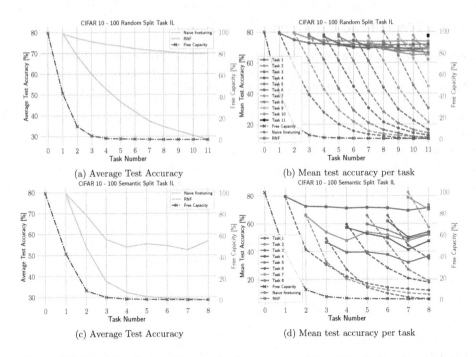

(a) Average Test Accuracy

(b) Mean test accuracy per task

(c) Average Test Accuracy

(d) Mean test accuracy per task

Fig. 3. CIFAR100 is split into 10 tasks for the random split and 8 tasks for the semantic split. In the semantic split setup, each task contains a different number of semantically similar classes. The left vertical axis shows the mean test accuracy over all already seen tasks. The horizontal axis shows task progression. The right vertical axis shows the model's free capacity. For (a), CIFAR100 was split randomly into 10 tasks, each containing 10 classes. The plot shows the average test accuracy over all previous tasks after introducing each new task on CIFAR100. (b) shows the mean test accuracy progression for each task on the random split. (c) shows the average test accuracy progression for both the naive finetuning and RNF on the semantic split of CIFAR100. (d) shows the mean test accuracy progression for each task on the semantic split.

4.5   Adience

The Adience benchmark dataset of unfiltered faces [9] is a dataset made up about 26,000 photos of human faces with binary gender- and eight different age group labels. The images are shot under real-world conditions, meaning different variations in pose, lighting conditions and image quality. We performed experiments on Adience in two scenarios:

- *Split*: The dataset is split into two tasks, each consisting of a four-class classification problem of different age groups. The classes are grouped in a mixed and an ordered setup (Adience-Mixed: Classes [0, 2, 4, 6] and [1, 3, 5, 7], Adience-Ordered: Classes [0–3] and [4–7]). As the classes are labeled with the corresponding age ranges increasing, the groups in the ordered mixing are expected to have more common features. The baseline is established by finetuning the model for three epochs per sequential task.

- *Entire Dataset*: In this scenario, the pretrained model was first pruned to retain the knowledge of ImageNet and then trained on the entire Adience dataset. As Adience is a very complex task, the pruning threshold is increased to 5% in order to increase the amount of free network capacity in the pretrained model. The model was finetuned on Adience for six epochs.

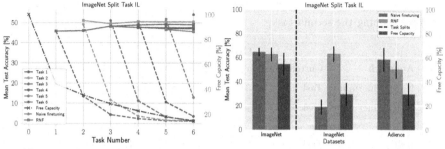

(a) Task mean test accuracy: ImageNet split   (b) Task mean test accuracy: ImageNet-Adience split

Fig. 4. (a): ImageNet split, where ImageNet is split into 6 sequential tasks consisting of 100 classes each. The left vertical axis shows the mean test accuracy over each already seen task. The horizontal axis shows the task progression. The dotted lines represent the test accuracy using the default naive finetuning approach. The solid lines show the results obtained using RNF. The right vertical axis shows the model's free capacity. (b): Task-IL protocol by sequentially training both ImageNet and Adience datasets. The left vertical axis shows the mean test accuracy of a specific task over five repetitions. The standard deviation is shown as a solid vertical black line at the top of the bar plot. The horizontal axis is split into two groups. The bar plots on the left side of the dotted line represent the mean test accuracy of the model on ImageNet. The bar plots on the right side show the test accuracy of ImageNet after training the Adience dataset and that of the Adience dataset. The test accuracy is computed for both the default baseline approach and RNF, the model's free capacity after each task training is shown on the right vertical axis.

Similarly to the experiments on the benchmark datasets, the results of the proposed method display a lower individual accuracy of the second task compared the naive finetuning baseline for both Split tasks, which can be seen in Table 1. Nevertheless, the baseline displays significant catastrophic forgetting, especially for the ordered setup with a drop of almost 46% for the previously learned task that is almost completely prevented when using RNF. The lower accuracy scores on task two can be explained by a stability-vs-plasticity dilemma: decreasing the plasticity of parts of the network can increase stability for already acquired knowledge, but can slow down learning of new tasks, so that with the same amount of training epochs, the task is not learned to the same degree. Even though the accuracy of the second task is lower, RNF still shows that the application of the method is advantageous for sequential tasks regardless of complexity of tasks and size of the model, as the mean accuracy over both tasks increases by about 18% compared to the naive finetuning. In both setups, the model retains about 30% of free capacity, making it possible to learn further tasks.

Table 1. Average test accuracy on Adience for both the ordered and the mixed setup in two tasks alongside the model's free capacity after each freezing stage. The first "test accuracy" column represents the accuracy of the respective task after training the first task. The second "test accuracy" column reports the test accuracy of the respective task after training the second task.

Ordered				
Approach	Current Task	Test Accuracy		Free Capacity
Naive Finetuning	Task 1	64.38(±3.4)	18.7(±6.24)	100
Naive Finetuning	Task 2	–	58.15(±9.76)	100
Average over both		38.43(±8)		
RNF	Task 1	62.52(±5.6)	62.87(±6.3)	54.53(±9.16)
RNF	Task 2	–	50.02(±7.23)	29.4(±9.4)
Average accuracy over both tasks		**56.45(±6.7)**		
Mixed				
Approach	Current Task	Test Accuracy		Free Capacity
Naive Finetuning	Task 1	68.3(±4.33)	52.71(±9.6)	100
Naive Finetuning	Task 2	–	80.42(±4.47)	100
Average accuracy over both tasks		66.57(±7)		
RNF	Task 1	70.89(±4.33)	70.11(±3.63)	47.57(±2.8)
RNF	Task 2	–	75.85(±4.24)	28.8(±3.76)
Average accuracy over both tasks		**72.98(±3.9)**		

Table 2. Average test accuracy on splitting ImageNet and Adience into two tasks alongside the model's free capacity after each pruning stage. The first "test accuracy" column represents the test accuracy after training ImageNet. The second column reports the test accuracy of ImageNet after training on Adience and the test accuracy on Adience.

Approach	Current Task	Test Accuracy		Free Capacity
Naive Finetuning	ImageNet	71.5(±0)	51.51(±3.42)	100
Naive Finetuning	Adience	–	51.02(±1.66)	100
Average accuracy over both tasks		51.27(±2.54)		
RNF	ImageNet	71.5(±0)	69.05(±0.16)	26.5(±0)
RNF	Adience	–	44.66(±1.61)	14.9(±0.02)
Average accuracy over both tasks		**56.86(±0.89)**		

As can be seen in Fig. 4b, preserving the knowledge of the ImageNet dataset in the pretrained model requires about 73% of the model's full capacity. The results of RNF on the Adience-ImageNet split can be found in Table 2: While the accuracy of task two after our method is again lower than the naive finetuning baseline, the accuracy of task one only drops about 2% compared to almost 20% in the baseline case, granting a mean accuracy increase of 5.6% over both tasks while still retaining about 15% free capacity in the model that can be used to learn further tasks.

4.6   Qualitative Results

Alongside the experiments, we also observe changes in the attribution heatmaps computed by LRP before and after the application of RNF. Figure 5a shows heatmaps computed w.r.t. different age groups as targets alongside the original image. Hot pixel color (red over yellow to white) denotes positive relevance towards the target class, i.e., marking features causing the model to predict in favor of the target class, while cold pixel color (blue over turquoise to white) denotes negative relevance. The model learns to associate different facial features with specific age groups, which is especially apparent in the first image, where positive relevance is assigned consistently to the glasses for the age group (48–53), while negative relevance is assigned in the upper area of the face when the target age group is (8–13). The effect of catastrophic forgetting on the relevance distribution can be observed in Fig. 5b. Computing the relevance after training the second task shows that using the naive finetuning baseline, the model now assigns negative relevance to the upper area of the face that was previously considered a positive class trait. RNF retains the initially learned features and prediction strategies of the model, which reflects in the relevance assignments which still focus on the glasses of the woman. Similar behavior is shown in the other images: while the assignment of relevance changes after training with the default baseline, the model that used RNF is still focusing on the areas that were relevant before training the second task and keeps the sign of the relevance consistent. Finally, the generated heatmaps are consistent with the test accuracy results displayed in the Figs. 4a and 4b, and illustrate how previously learned features are preserved despite the introduction of new classes when RNF is employed.

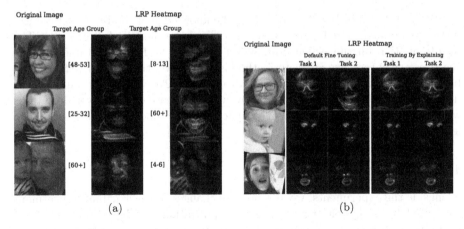

Fig. 5. (a): Images from the Adience dataset alongside their explanations. The hot colored regions in the heatmaps mark relevant features used by the model for recognizing the chosen class, while coldly colored regions show negative relevance, marking contradicting evidence. The relevance is computed w.r.t. the target class labels indicated on the left of each heatmap. Choosing different target classes produces different explanations, as different class outputs of the model utilize the presented input information differently. (b): Original images from the Adience dataset alongside their explanations for the true class, after either using naive finetuning over several tasks, or RNF. The figure shows samples from task 1 before and after learning task 2 in the ordered split experiment. This demonstrates that RNF prevents catastrophic forgetting, i.e., a drift in the reasoning of the model that occurs during naive finetuning under continued training for tasks already optimized. (Color figure online)

5   Conclusion

Overcoming catastrophic forgetting is one key obstacle towards achieving reliable lifelong learning. Retraining the model from scratch every time new data or tasks are added is sometimes possible, but very inefficient. In order to prevent the model from forgetting previously learned information, the plasticity of important neurons can be lowered for further training, so that they retain the ability to solve earlier tasks. We present an effective algorithm that uses LRP to identify the neurons that contribute the most to a given task, showing a significant increase in accuracy compared to naive finetuning, while only introducing minimal additional computation cost and requirements on data availbility for previously optimized tasks. Evaluation of the proposed method on the CIFAR10-100 split achieved an increase in accuracy of about 40% compared to the baseline, which could also be achieved in a semantic split setting. The success of our method was also demonstrated on larger datasets, achieving a 35% increase of accuracy across all sequential tasks when trained on ImageNet. In addition to MNIST, CIFAR and ImageNet, we further evaluated three scenarios on the Adience dataset. We were able to show that our method not only performs favourably with difficult and unbalanced data, but also in a multi-dataset scenario. Retaining the knowledge after training on ImageNet to learn the entire Adience dataset conserved 18% of accuracy compared to the naive finetuning baseline, achieving a net gain of about 11% in accuracy over both tasks. We were able to show that *RNF* is scalable, efficient, and effective for rendering neural networks resilient against catastrophic forgetting in sequential learning setups. Our technique additionally allows for the functional annotation of neural networks. After identifying the relevant parts of the model for a specific task, the obtained learning rate mask could be shared with other researchers, allowing them to employ transfer learning that benefits from existing knowledge through feature reuse, leveraging free network capacity for new tasks while not losing the ability to solve already learned ones. As LRP is applicable to a wide range of network architectures, our RNF technique can also be applied beyond the image domain, e.g., for natural language processing, or in Reinforcement Learning. In the latter especially, the impact of non-i.i.d. data is significant, and existing solutions, e.g., experience replay, are highly inefficient. Here, exploring RNF-based solutions is an interesting prospect for future work.

Acknowledgment. This work was supported by the German Ministry for Education and Research as BIFOLD (ref. 01IS18025A and ref. 01IS18037A), the European Union's Horizon 2020 programme (grant no. 965221 and 957059), and the Investitionsbank Berlin under contract No. 10174498 (Pro FIT programme).

A   Appendix

A.1   MNIST-Split

The model architecture was taken from [40], which compares multiple methods for mitigating catastrophic forgetting. It consists of two hidden layers with 400 neurons each and ReLU activations. As for most experiments except the real-world dataset, the pruning threshold is set to 2%, meaning that the accuracy can drop by up to 2% before the pruning procedure is halted. We use the Adam optimizer with a learning rate of 0.001, $\beta_1 = 0.9$ and $\beta_2 = 0.999$ with a batch size of 128.

A.2   MNIST-Permuted

For this experiment, the architecture from [34] was adapted by increasing the number of hidden layer units to 1000 per layer to match the increased complexity of the task. Additionally, the learning rate was decreased to 0.0001 and the model was trained for ten instead of four epochs per task.

A.3   CIFAR10 and CIFAR100

In this experiment, we adopted architecture and experimental setup from [46].

A.4   ImageNet Split

Here, we replicate the conditions from [43] but establish our baseline after ten instead of 70 epochs, which we also use when applying RNF.

A.5   Adience

As is state-of-the art for this dataset [9], we normalize to zero mean and unit standard deviation during training and apply data augmentation for the training data by randomly cropping to 224×224 as well as horizontal flipping. For testing, each sample is cropped five times to 224×224 (four corner crops and one center crop), where each crop is additionally mirrored horizontally. The ground truth is then compared to the mean of the Softmax activations of the ten samples. Only the center crops are used in the reference data. As the dataset is strongly imbalanced, we additionally employ a resampling strategy during training that undersamples classes with a high number of samples and oversamples classes with a low number of samples by computing the class probabilities and then sampling from a multinomial distribution.

In this experiment, we employ a VGG-16 network architecture [38] that has been pretrained on ImageNet (from the PyTorch [29] model zoo), as well as an Adam optimizer and L2 regularization.

- *Split*: We used a learning rate of 0.0001, L2 regularization with $\lambda = 0.01$ and a batch size of 32.
- *Entire Dataset*: The model was trained with a learning rate of 0.00001 and $\lambda = 0.001$.

References

1. Alber, M., et al.: iNNvestigate neural networks! J. Mach. Learn. Res. **20**(93), 1–8 (2019)
2. Anders, C.J., Neumann, D., Samek, W., Müller, K.-R., Lapuschkin, S.: Software for dataset-wide XAI: from local explanations to global insights with Zennit, CoRelAy, and ViRelAy. arXiv preprint arXiv:2106.13200 (2021)
3. Bach, S., Binder, A., Montavon, G., Klauschen, F., Müller, K.-R., Samek, W.: On pixel-wise explanations for non-linear classifier decisions by layer-wise relevance propagation. PLoS ONE **10**(7), e0130140 (2015)
4. Baehrens, D., Schroeter, T., Harmeling, S., Kawanabe, M., Hansen, K., Müller, K.-R.: How to explain individual classification decisions (2010)
5. Becking, D., Dreyer, M., Samek, W., Müller, K., Lapuschkin, S.: ECQ^x: explainability-driven quantization for low-bit and sparse DNNs. In: Holzinger, A., Goebel, R., Fong, R., Moon, T., Müller, K.R., Samek, W. (eds.) xxAI 2020. LNCS, vol. 13200, pp. 271–296. Springer, Cham (2022). https://doi.org/10.1007/978-3-031-04083-2_14
6. Chereda, H., et al.: Explaining decisions of graph convolutional neural networks: patient-specific molecular subnetworks responsible for metastasis prediction in breast cancer. Genome Med. **13**(1), 1–16 (2021). https://doi.org/10.1186/s13073-021-00845-7

7. Deng, J., Dong, W., Socher, R., Li, L.J., Li, K., Fei-Fei, L.: ImageNet: a large-scale hierarchical image database. In: 2009 IEEE Conference on Computer Vision and Pattern Recognition, pp. 248–255 (2009)
8. Deng, L.: The MNIST database of handwritten digit images for machine learning research. IEEE Sig. Process. Mag. **29**(6), 141–142 (2012)
9. Eidinger, E., Enbar, R., Hassner, T.: Age and gender estimation of unfiltered faces. IEEE Trans. Inf. Forensics Secur. **9**(12), 2170–2179 (2014)
10. Erhan, D., Bengio, Y., Courville, A., Vincent, P.: Visualizing higher-layer features of a deep network. Technical report, Univeristé de Montréal, January 2009
11. Evans, T., et al.: The explainability paradox: challenges for xAI in digital pathology. Future Gener. Comput. Syst. **133**, 281–296 (2022)
12. Farquhar, S., Gal, Y.: Towards robust evaluations of continual learning. arXiv preprint arXiv:1805.09733 (2018)
13. Fong, R.C., Vedaldi, A.: Interpretable explanations of black boxes by meaningful perturbation. In: 2017 IEEE International Conference on Computer Vision (ICCV) (2017)
14. French, R.: Catastrophic forgetting in connectionist networks. Trends Cogn. Sci. **3**, 128–135 (1999)
15. Guyon, I., Elisseeff, A.: An introduction to variable and feature selection. J. Mach. Learn. Res. **3**, 1157–1182 (2003)
16. Guyon, I., Weston, J., Barnhill, S., Vapnik, V.: Gene selection for cancer classification using support vector machines. Mach. Learn. **46**, 389–422 (2002). https://doi.org/10.1023/A:1012487302797
17. Hohman, F., Park, H., Robinson, C., Chau, D.H.: Summit: scaling deep learning interpretability by visualizing activation and attribution summarizations. arXiv preprint arXiv:1904.02323 (2019)
18. Hägele, M., et al.: Resolving challenges in deep learning-based analyses of histopathological images using explanation methods. Sci. Rep. **10**, 6423 (2020)
19. Kim, B., et al.: Interpretability beyond feature attribution: quantitative testing with concept activation vectors (TCAV). In: International Conference on Machine Learning, pp. 2668–2677. PMLR (2018)
20. Kirkpatrick, J., et al.: Overcoming catastrophic forgetting in neural networks (2017)
21. Kohlbrenner, M., Bauer, A., Nakajima, S., Binder, A., Samek, W., Lapuschkin, S.: Towards best practice in explaining neural network decisions with LRP. In: 2020 International Joint Conference on Neural Networks (IJCNN), pp. 1–7 (2020)
22. Krizhevsky, A.: Learning multiple layers of features from tiny images. Master's thesis, University of Toronto, Department of Computer Science (2009)
23. Lange, M.D., et al.: Continual learning: a comparative study on how to defy forgetting in classification tasks. arXiv preprint arXiv:1909.08383 (2019)
24. Lee, J., Yoon, J., Yang, E., Hwang, S.J.: Lifelong learning with dynamically expandable networks. arXiv preprint arXiv:1708.01547 (2017)
25. Montavon, G., Lapuschkin, S., Binder, A., Samek, W., Müller, K.-R.: Explaining nonlinear classification decisions with deep Taylor decomposition. Pattern Recogn. **65**, 211–222 (2017)
26. Nguyen, A.M., Dosovitskiy, A., Yosinski, J., Brox, T., Clune, J.: Synthesizing the preferred inputs for neurons in neural networks via deep generator networks. arXiv preprint arXiv:1605.09304 (2016)
27. Olah, C., Mordvintsev, A., Schubert, L.: Feature visualization. Distill **2**(11), e7 (2017)

28. Oren, G., Wolf, L.: In defense of the learning without forgetting for task incremental learning. In: Proceedings of the IEEE/CVF International Conference on Computer Vision Workshops, pp. 2209–2218 (2021)
29. Paszke, A., Gross, S., Massa, F., Lerer, A., Bradbury, J., et al.: PyTorch: an imperative style, high-performance deep learning library. In: Advances in Neural Information Processing Systems, pp. 8024–8035 (2019)
30. Radford, A., et al.: Learning transferable visual models from natural language supervision (2021)
31. Ribeiro, M.T., Singh, S., Guestrin, C.: "Why should I trust you?": Explaining the predictions of any classifier. In: Proceedings of the 22nd ACM SIGKDD International Conference on Knowledge Discovery and Data Mining, pp. 1135–1144 (2016)
32. Samek, W., Binder, A., Montavon, G., Bach, S., Müller, K.-R.: Evaluating the visualization of what a deep neural network has learned. IEEE Trans. Neural Netw. Learn. Syst. **28**(11), 2660–2673 (2017)
33. Samek, W., Montavon, G., Lapuschkin, S., Anders, C.J., Müller, K.-R.: Explaining deep neural networks and beyond: a review of methods and applications. Proc. IEEE **109**(3), 247–278 (2021)
34. Samek, W., Wiegand, T., Müller, K.-R.: Explainable artificial intelligence: understanding, visualizing and interpreting deep learning models. ITU J. ICT Discov. **1**(1), 39–48 (2018)
35. Schuhmann, C., et al.: LAION-400M: open dataset of clip-filtered 400 million image-text pairs. arXiv preprint arXiv:2111.02114 (2021)
36. Serrà, J., Surís, D., Miron, M., Karatzoglou, A.: Overcoming catastrophic forgetting with hard attention to the task. arXiv preprint arXiv:1801.01423 (2018)
37. Silver, D., Huang, A., Maddison, C., Guez, A., Sifre, L., Driessche, G., et al.: Mastering the game of go with deep neural networks and tree search. Nature **529**, 484–489 (2016)
38. Simonyan, K., Zisserman, A.: Very deep convolutional networks for large-scale image recognition. In: 3rd International Conference on Learning Representations (ICLR) (2015)
39. Sundararajan, M., Taly, A., Yan, Q.: Axiomatic attribution for deep networks. In: International Conference on Machine Learning, pp. 3319–3328. PMLR (2017)
40. van de Ven, G.M., Tolias, A.S.: Three scenarios for continual learning. arXiv preprint arXiv:1904.07734 (2019)
41. Wilm, F., Benz, M., Bruns, V., Baghdadlian, S., Dexl, J., Hartmann, D., et al.: Fast whole-slide cartography in colon cancer histology using superpixels and CNN classification. J. Med. Imaging **9**(2), 027501 (2022)
42. Wortsman, M., et al.: Supermasks in superposition. arXiv preprint arXiv:2006.14769 (2020)
43. Wu, Y., et al.: Large scale incremental learning. arXiv preprint arXiv:1905.13260 (2019)
44. Yeom, S.K., et al.: Pruning by explaining: a novel criterion for deep neural network pruning. Pattern Recogn. **115**, 107899 (2021)
45. Zeiler, M.D., Fergus, R.: Visualizing and understanding convolutional networks. arXiv preprint arXiv:1311.2901 (2013)
46. Zenke, F., Poole, B., Ganguli, S.: Improved multitask learning through synaptic intelligence. arXiv preprint arXiv:1703.04200 (2017)
47. Zintgraf, L.M., Cohen, T.S., Adel, T., Welling, M.: Visualizing deep neural network decisions: prediction difference analysis. arXiv preprint arXiv:1702.04595 (2017)

Approximation of SHAP Values
for Randomized Tree Ensembles

Markus Loecher[1]([✉]) [ID], Dingyi Lai[2] [ID], and Wu Qi[2] [ID]

[1] Berlin School of Economics and Law, 10825 Berlin, Germany
markus.loecher@hwr-berlin.de
[2] Department of Statistics, Humboldt University, Berlin, Germany

Abstract. Classification and regression trees offer straightforward methods of attributing importance values to input features, either globally or for a single prediction. Conditional feature contributions (CFCs) yield *local*, case-by-case explanations of a prediction by following the decision path and attributing changes in the expected output of the model to each feature along the path. However, CFCs suffer from a potential bias which depends on the distance from the root of a tree. The by now immensely popular alternative, *SHapley Additive exPlanation* (SHAP) values appear to mitigate this bias but are computationally much more expensive. Here we contribute a thorough, empirical comparison of the explanations computed by both methods on a set of 164 publicly available classification problems in order to provide data-driven algorithm recommendations to current researchers. For random forests and boosted trees, we find extremely high similarities and correlations of both local and global SHAP values and CFC scores, leading to very similar rankings and interpretations. Unsurprisingly, these insights extend to the fidelity of using global feature importance scores as a proxy for the predictive power associated with each feature.

Keywords: SHAP values · Saabas value · Variable importance · Random forests · Boosting · GINI impurity

1 Interpreting Model Predictions

Tree-based algorithms such as random forests and gradient boosted trees continue to be among the most popular and powerful machine learning models used across multiple disciplines.

While variable importance is not easily defined as a concept [6], the conventional wisdom of estimating the impact of a feature in tree based models is to measure the *node-wise reduction of a loss function*, which (i) yields only global importance measures and (ii) is known to suffer from severe biases. Nevertheless, variable importance measures for random forests have been receiving increased attention in bioinformatics, for instance to select a subset of genetic

© IFIP International Federation for Information Processing 2022
Published by Springer Nature Switzerland AG 2022
A. Holzinger et al. (Eds.): CD-MAKE 2022, LNCS 13480, pp. 19–30, 2022.
https://doi.org/10.1007/978-3-031-14463-9_2

markers relevant for the prediction of a certain disease. They also have been used as screening tools [4,14] in important applications, highlighting the need for reliable and well-understood feature importance measures.

The default choice in most software implementations [8] of random forests [1] is the *mean decrease in impurity (MDI)*. The MDI of a feature is computed as a (weighted) mean of the individual trees' improvement in the splitting criterion produced by each variable. A substantial shortcoming of this default measure is its evaluation on the in-bag samples, which can lead to severe overfitting and bias [7,9,10,18].

1.1   Conditional Feature Contributions (CFCs)

The conventional wisdom of estimating the impact of a feature in tree based models is to measure the **node-wise reduction of a loss function**, such as the variance of the output Y, and compute a weighted average of all nodes over all trees for that feature. By its definition, such a *mean decrease in impurity* (MDI) serves only as a global measure and is typically not used to explain a *per-observation, local impact*. Saabas [17] proposed the novel idea of explaining a prediction by following the decision path and attributing changes in the expected output of the model to each feature along the path.

Let f be a decision tree model, x the instance we are going to explain, $f(x)$ the output of the model for the current instance, and $f_x(S) \approx E[f(x) \mid x_S]$ the estimated expectation of the model output conditioned on the set S of feature values, then -following [12]- we can define the *conditional feature contributions* (CFCs)[1] for the i'th feature as

$$\phi_i^s(f, x) = \sum_{j \in D_x^i} f_x(A_j \cup j) - f_x(A_j), \tag{1}$$

where D_x^i is the set of nodes on the decision path from x that split on feature i, and A_j is the set of all features split on by ancestors of j. These feature attributions sum up to the difference between $E(\hat{(y)})$ (the expected output of the model) and the specific prediction $\hat{(y_i)}$; an appealing property shared with the SHAP values defined below. For ensemble models such as bagged/boosted trees or random forests, the CFCs for the ensemble model are naturally defined as the average of the CFCs over all trees.

In the light of wanting to explain the predictions from tree based machine learning models, these *conditional feature contributions* are rather appealing, because

1. The positive and negative contributions from nodes convey directional information unlike the strictly positive purity gains.
2. By combining many local explanations, we can represent global structure while retaining local faithfulness to the original model.

---

[1] Synonymous with *Saabas value*.

3. The expected value of every node in the tree can be estimated efficiently by averaging the model output over all the training samples that pass through that node.
4. The algorithm has been implemented and is easily accessible in a python [17] and R [19] library.

1.2 SHAP Values

However, Lundberg et al. [11] pointed out that it is strongly biased to alter the impact of features based on their distance from the root of a tree. This causes CFC scores to be inconsistent, which means one can modify a model to make a feature clearly more important, and yet the CFC attributed to that feature will decrease. As a solution, the authors developed an algorithm ("TreeExplainer") that computes local explanations based on exact Shapley values in polynomial time[2]. As explained in [12], Shapley values are a method to spread credit among players in a "coalitional game" which translates to a statistical model as follows: the difference in conditional expectation function of the model's output is averaged over all possible feature orderings with and without the feature of interest. Shapley values are known to be a unique solution satisfying the following three desirable properties: *local accuracy, consistency,* and *missingness*:

$$\phi_i(f, x) = \sum_{R \in \mathcal{R}} \frac{1}{M!} \left[f_x(P_i^R \cup i) - f_x(P_i^R) \right] \tag{2}$$

where $\mathcal{R}$ is the set of all feature orderings, P_i^R is the set of all features that come before feature i in ordering R, and M is the number of input features for the model.

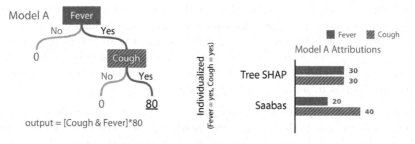

Fig. 1. Supplementary Figure 5(A) from [11] used to demonstrate inconsistencies in the Saabas method: The cough feature is assigned a larger CFC than fever whereas SHAP allocates credit equally, at least for the local attribution assigned to the instance (fever= True, cough = True).

One should not forget though that the same idea of adding *conditional feature contributions* lies at the heart of *TreeExplainer* with one important difference.

---

[2] A python library is available at https://github.com/slundberg/shap.

While SHAP values average the importance of introducing a feature over all possible feature orderings, CFC scores only consider the single ordering defined by a tree's decision path. Lundberg et al. [12] warn: "But since they do not average over all orderings, they do not match the SHAP values, and so must violate consistency." and show a simple example of a hypothetical "cough/fever" dataset on which a single tree is fitted - reproduced in Fig. 1. The data generating process is symmetrical in both features but the local Saabas values are different depending on their position in the tree path whereas SHAP allocates credit equally.

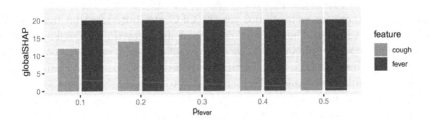

Fig. 2. Generalizing the two-way-AND data generation process as in Fig. 1 for unbalanced data sets with focus on global SHAP scores. Note that the global CFC scores are always equal to $(20, 20)$ for this range of marginal proportions of fever symptoms. We fix $p_{cough} = 0.5$ and observe credit allocation inconsistent with the true underlying model.

We point out that this conclusion does not hold for the **global** importance scores: (i) global CFC scores (i.e. the sum of the absolute values of the local scores) are equal, whereas (ii) global SHAP values do **not** allocate credit equally for unbalanced data sets! This is illustrated in Fig. 2 where we repeat the pure interaction data generating process from Fig. 1 but vary the marginal proportion of fever symptoms between $0.1, \ldots 0.5$ while fixing the marginal proportion of cough symptoms to 0.5 (maintaining perfect independence between features).

Upon further reflection, this chosen example suffers from the additional following limitations to serve as a general point in case:

1. For this choice of perfectly balanced data, the impurity reduction in the first split is equal for both features and hence the feature is chosen randomly by the tree; both choices lead to identical model performance. In such special/degenerate cases, a sensible model interpreter should probably average over all equivalent split scenarios.
2. For the general case of unbalanced data even the local SHAP scores are no longer equal.
3. The data generating process only depends on pure interaction terms but we are requiring **additive** explanations, which seems to be a mismatch.
4. One could argue that such a specific data set merely highlights the consequences of model misspecification than truly general shortcomings of e.g. CFC scores.

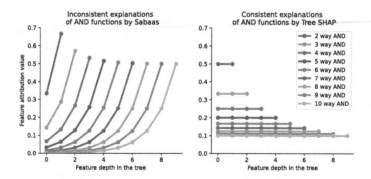

Fig. 3. Supplementary Figure 4(A-B) from [11] reproduced: Demonstration of the superiority of SHAP over CFC values in terms of consistency in single trees. (multiway AND for p binary predictors: $y = 1$ iff $\forall_j x_j = 1$ else $y = 0$) Left panel: CFCs ("Saabas values") give little credit to features near the root leading to inconsistencies. Right panel: Tree SHAP evenly distributes credit among all the features involved in the AND function

We are now in a position to understand why randomization in algorithms might alleviate most of the described issues.

Supplementary Figure 4 (A-B) in [11] showcases the inconsistency of CFCs and respective consistent credit allocation of SHAP values for the very special case of multiway AND functions. (The output y is zero unless all p binary features are equal to 1.) Clearly -this being a generalization of the fever/cough example-, all features contribute equally to the outcome and should receive equal credit which is not the case for CFCs. For the reader's convenience, we reproduce these experiments in Fig. 3 and point out that these extreme results only hold for a **single** tree.

These discrepancies almost disappear in randomized ensembles such as forests or boosted trees as illustrated in Fig. 4 for random forests ($ntree = 100$), where the dependence of CFCs on the distance from the root is almost absent.

We close this section by noting that Theorem 1 in the supplement of [11] already states that *"In the limit of an infinite ensemble of totally random fully developed trees on binary features the Saabas values equal the SHAP values"*. However, these extreme assumptions (infinite number of trees and data, mtry= 1, only categorical features, etc..) are quite far from realistic settings in which random forests are usually deployed.

2   SHAP Versus CFCs

The main contribution of this paper is a thorough, direct, empirical comparison of CFC and SHAP scores for two tree ensemble methods: random forests (RF) and gradient boosted trees (XGboost). (All our simulations utilize the *sklearn* library, in particular the methods *RandomForestClassifier* and *XGBClassifier*.) Our initial focus in Sect. 2.2 are the local explanations, while Sect. 2.3 provides a different perspective on the global discriminative power.

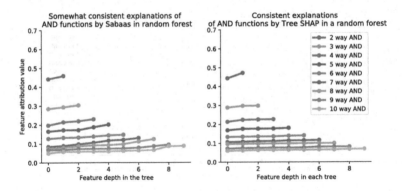

Fig. 4. Same numerical experiment as detailed in Fig. 3 but with random forests instead of a single tree. ($ntree = 100, mtry = \sqrt{p}$) The inconsistencies for Saabas/CFCs are much less pronounced and seem to disappear entirely for higher order AND functions.

2.1   Data Investigated

We are first going to describe the data sets used in our study.

Classification. The algorithms were compared on 164 supervised classification datasets from the *Penn Machine Learning Benchmark (PMLB)* [15]. PMLB is a collection of publicly available classification problems that have been standardized to the same format and collected in a central location with easy access via Python[3]. Olson et al. [15] compared 13 popular ML algorithms from scikit-learn[4] and found that Random Forests and boosted trees ranked consistently at the top.

Regression. In addition to the classification tasks described , we now include 10 additional datasets intended for regression from the UCI Machine Learning Repository [5], summaries of which are provided in Table 1. As an additional regression model, we generated synthetic data as defined in [13]:

$$Y = 10\sin(\pi X_1 X_2) + 20(X_3 - 0.05)^2 + 10X_4 + 5X_5 + \epsilon$$

where features X_j are sampled independently from Unif$(0, 1)$. In our simulations we choose sample sizes of $n = 500$ and a signal-to-noise ratio of SNR$= 3.52$.

---

[3] URL: https://github.com/EpistasisLab/penn-ml-benchmarks.

[4] The entire experimental design consisted of over 5.5 million ML algorithm and parameter evaluations in total.

Table 1. Summary of real-world data utilized.

Dataset	p	n
Abalone age [abalone]	8	4177
Bike sharing [bike]	11	731
Boston housing [boston]	13	506
Concrete compressive strength [concrete]	8	1030
CPU Performance [cpu]	7	209
Conventional and social movie [csm]	10	187
Facebook metrics [fb]	7	499
Parkinsons telemonitoring [parkinsons]	20	5875
Servo system [servo]	4	167
Solar flare [solar]	10	1066

2.2   Comparative Study: Local Explanations

Out of the 164 datasets in PMLB collection, one dataset *'analcatdata japan-solvent'* is chosen here to illustrate the local variable importance comparison between SHAP and CFC.

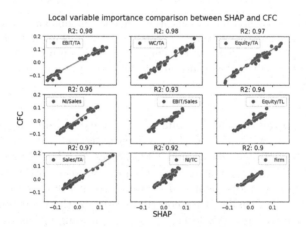

Fig. 5. Scatter plots of local SHAP and CFC values for dataset 'analcatdata japan-solvent', a linear regression is fitted to each variable and the corresponding R-Squared values are reported

It is a binary classification task, the dataset consists of 9 numerical variables and 52 observations. A more detailed description about each dataset is provided by the authors[5]. The scatter plots shown in Fig. 5 demonstrate the strong correlation between the variable importance calculated by SHAP and CFC methods

---

[5] https://epistasislab.github.io/pmlb/.

for random forests. The plots are ranked by the (global) SHAP importance values, i.e. the first variable EBIT/TA on the top left graph is the most important variable for the classification task according to SHAP, and the second variable WC/TA is the second important, and so forth. Additionally, a linear regression is fitted to each scatter plot, the corresponding R-Squared value of each linear regression is shown above each scatter plot for each variable, which confirms the strong linear correlation between SHAP and CFC computed variable importance.

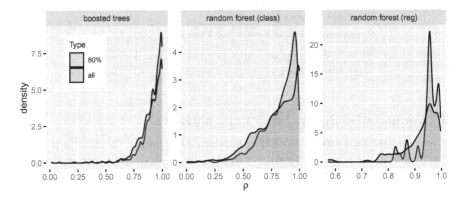

Fig. 6. Left panel is for boosted classification while the middle and right panel depict random forests for classification and regression, respectively. Note that the total number of regression data sets is much lower leading to a less smooth density plot. Filtered (80%) stands for variables that contribute to 80% of the total importance for each dataset.

In addition to this anecdotal example of the local variable importance comparison between SHAP and CFC, Fig. 6 depicts the general distribution of all local correlation between SHAP and CFC scores in all PMLB datasets. For both RF and XGboost very strong correlations between SHAP and CFC are observed. The density plots display all correlations as well as a filtered distribution, including only variables that contribute 80% to the total importance for each dataset. We consider this filtering criterion meaningful, as the correlation calculation of SHAP and CFC might not be highly representative for variables with a low importance score[6]. We mention in passing that about half of the variables are filtered out. Those "unimportant" features seem to show less similarity between their CFC and SHAP version. We also observe that there's a peak correlation value 1 in the total correlation distribution plot, which is due to the bias of low/zero importance variables. At this point we do not have a satisfactory explanation for the observed differences between boosted trees and random forests.

---

[6] Although a more systematic approach would be to perform statistical tests on the feature scores, see [2], and references therein.

2.3   Predictive Power of Feature Subsets

In the previous section, we compared the local explanations provided by two rather different algorithms. Since these data sets are not generated by simulations, we do not really have any ground truth on the "importance" or relevance of each feature's contribution to the outcome. In fact, Loecher [9] showed that even SHAP and CFC scores suffer from the same bias observed in the original MDI. Hence, we cannot really quantify which score provides more meaningful or less misleading measures. At least for the global scores, we can inquire whether the sum of a method's importance scores is a reliable proxy for the predictive power of subsets of features. We chose to replicate the approach taken in [3]: "We generated several thousand feature subsets for each dataset, re-trained models for each subset, and then measured the correlation between the model loss and the total importance of the included features". In particular, the feature subsets were independently sampled as follows: sample the subset's cardinality $k \in 0, 1, \ldots, d$ uniformly at random, then select k elements from $1, \ldots, d$ uniformly at random. We computed both SHAP/CFC and the retrained models' loss (either negative log loss or rmse, respectively) on a test set. This strategy yields one correlation coefficient for SHAP and CFC for each of the data sets described in Sect. 2.1.

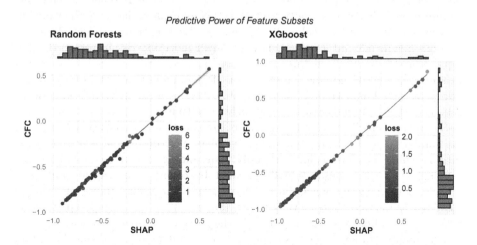

Fig. 7. For both CFC and SHAP we measured the correlation between the model loss and the total importance of the included features. Here, loss is measured either as negative log loss for classification or as (rescaled) rmse for regression. The left panel shows results for random forests, the right panel for XGboost. Note that the latter one does not include results for regression.

Figure 7 depicts these correlations for random forests and boosted trees as a scatterplot with marginal distributions [16]. In order to show both log loss and rmse on the same graph, we rescaled the latter on the same range. Across all data sets, the measured correlations are almost identical for the global SHAP and CFC scores. The strength of this association surprised us and suggests that the global CFC scores measure the predictive information content of a features with as much fidelity as the global SHAP values. We mention in passing the (albeit few) surprisingly positive correlations between the model loss and feature importances suggesting that sometimes including features with high importance scores can hurt model performance. We speculate that this counter-intuitive phenomenon is due to the overfitting/bias of SHAP and MDI scores reported in [9].

3   Discussion

The rapidly rising interest in explainable AI and interpretable machine learning has made it very clear that local explanations for complex models have a significant impact for most applications in general and particularly so for highly regulated domains, such as healthcare, finance and public services. Explanations can provide transparency to customers and overseeing agencies, help find biases in algorithmic decision making, facilitate human-AI collaboration, and could even speed up model development and debugging. The increasingly popular SHAP scores play a vital role in this context since they fulfil certain optimality requirements borrowed from game theory and -for trees only- offer somewhat efficient computational shortcuts. Lundberg et al. [11] convincingly showcase the value of local explanations in the realm of medicine. SHAP values are used to (1) identify high-magnitude but low-frequency nonlinear mortality risk factors in the US population, (2) highlight distinct population subgroups with shared risk characteristics, (3) identify nonlinear interaction effects among risk factors for chronic kidney disease. Furthermore, "they enable doctors to make more informed decisions rather than blindly trust an algorithm's output".

The main contribution of this paper is to demonstrate empirically the effective similarity of SHAP scores and the much simpler conditional feature contributions which are defined for trees only. Our findings are significant not only because CFCs are computationally cheaper by orders of magnitude but also because of their close connection to the conventional feature importance measure MDI as derived in [10]. Why is this important? It could help explain why SHAP scores suffer from the same bias in the presence of predictors with greatly varying number of levels. For example, Loecher [9] demonstrated that both the CFCs as well as the SHAP values are highly susceptible to "overfitting" to the training data and proposed a correction based on out-of-bag (oob) data. In that light, debiasing SHAP values could be achieved using similar ideas.

Our empirical findings complement the recent theoretical work by Sutera et al. [20] who derived conditions under which global Shapley values are identical to global MDI (i) in the limit of infinite, fully randomized tree ensembles and (ii) for impurity defined as cross entropy. The authors further defined local MDI values in trees as local node-wise impurity reductions which also tend to converge to local Shapley values under similar assumptions, but with a modified characteristic function. We did not investigate the dependence of the similarity between SHAP and CFC on the number of subsampled columns in random forests (parameter *mtry*). Instead, we choose to tune the ensembles and use the optimal value for *mtry* which we believe to be more relevant for applied modelling.

Acknowledgements. We thank [15] for providing the complete code (https://github. com/rhiever/sklearn-benchmarks) required both to conduct the algorithm and hyperparameter optimization study, as well as access to the analysis and results. We also thank the authors of [3] for providing us details on their simulations.

References

1. Breiman, L.: Random forests. Mach. Learn. **45** (2001). https://doi.org/10.1023/A:1010933404324
2. Coleman, T., Peng, W., Mentch, L.: Scalable and efficient hypothesis testing with random forests. arXiv preprint arXiv:1904.07830 (2019)
3. Covert, I., Lundberg, S.M., Lee, S.I.: Understanding global feature contributions with additive importance measures. In: Larochelle, H., Ranzato, M., Hadsell, R., Balcan, M.F., Lin, H. (eds.) Advances in Neural Information Processing Systems. vol. 33, pp. 17212–17223. Curran Associates, Inc. (2020). https://proceedings.neurips.cc/paper/2020/file/c7bf0b7c1a86d5eb3be2c722cf2cf746-Paper.pdf
4. Díaz-Uriarte, R., De Andres, S.A.: Gene selection and classification of microarray data using random forest. BMC Bioinform. **7**(1), 3 (2006)
5. Dua, D., Graff, C.: UCI machine learning repository (2017). http://archive.ics.uci.edu/ml
6. Grömping, U.: Variable importance assessment in regression: linear regression versus random forest. Am. Stat. **63**(4), 308–319 (2009)
7. Kim, H., Loh, W.Y.: Classification trees with unbiased multiway splits. J. Am. Stat. Assoc. **96**(454), 589–604 (2001)
8. Liaw, A., Wiener, M.: Classification and regression by randomforest. R. News **2**(3), 18–22 (2002). https://CRAN.R-project.org/doc/Rnews/
9. Loecher, M.: From unbiased MDI feature importance to explainable AI for trees. arXiv preprint arXiv:2003.12043 (2020)
10. Loecher, M.: Unbiased variable importance for random forests. Commun. Stat. Theory Methods 51, 1–13 (2020)
11. Lundberg, S.M., et al.: From local explanations to global understanding with explainable AI for trees. Nat. Mach. Intell. **2**(1), 56–67 (2020)
12. Lundberg, S.M., Erion, G.G., Lee, S.I.: Consistent individualized feature attribution for tree ensembles. arXiv preprint arXiv:1802.03888 (2018)
13. Mentch, L., Zhou, S.: Randomization as regularization: a degrees of freedom explanation for random forest success. J. Mach. Learn. Res. **21**(171) (2020)
14. Menze, B.R., et al.: A comparison of random forest and its GINI importance with standard chemometric methods for the feature selection and classification of spectral data. BMC Bioinform **10**(1), 213 (2009)
15. Olson, R.S., La Cava, W., Mustahsan, Z., Varik, A., Moore, J.H.: Data-driven advice for applying machine learning to bioinformatics problems. arXiv preprint arXiv:1708.05070 (2017)
16. Patil, I.: Visualizations with statistical details: the 'ggstatsplot' approach. J. Open Sour. Softw.**6**(61), 3167 (2021)

17. Saabas, A.: Treeinterpreter library (2019). https://github.com/andosa/treeinterpreter
18. Strobl, C., Boulesteix, A.L., Zeileis, A., Hothorn, T.: Bias in random forest variable importance measures: illustrations, sources and a solution. BMC Bioinform. **8** (2007). https://doi.org/10.1186/1471-2105-8-25
19. Sun, Q.: tree.interpreter: random forest prediction decomposition and feature importance measure (2020). https://CRAN.R-project.org/package=tree.interpreter, r package version 0.1.1
20. Sutera, A., Louppe, G., Huynh-Thu, V.A., Wehenkel, L., Geurts, P.: From global to local MDI variable importances for random forests and when they are shapley values. In: Advances in Neural Information Processing Systems, vol. 34 (2021)

Color Shadows (Part I): Exploratory Usability Evaluation of Activation Maps in Radiological Machine Learning

Federico Cabitza[1,2(✉)], Andrea Campagner[1], Lorenzo Famiglini[1], Enrico Gallazzi[3], and Giovanni Andrea La Maida[3]

[1] Universitá degli Studi di Milano-Bicocca, Milan, Italy
`federico.cabitza@unimib.it`
[2] IRCCS Istituto Galeazzi Milano, Milan, Italy
[3] Istituto Ortopedico Gaetano Pini — ASST Pini-CTO, Milan, Italy

Abstract. Although deep learning-based AI systems for diagnostic imaging tasks have virtually showed superhuman accuracy, their use in medical settings has been questioned due to their "black box", not interpretable nature. To address this shortcoming, several methods have been proposed to make AI eXplainable (XAI), including Pixel Attribution Methods; however, it is still unclear whether these methods are actually effective in "opening" the black-box and improving diagnosis, particularly in tasks where pathological conditions are difficult to detect. In this study, we focus on the detection of thoraco-lumbar fractures from X-rays with the goal of assessing the impact of PAMs on diagnostic decision making by addressing two separate research questions: first, whether activation maps (as an instance of PAM) were perceived as useful in the aforementioned task; and, second, whether maps were also capable to reduce the diagnostic error rate. We show that, even though AMs were not considered significantly useful by physicians, the image readers found high value in the maps in relation to other perceptual dimensions (i.e., pertinency, coherence) and, most importantly, their accuracy significantly improved when given XAI support in a pilot study involving 7 doctors in the interpretation of a small, but carefully chosen, set of images.

Keywords: eXplainable AI · Medical machine learning · Activation maps · Thoracolumbar fractures · X-rays

1 Introduction

In recent years, the development of Machine Learning (ML) and Deep Learning (DL) models in image-centered medical areas such as radiology, pathology and, to a lesser extent, oncology and cardiology has significantly grown, showing results of increasing, although not complete, reliability and robustness [41]. As

© IFIP International Federation for Information Processing 2022
Published by Springer Nature Switzerland AG 2022
A. Holzinger et al. (Eds.): CD-MAKE 2022, LNCS 13480, pp. 31–50, 2022.
https://doi.org/10.1007/978-3-031-14463-9_3

demonstrated by some studies [9], including an AI-based decision support system (DSS) in medical tasks like diagnosis, planning and prognosis might result in an increase in the accuracy of the medical team, especially in case of less expert diagnosticians [23]. However, although AI systems that classify diagnostic images by applying deep learning techniques have shown almost superhuman accuracy [26], they have also raised concerns about their real applicability in medical contexts since they do not provide any support for understanding their advice, i.e. they are so-called inscrutable "black boxes" [20]. To overcome this apparent shortcoming, several methods have been introduced to make AI eXplainable (XAI) [21], including a particular type of feature attribution methods applied to images, namely Pixel attribution methods (PAM) [30]. These latter produce color maps (or, heatmaps) highlighting regions that were relevant for certain image classification: while PAMs take different names depending on the techniques used to generate them, such as saliency maps and activation maps (including class activation maps, or CAMs)[1], they all produce as output, as said above, essentially a visual representations that highlights feature points (pixels and pixel regions) of interest in the image that are related to the automatic classification (more details in Sect. 2), allowing to answer questions like "How would the model's classification be affected if the value of those pixel regions changed?". A still open question regards whether these methods are actually effective in "opening" the black-box and are of actual support in real-world diagnostic tasks, especially in tasks where it is still difficult to detect pathological conditions (e.g. [4]).

One of these tasks is the detection of traumatic thoracolumbar (TL) fractures from vertebral X-rays; in this application domain a relatively high proportion of missed diagnoses is still observed, with a false negative rate that is close to 30% [17]. Since the introduction of AI-based DSS and XAI methods to reduce such error rate seems plausible (as also shown by some recent systematic reviews in the radiological domain [1,41]), the present study focuses on the detection of TL fractures from X-rays, and is aimed at assessing the impact of AI-generated activation maps (AM) on diagnostic decision making.

In particular, we will consider a particular human - AI collaboration protocol, the *human-first, strict second-opinion* protocol, that is depicted in Fig. 1: according to this protocol, the human diagnostician is required to provide a tentative diagnosis by considering the diagnostic image *before* receiving the machine's advice and being influenced by it; then the machine processes the image and yields its best diagnostic advice, all together with an AM; finally, the human doctor is required to reconsider the case following the machine advice and map and consolidate their final decision in the report. This protocol is conjectured to be at low risk of automation bias [29] and deskilling [12], and was chosen because

---

[1] It is worth of note that activation maps are different from saliency maps, although the two terms are often used interchangeably. In fact, the two approaches rely on different methods to compute heatmaps: saliency maps are usually generated by means of back-propagation w.r.t. to the input of the network [35], while activation maps are obtained by means of the feature maps obtained at a specific layer of the network [42].

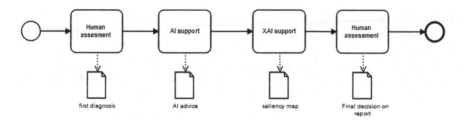

Fig. 1. BPMN diagram of the human-AI collaboration protocol adopted in this study.

it allows to assess the *diagnostic benefit*, in terms of accuracy difference between the first (pre-AI) and the final (post-AI) human decision.

In this exploratory study, we aim to tackle two related but distinct research questions: first, to assess whether activation maps (as an example of PAMs) were perceived as useful (H_1) in the radiological task of vertebral fracture detection; and, more importantly yet, whether maps were also effective (H_2), that is they improved the accuracy of the decision makers involved. These hypotheses, which can be traced back to the assessment of the usability of activation maps, broadly speaking, and of their actual and perceived utility, are seldom considered in the field at the intersection of medical AI and eXplainable AI [22] (XAI). For instance, Han et al. [19] studied the capability of saliency maps to guide radiologists to the correct pathology, considering both thoracic and knee imaging, showing promising results. By contrast, Arun et al. [3] studied the trustworthiness of saliency maps, showing that commonly employed saliency map techniques fail to be trustworthy compared to commonly adopted baselines (e.g., segmentation models for lesion localization), and therefore argued against the use of saliency map-like support in high-risk medical domains. A similar finding was also reported by Alqaraawi et al. [2]: while the authors report that saliency maps helped radiologists in identifying features to which the AI models were sensitive, these maps did not provide helpful support in terms of predicting or understanding the classifications of the AI system. Finally, and more akin to the aims of this paper, Bansal et al. [7] reported how, even though human diagnostic performance improved when provided with AI support, explanations (though generated by means of LIME [34], a feature attribution method, rather than saliency or activation maps) did not have a beneficial effect on human performance, rather they increased the acceptance of AI classifications irrespective of whether they were correct or not, thus suggesting a potential for increased automation bias.

The exploratory nature of the study lies in the fact that we were not only interested in detecting significant differences between the various experimental conditions, namely unsupported diagnosis vs. PAM-supported analysis; in fact, due to the relatively small number of images considered, we expected this result could be difficult to obtain or generalize. Rather, we also want to assess the *size effect* of the use of activation maps in diagnostic radiological decision making. Indeed, estimates of size effect could be used in reliable power analyses for other more in-depth studies regarding this relevant method to achieve and guarantee explainability of AI-based decision support. To our knowledge, this is the first study specifically aimed at evaluating effect size for the perceived utility and observed effectiveness of activation maps.

2   Methods

To conduct this user study we: 1) first, trained a classification model to detect thoracolumbar (TL) fractures in vertebral X-rays; 2) then, we had an expert orthopedist select a number of images (from the test set) that could represent interesting cases of both positive (fracture) and negative cases (no fracture); 3) then, we produced an activation map for each of these images; 4) then, we followed the interaction protocol depicted in Fig. 1 and 4a) asked a sample of orthopedists to classify the above images before being supported by the AI; 4b) then we showed these orthopedists the advice produced by the model built at the first step (see above) and the activation maps developed at step 2; and 4c) lastly, we collected their definitive diagnosis regarding the images considered and their perceptions about some usability dimensions regarding the maps by means of a short psychometric questionnaire.

In what follows, we will report about the ML training, the production of the activation maps and the definition of the psychometric scales considered in this study. In Sect. 3 we will show the results regarding the users' responses collected at step 4 above, while in Sect. 4 we will comment on the main findings.

2.1   Data

As said above we considered a binary classification task: 'TL fracture present' vs 'no TL fracture present'. To this aim, we considered a total of 630 vertebral cropped X-rays, which had been collected at the Spine Surgery Centre of the Niguarda General Hospital of Milan (Italy) from 2010 to 2020, from 151 trauma patients over 18 years old, split into 328 no-fracture images (52%) and 302 fracture ones (fractures associated with other conditions than trauma, such as osteoporosis or neoplasms, were excluded from the dataset).

For the ground truthing of the training dataset, 3 experienced spine surgeons annotated the 630 images mentioned above in conjunction with Gold Standard CT and MRI images. Each annotator selected and cropped individual vertebra through an image cropping software, specifically developed to create this dataset (using the C++ programming language): therefore, multiple images were obtained from the same X-ray image, so as to increase the sample size. The available data was split into 80% training, 10% validation, and 10% test set to provide sound evaluations in terms of over/under-fitting.

Data Augmentation [15] was applied only to the training batch. Specifically, we applied the following augmentation methods: random horizontal flip (p = .6), random vertical flip (p = .5) and random rotation within the [−5, 5] degree range. Furthermore, as additional pre-processing steps, the input images were resized to 224 × 224 pixels, and normalized (mean = [0.485, 0.456, 0.406], standard deviation = [0.229, 0.224, 0.225]; one value for each of the RGB channels.

2.2   Model Development

For the purpose of this research, we adopted a transfer-learning approach. To this aim, we selected a ResNeXt–50 [40] model pre-trained on images extracted from Imagenet. The choice to use a pre-trained model is justified by the fact that our training sample is relatively small, thus, a transfer learning-based approach (even though based on images from different application domains than the one considered in this study) could have beneficial effects in terms of generalization [25]. On the other hand, the choice of the ResNeXt-50 architecture is based on promising results of this network reported in the literature [40]. The above architecture was slightly modified to adapt it to the binary classification task: the number of neurons in the last dense layer was modified from 1000 to 2 to become our classification layer.

We chose softmax as the final activation function and cross-entropy as the corresponding loss. We then trained the weights of the fourth block of ResNeXt-50 and the last dense layer (previously added). We chose to contextualize only the last layer of the network's backbone because the original model was trained on data from a different domain. Following state-of-the-art practices [8] w.r.t. the contextualization of imaging models, we adopted different learning rates for the network's backbone and the classification layer: specifically, a learning rate of $lr_g = 2e − 3$ was adopted for the dense classification layer, while the learning rate for the fourth block mentioned above was $lr = 1/6 * lr_g$. The number of training epochs was set to 200. The final model's weights were selected based on the minimum loss value reported on the validation set.

2.3   Activation Maps

As said above, the main focus of this study does not regard the classification performance of the model, but, rather, whether a ML model could meaningfully suggest regions of interest in an X-ray to the human reader called to report on that exam. To provide such visual indications, we adopted the class activation map (CAM) method [42]. To the users involved in our user study we introduced the term activation map as a "particular heat map that, juxtaposed to the diagnostic image, highlights the areas that the classification model considers most informative (and hence predictive) in order to reach a correct diagnosis (presence or absence of fracture)". Technically speaking, CAM is a method to generate weight matrices associated with an image, with the understanding that weights of greater magnitude denote an area of particular importance for the model's classification. An example of an x-ray, with the corresponding activation map, is shown in Fig. 2.

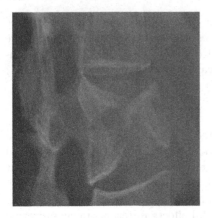 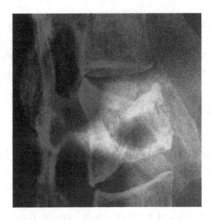

Fig. 2. An example of an x-ray and the corresponding activation map.

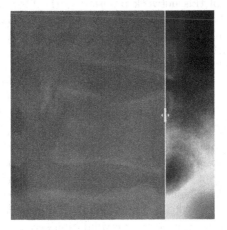

Fig. 3. The image comparator used to consult both the x-ray and the activation map. Both images are 800 × 800 pixel wide. Users could scroll a kind of slider by moving a handle (at the center of the comparator) horizontally so as to superimpose one image on another and better appreciate the underlying anatomical structures.

Figure 3 shows an X-ray given as input to the model and the corresponding activation map, generated through the CAM method. Two authors, both board-certified orthopaedic specialists and spine surgeons, selected 12 x-rays from a random sample of the test set, by selecting images that: 1) would be *clear* enough to be interpreted notwithstanding the relatively low resolution (800 × 800) in comparison with typical original x-rays[2]; and, 2) respected the constraints of a balanced sample (i.e. 6 with fracture and 6 without fracture) on which the AI exhibited approximately the same accuracy as the lower bound of model's performance on the test set (i.e., 8 right and 4 wrong)[3].

2.4   Utility and Other Subjective Dimensions

After having generated 12 activation maps from as many X-ray images (see above), we presented them to 7 orthopedists, with varying expertise, involved in the study, by means of an online multi-page questionnaire, developed on the Limesurvey platform[4]. We asked the orthopedists to consider each x-ray and corresponding map in a page that displayed a Javascript comparator component[5], as the one shown in Fig. 3. Immediately below this comparator, we asked the physicians to evaluate the map along the three qualitative dimensions mentioned above, *coherence*, *pertinency* and *utility*, in this order, on a 6-value ordinal scale, a semantic differential from 1 (not at all) to 6 (totally)[6].

Intuitively, pertinency regards meaningfulness and informativeness in regard to the diagnostic question; coherence with the advice given by the machine; utility regards practical support in reaching a correct diagnosis. In the first page of the online questionnaire, we presented these concepts in the following terms: *Coherence* (between the map and the suggestion provided by the machine) was described as the extent "the map highlights the structures that present a fracture or those in which fractures are excluded according to the machine's suggestion". *Pertinency* (of the map with respect to the diagnosis considered correct), defined as the extent the map highlights structures that have lesions or demonstrate their absence depending on the diagnosis considered correct (and regardless of the machine's suggestion). Lastly, *Utility* was defined as "the extent the map highlights structures or findings that allow for the correct diagnosis to be reached more easily than if the map was not seen". The rationale for considering the above three dimensions

---

[2] More precisely, both specialists had to convene that the images were at least of level 3 ("sufficient image quality: moderate limitations for clinical use but no substantial loss of information") on the absolute Visual Grading Analysis (VGA) scale [28], which is a 5-value ordinal scale from 1 ("excellent image quality: no limitations for clinical use") to 5 ("poor image quality: image not usable, loss of information, image must be repeated").

[3] Balance in model accuracy was also guaranteed at class level: in each group (fractured vs non-fractured), 2 images were associated with a misdiagnosis by the model while 4 were correctly classified.

[4] https://www.limesurvey.org/.

[5] https://juxtapose.knightlab.com/.

[6] *Clarity* would be a fourth relevant dimension of heatmaps for radiological use or XAI settings, as it relates to the accurate presentation of anatomical or pathological structures. However, this dimension was not assessed by the sample of readers involved, as images had already been selected to be of optimal clarity. Moreover, correlation between clarity and utility has been conjectured to be obvious and not worthy of investigation.

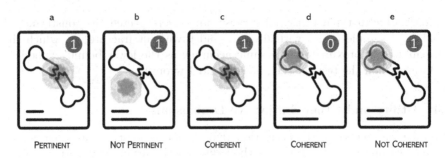

Fig. 4. Five possible cases of activation maps: a and b, associated with the machine advice reported in the top right-most angle of the X-ray, regard pertinency (with respect to the case at hand); c, d and e, associated with the machine advice reported in the top right-most angle of the X-ray, regard coherence (with respect to the advice given).

was motivated by the need to determine their collinearity with respect to a latent construct of *"usability"* of radiological heatmaps, and determine the size of correlation between the above dimensions. In particular, we shared the intuition that a map that is both coherent and pertinent map is likely to be also practically useful, but, conversely, an useful map is not necessarily also coherent and pertinent (because its low pertinency or coherence could rightly suggest the reader to be wary of the machine' advice). Figure 4 shows these categories in visual form.

We highlight the fact that coherence and pertinency are not equivalent concepts. Nonetheless, they are both related to explainability, in that maps give clues on what has been considered informative by the model: if doctors cannot detect anything relevant in those areas (pertinency), or fail to recognize an area that is associated with a specific advice (coherence), either things could be interpreted as a sign that the model is producing the wrong advice.

Aside from a global analysis among the above mentioned perceptual dimensions, we also performed a comparative analysis, through which we analyzed the distributions of the perceptions across three possible stratification criteria: by agreement (i.e., cases on which the human readers agreed with the classification of the AI, against cases on which such an agreement was not observed); by class type (i.e., fractured cases, against non-fractured cases); by correctness (i.e., cases on which the AI classification was correct, against cases on which the AI classification was wrong); and by expertise of the human readers.

The 7 readers considered 12 cases (and corresponding X-rays and activation maps), providing a total of 84 evaluations.

3   Results

The results reported by the ML model on the test set are reported in Table 1 and Fig. 5, which also reports the accuracy of the human readers both with and without the support of the AI.

The risk of overfitting can be ruled out on the basis of a hypothesis test on the null hypothesis that the accuracy scores observed on the validation and test set are not significantly different (two proportion test: $z = -.7432$, p-value $= .45$).

Table 1. Average performance scores (and their binomial confidence Intervals at 95% confidence level) of the AI model calculated on the test set. AUROC stands for area under the ROC curve.

Metric	*Accuracy*	*Sensitivity*	*F1-score*	*AUROC*
	.77 [.66, .88]	.76 [.65, .87]	.77 [.66, .88]	.77 [.66, .88]

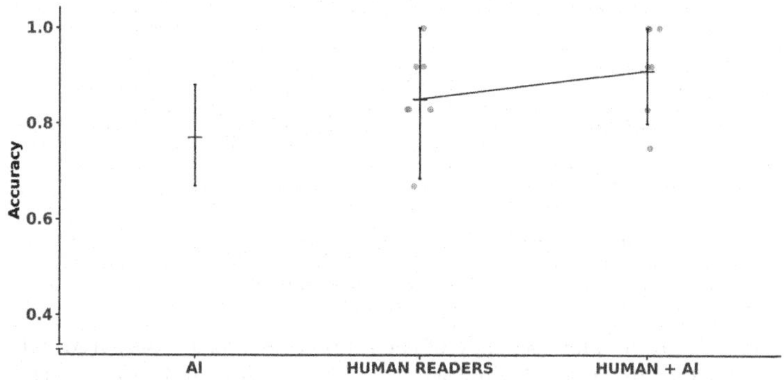

Fig. 5. Diagnostic accuracy of the AI and the human readers, before and after having been supported by the AI (and AMs), with their binomial confidence intervals at 95% confidence level. The AI performance has been evaluated on the test set (N = 52), while the human readers on the 12-image set used in this study.

The diagnostic accuracy of human readers before and after having received the support of the AI is reported in Table 2, as well as in the form of a benefit diagram in Fig. 6: this latter diagram (first proposed in [39]) reports on the benefit of the interaction with AI by representing the difference in human readers' performance against their baseline accuracy.

Table 2. Accuracy of human readers pre and post AI support, with their binomial confidence interval at 95% confidence level. Group comparison has been evaluated through a McNemar test; effect size computed through odds ratio [14] (1st group human readers, 2nd human + AI, (N = 84).

	Accuracy (95%CI)	P-value	Effect size
Human readers	.82 [.73, .90]	.021	.28
Human + AI	.92 [.86, .97]		

The distribution of readers' perceptions about the use of activation maps, in terms of the considered perceptual dimensions is reported in Fig. 7. The (Spearman) correlation among the considered perceptual dimensions is reported in Table 3 the figure highlights a moderate and significant correlation between utility and pertinency, while the correlation among the other dimensions were weak and not significant (p-values > .05).

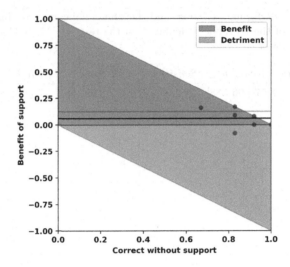

Fig. 6. Benefit diagram. The accuracy of each human reader is compared with respect to the gain obtained by the AI support. The blue region indicates whether the AI support helps to improve human Accuracy, while the red region indicates a negative impact by the AI support. The black line indicate the average benefit, with its confidence intervals. If this interval is entirely above the 0 line, then the support is significantly beneficial. (Color figure online)

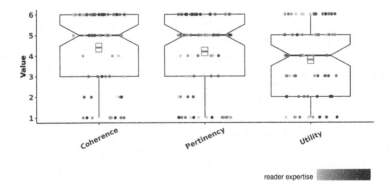

Fig. 7. Readers' perceptions about the use of activation maps in the diagnostic user study.

Table 3. Spearman correlation scores (at 95% confidence level) between the perceptual dimensions identified in this study.

Correlation btw	Value	P-value
Utility - pertinency	.40	<.001
Utility - coherence	−.10	.37
Pertinency - coherence	.07	.54

The distributions of perceptual dimension stratified by the four previously described categories (agreement, class, expertise and correctness) are reported in Figs. 8, 9, 10 and 11. In Table 4 we report about the effect sizes for the AM dimensions across different groups and conditions.

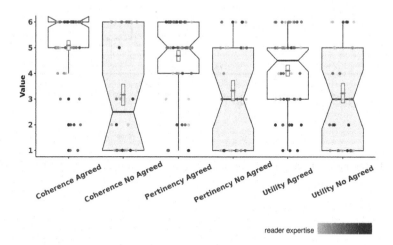

Fig. 8. Rater perceptions about the use of activation maps, stratified according to whether humans and AI agreed or not (regardless of the correctness of their diagnosis).

4 Discussion

First of all, we should comment on the fact that the accuracy of the diagnostic model seems relatively low and it is actually lower than the average accuracy observed in the readers' team. To this regard, Fig. 5 shows the average accuracy exhibited by the AI system (on the test set) and the team of 7 medical doctors involved in the study (on the 12 X-ray images). However, beyond appearances, the AI and humans accuracy did not significantly differ (two proportion test: $z = .37$, $p = .70$) as also the overlapping confidence intervals suggest. Moreover, both the observed accuracy of the AI and the humans is in accordance with other studies, of fracture diagnostic systems [26,32] and of human diagnostic errors [17,26]. Also, the seemingly lower accuracy of the AI with respect to the average human accuracy should not be taken as a sign of useless, or worse yet harmful, diagnostic support.

Indeed, as we analytically demonstrated in [10], the contribution of an AI system "adjuncted" into a medical team can significantly improve their performance even if its average accuracy is lower than the average accuracy of the medical team (see Fig. 12). Even AIs exhibiting an accuracy lower than 80% can then help achieve an average team accuracy that is several percentage points higher than the accuracy exhibited by the unsupported team (see [10] for more details).

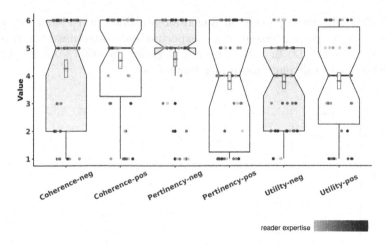

Fig. 9. Rater perceptions about the use of activation maps, stratified by class (neg vs pos).

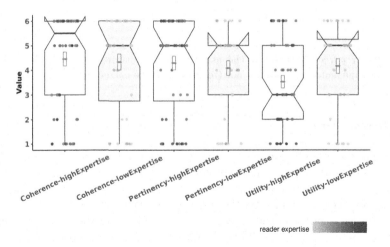

Fig. 10. Rater perceptions about the use of activation maps, stratified by the expertise level of the human readers.

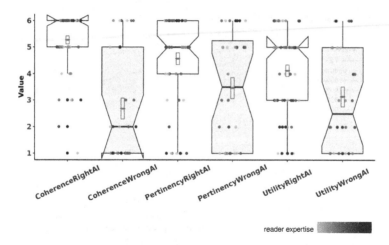

Fig. 11. Rater perceptions about the use of activation maps, stratified according to whether AI was correct or not (regardless of the agreement between human and AI).

Table 4. (Standardized) effect sizes, in terms of Cliff's Delta, for the inter-strata differences observed in the study (i.e., agreement vs no agreement, positive class vs negative class, low vs high expertise and AI correctness). Minimum sample sizes computed on T distribution to obtain a power of .8.

		Effect size [95% CI] (minimum sample size)
Agreement	*Coherence*	.44 [.1, .6] (165)
	Pertinency	.35 [.15, .65] (260)
	Utility	.25 [−.03, .5] (500)
Class	*Coherence*	.05 [−.18, .28] (12500)
	Pertinency	.2 [−.42, .05] (790)
	Utility	.02 [−.22, .25] (80,000)
Expertise	*Coherence*	−.06 [−.29, .17] (8,600)
	Pertinency	−.12 [−.35, .11] (1,100)
	Utility	.24 [−.05, .43] (550)
Correctness	*Coherence*	−.63 [−.8, −.36] (80)
	Pertinency	−.27 [−.5, 0] (430)
	Utility	−.29 [−.53, 0] (380)

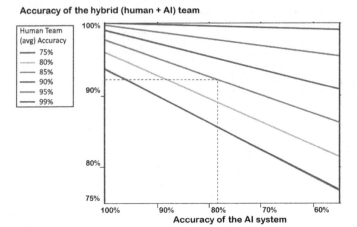

Fig. 12. Nomogram that allows to determine the theoretical accuracy that an hybrid team can achieve, once the AI accuracy and the average accuracy of the human (unsupported) team are known.

Coming back to our research questions, we will first consider the perceived dimensions, among which perceived utility, and then effectiveness. In particular, we considered three dimensions, to qualitatively gauge by means of semantic differentials, chosen to comprehensively represent the information needs of AM consumers: indeed, the Cronbach's alpha is low (.37) and this can hint at that the dimensions are orthogonal (and by excluding utility, which in Table 3 we see being moderately correlated with pertinency, correlation rises only slightly to .40, as expected).

In [36], the authors address similar aspects to compare saliency maps generated at different levels of abstractions by using the terms 'justification' and 'understanding': respectively, the extent maps "help explain why [a specific] diagnosis was put forward" (which we trace back to coherence); and the extent "the network appears to have a correct understanding of which parts of the X-ray are important for the diagnosis" (which we trace back to pertinency).

Thus, in regard to our first hypothesis (H_1), if AMs are useful aids, we cannot state if this is true in general. In fact, the AMs that we used in our study were not found to be significantly useful (nor useless) in supporting the detection of TL fractures by the sample of doctors involved: more precisely, on the 6-value scale, the median value of the perceived utility was 4, that is just above the middle point. Moreover, the proportion of judgments in the higher half of the scale (4, 5 and 6) was not significantly different from .5 (2-proportion test, p = .33); and even if we do not consider the undecided respondents (who rated either 3 or 4) the difference is still not significant (2-proportion test, p = .10). The null hypothesis could not thus be rejected at a confidence level of .95. We estimate, by setting the power to 80%, that slightly less than 200 judgments are required to identify a significant difference to this respect.

Interestingly though, AMs were instead found to be significantly pertinent (median value: 6, and 79% of judgments above middle point, for a p-value of .007 after a one-sample proportion binomial test), and also quite coherent (median value: 6, although not significantly so, 71%, p-value = .064). Moreover, the correlation between pertinency and utility (i.e., .4) was both moderate [37] and statistically significant, while the same does not hold between coherence and utility, even if also coherence was found to be high across the image set (median = 6). This suggests that readers found more useful a map that would highlight relevant areas of the radiological image, irrespective of the advice given by the machine. This could also suggest that the advice and the AM are seen as two distinct aids, which trigger different but complementary reasonings.

As shown in Sect. 3, we also investigated different effects under various conditions, namely: if AMs for positive (fracture) images were found to be more useful (or pertinent or coherent) than negative fractures, as shown in Fig. 9, that is whether perceptions were affected by types (i.e., classes) of images; if AMs associated with correct advice were found to be more useful (or pertinent or coherent) than those associated with wrong labels, as shown in Fig. 11, that is whether perceptions were affected by the AI accuracy; if readers with lower expertise found the AMs more useful (or pertinent or coherent) than the more expert readers did, as shown in Fig. 10, that is whether perceptions were affected by expertise; if agreeing with the AI advice changed how raters perceived the AM regarding utility (or pertinent or coherent), as shown in Fig. 8, that is whether perceptions were affected by human-AI agreement.

Thus, we observed that when readers agree with the AI, the perceived utility is higher, as if PAMs exert a more reassuring function, than an explanatory one. In the rare cases when the readers and AI disagreed (8%) *and* the latter one was right, median utility was pretty high (5) and the number of times the utility was appreciated was twice as high as the number of times it was considered totally unnecessary, thus suggesting that readers appreciated the corrective potential of the machine. It was also expected that the utility of activation maps was considered significantly higher when the AI advice was right, and hence contributing to a correct diagnosis, as well as the pertinency and coherence scores that collapse when AI gets the diagnosis wrong, suggesting that through PAMs the machine reveals its inadequacy, by emphasizing anatomical structures that have no clinical relevance. Conversely, a quite unexpected (and still relevant) finding regards the absence of significant differences in perceived pertinency and utility depending on the class type: one could expect that signs pointing out fractured structures would be considered more clear (at least for their confirmatory value) than the highlighting of areas that do *not* show anything that is clinically relevant, but this was not the case. This might suggest that PAMs can highlight areas where one could *expect* signs of fractures, having readers look at the right spots to rule them out. According to the observed effect size, a sample size of slightly less than 800 evaluations would be needed to detect any significant difference in this respect, which could be pragmatically relevant in diagnostic tasks where sensitivity, i.e., avoiding false negatives, is crucial. Finally, it is noteworthy to observe that the less expert the reader, the more useful the AMs were perceived to be, with a small-to-medium effect (effect size .22) for this condition. As shown in Fig. 10, this difference is significant and indicates the (perceived) educational potential of PAMs in improving novices' and trainees' skills.

In regard to our second, and perhaps more practically relevant, research question (H_2), that is whether AMs can improve decision accuracy: from a qualitative point of view (see Fig. 5) the confidence intervals for the pre-AI and post-AI groups overlap; that notwithstanding, we observe a significant difference in the accuracy of the two groups of approximately 10% (p-value = .021, effect size = .28), as reported in Table 2. With a power of 80%, the required sample size to detect a significant benefit is approximately of 90 ratings. The same effect can also be observed from the benefit diagram depicted in Fig. 6), where for almost all of the involved readers, the AI and XAI support had a beneficial effect in terms of diagnostic effectiveness. Thus, these results allow us to give a positive, albeit tentative answer to our second research question: providing human readers with both AI support and PAMs (as XAI support) could be effective in reducing the rate of diagnostic errors.

On the other hand, it is noteworthy to report that 9 diagnosis (out of 84) were changed and corrected by 4 physicians (out of 7) after having observed the activation map, thus receiving a relevant benefit by exploiting the AI+PAM support; however, in 1 case a physician changed his right diagnosis and made a mistake, see also Fig. 6 where one circle (i.e., physician) dropped in the red region of detrimental aid. On a more qualitative side, we also comment on a finding that emerged also in a previous study of ours on AI-supported ECG reading [31] and that we discussed upon in the informal talks we had with the X-ray readers after the experimental session. The almost null correlation between pertinency and coherence could indicate that activation maps, as well as any other methods that inform human decision makers about where the model "looked at", or what regions it considered more informative and predictive, in order to propose a specific advice, follow lines of reasoning that are hardly traceable back (or assimilable) to traditional medical methods and radiological semiotics. Even when images are not found to be coherent with the case at hand, the advice can nevertheless be correct, but low understandability of explanations could induce an unjustified wariness of the machine' advice, and lead to a form of algorithmic aversion [11,27] and a paradoxical form of automation bias, which we called *white-box paradox* [13]; this occurs when the explanation undermines trust and misleads the decision maker into rejecting an advice that is correct, so as to potentially misdiagnosing the case[7].

On one hand, these shortcomings have led some observers to critically reconsider the usefulness and meaningfulness of visual maps in radiology diagnosis (see, e.g., [18]); but, on the other hand, this also suggests the importance to train doctors in a sort of *machine semiotics* [24]. Although this "discipline" could be undermined by the current lack of standardization in radiological heatmaps and

---

[7] It is noteworthy that the white-box paradox can also mislead doctors when the advice is wrong, in that it can convince them of the opposite, as it has been reported in [7].

the fact that their manifestations could relevantly differ depending on the PAM methods adopted, pre-trained models, layers of choice, network architectures and diagnostic task, we believe it is worthy of further investigation to see if human readers can become proficient in reading seemingly misleading visual indications (the color shadows hinted at in the title), so as to exploit the additional information that activation maps and similar methods can provide them with to interpret the advice and the associated confidence scores in a more knowledgeable and informed manner: in so doing, these competent map readers would not only identify machine's blunders (like those reported in [5]) but also potentially valuable machine "intuitions"[8].

5   Conclusion

In this study, we focused on the effect of providing explainable AI support, in the form of Pixel Attribution Methods (PAM) and in particular Activation Maps (AM), as an aid to human readers in the task of TL fracture detection from x-rays. Through a pilot study in which we involved 7 doctors of varying expertise in the interpretation of a small, but carefully chosen, set of images with the aid of an AI system, we showed that, even though AMs were not considered particularly useful (nor useless) by the human readers (since no statistically significant difference was found in either directions), the readers did find higher value in AMs that were pertinent (i.e., which highlighted relevant areas of the diagnostic image, irrespective of the advice given by the machine and of its correctness) and, most importantly, their diagnostic accuracy significantly improved when given XAI support (as compared with their baseline diagnostic error rate).

We believe that these findings can be of interest for XAI scholars and practitioners, as they suggest that PAMs can improve diagnostic decision making and trigger different forms of reasoning from categorical (or probabilistic) output.

Moreover, with this pilot study, we provide interested researchers with effect size estimates that can be used as reference points for reliable power analysis (and sample size analysis) [6] that future studies investigating the role of PAM in medical decision making could adopt for a more efficient experimental design.

As the title of this work suggests, this contribution is only preparatory to further research. In this strand of research, interesting future work can regard: 1)

---

[8] It should be noted that "the implicit assumption [...] that the specific (diagnostic) message of the X-ray images resided inside them from the beginning, and that it is obscured either by technological or epistemological problems [is contestable as too naive]". Conversely it has been argued [33] that "the specific content of the images was shaped by the activities of X-ray workers within the context of medical developments of the time" when x-ray imaging was introduced in medical practice at the beginning of the 20th century. In these days, we could be witnessing the same phenomenon, in which radiologists, specialists, data scientists and ML developers could participatorily co-develop a machine semiotics for specific diagnostic tasks, if they are willing.

collecting readers' confidence on their decisions, both before and after the XAI support, to understand whether the main function of PAMs could regard more corroborating beliefs and making readers more self-confident in their final decision rather then explaining the machines advice (or, conversely, make them more doubtful and more likely to ask for a colleague's second opinion); 2) decoupling the effect of advice provision from the XAI effect, by collecting an intermediate decision (after the AI advice and before the XAI support in Fig. 1) and producing two benefit analyses (like the one presented in Fig. 6), comparing human accuracy before and after consulting the visual aid; this would allow to get better estimates of phenomena such PAM-induced automation bias and white-box paradoxes; 3) comparing pertinency and utility of either different methods of PAM production, or PAMs generated from data of different network layers (i.e. levels of abstraction), as done in [36]; 4) investigating the potential increase of inattentional blindness episodes and incidental findings [38] due to PAM-based support; 5) investigating PAM semiotics and opportunities to standardize PAM presentation for relevant radiological tasks, so as to learn how to interpret the color shadows [16] that these methods can enrich diagnostic images with, so as to make their reading and classification more efficient and accurate, and to positively contribute to the better understanding of the clinical case at hand by the treating physicians.

References

1. Aggarwal, R., et al.: Diagnostic accuracy of deep learning in medical imaging: a systematic review and meta-analysis. NPJ Digit. Med. **4**(1), 1–23 (2021)
2. Alqaraawi, A., Schuessler, M., Weiß, P., Costanza, E., Berthouze, N.: Evaluating saliency map explanations for convolutional neural networks: a user study. In: Proceedings of the 25th International Conference on Intelligent User Interfaces, pp. 275–285 (2020)
3. Arun, N., et al.: Assessing the trustworthiness of saliency maps for localizing abnormalities in medical imaging. Radiol. Artif. Intell. **3**(6), e200267 (2021)
4. Ayhan, M.S., et al.: Clinical validation of saliency maps for understanding deep neural networks in ophthalmology. Med. Image Anal. **77**, 102364 (2022)
5. Badgeley, M.A., et al.: Deep learning predicts hip fracture using confounding patient and healthcare variables. NPJ Digit. Med. **2**(1), 1–10 (2019)
6. Balki, I., et al.: Sample-size determination methodologies for machine learning in medical imaging research: a systematic review. Can. Assoc. Radiol. J. **70**(4), 344–353 (2019)
7. Bansal, G., et al.: Does the whole exceed its parts? The effect of AI explanations on complementary team performance. In: Proceedings of the 2021 CHI Conference on Human Factors in Computing Systems, pp. 1–16 (2021)
8. Becherer, N., Pecarina, J., Nykl, S., Hopkinson, K.: Improving optimization of convolutional neural networks through parameter fine-tuning. Neural Comput. Appl. **31**(8), 3469–3479 (2017). https://doi.org/10.1007/s00521-017-3285-0
9. Brynjolfsson, E., Mitchell, T.: What can machine learning do? Workforce implications. Science **358**(6370), 1530–1534 (2017)

10. Cabitza, F., Campagner, A., Del Zotti, F., Ravizza, A., Sternini, F.: All you need is higher accuracy? On the quest for minimum acceptable accuracy for medical artificial intelligence. In: e-Health 2020, Proceedings of the 12th International Conference on e-Health, pp. 159–166 (2020)
11. Cabitza, F.: Biases affecting human decision making in AI-supported second opinion settings. In: Torra, V., Narukawa, Y., Pasi, G., Viviani, M. (eds.) MDAI 2019. LNCS (LNAI), vol. 11676, pp. 283–294. Springer, Cham (2019). https://doi.org/10.1007/978-3-030-26773-5_25
12. Cabitza, F., Campagner, A., Cavosi, V.: Assessing the impact of medical AI: a survey of physicians' perceptions. In: 2021 5th International Conference on Medical and Health Informatics, pp. 225–231 (2021)
13. Cabitza, F., Campagner, A., Simone, C.: The need to move away from agential-AI: empirical investigations, useful concepts and open issues. Int. J. Hum Comput Stud. **155**, 102696 (2021)
14. Chinn, S.: A simple method for converting an odds ratio to effect size for use in meta-analysis. Stat. Med. **19**(22), 3127–3131 (2000)
15. Chlap, P., Min, H., Vandenberg, N., Dowling, J., Holloway, L., Haworth, A.: A review of medical image data augmentation techniques for deep learning applications. J. Med. Imaging Radiat. Oncol. **65**(5), 545–563 (2021)
16. Croskerry, P., Cosby, K., Graber, M.L., Singh, H.: Diagnosis: Interpreting the Shadows. CRC Press, Boca Raton (2017)
17. Delmas, P.D., et al.: Underdiagnosis of vertebral fractures is a worldwide problem: the IMPACT study. J. Bone Miner. Res. **20**(4), 557–563 (2005)
18. Ghassemi, M., Oakden-Rayner, L., Beam, A.L.: The false hope of current approaches to explainable artificial intelligence in health care. Lancet Digit. Health **3**(11), e745–e750 (2021)
19. Han, T., et al.: Advancing diagnostic performance and clinical usability of neural networks via adversarial training and dual batch normalization. Nat. Commun. **12**(1), 1–11 (2021)
20. Handelman, G.S., et al.: Peering into the black box of artificial intelligence: evaluation metrics of machine learning methods. Am. J. Roentgenol. **212**(1), 38–43 (2019)
21. Holzinger, A., Saranti, A., Molnar, C., Biecek, P., Samek, W.: Explainable AI methods - a brief overview. In: Holzinger, A., Goebel, R., Fong, R., Moon, T., Müller, K.R., Samek, W. (eds.) xxAI 2020. LNCS, vol. 13200, pp. 13–38. Springer, Cham (2022). https://doi.org/10.1007/978-3-031-04083-2_2
22. Holzinger, A.T., Muller, H.: Toward human-AI interfaces to support explainability and causability in medical AI. Computer **54**(10), 78–86 (2021)
23. Hwang, E.J., et al.: Development and validation of a deep learning-based automated detection algorithm for major thoracic diseases on chest radiographs. JAMA Netw. Open **2**(3), e191095 (2019)
24. Jha, S., Topol, E.J.: Adapting to artificial intelligence: radiologists and pathologists as information specialists. JAMA **316**(22), 2353–2354 (2016)
25. Ke, A., Ellsworth, W., Banerjee, O., Ng, A.Y., Rajpurkar, P.: CheXtransfer: performance and parameter efficiency of ImageNet models for chest X-Ray interpretation. CoRR abs/2101.06871 (2021)
26. Liu, X., et al.: A comparison of deep learning performance against health-care professionals in detecting diseases from medical imaging: a systematic review and meta-analysis. Lancet Digit. Health **1**(6), e271–e297 (2019)
27. Lohoff, L., Rühr, A.: Introducing (machine) learning ability as antecedent of trust in intelligent systems. In: ECIS 2021 Research Papers, vol. 23 (2021)

28. Ludewig, E., Richter, A., Frame, M.: Diagnostic imaging-evaluating image quality using visual grading characteristic (VGC) analysis. Vet. Res. Commun. **34**(5), 473–479 (2010). https://doi.org/10.1007/s11259-010-9413-2

29. Lyell, D., Coiera, E.: Automation bias and verification complexity: a systematic review. J. Am. Med. Inform. Assoc. **24**(2), 423–431 (2017)

30. Nandi, A., Pal, A.K.: Detailing image interpretability methods. In: Nandi, A., Pal, A.K. (eds.) Interpreting Machine Learning Models, pp. 271–293. Springer, Cham (2022). https://doi.org/10.1007/978-1-4842-7802-4_12

31. Neves, I., et al.: Interpretable heartbeat classification using local model-agnostic explanations on ECGs. Comput. Biol. Med. **133**, 104393 (2021)

32. Olczak, J., et al.: Artificial intelligence for analyzing orthopedic trauma radiographs. Acta Orthop. **88**(6), 581–586 (2017)

33. Pasveer, B.: Knowledge of shadows: the introduction of X-ray images in medicine. Sociol. Health Illn. **11**(4), 360–381 (1989)

34. Ribeiro, M.T., Singh, S., Guestrin, C.: Model-agnostic interpretability of machine learning. arXiv preprint arXiv:1606.05386 (2016)

35. Simonyan, K., Vedaldi, A., Zisserman, A.: Deep inside convolutional networks: visualising image classification models and saliency maps. In: Workshop at International Conference on Learning Representations (2014)

36. Spinks, G., Moens, M.F.: Justifying diagnosis decisions by deep neural networks. J. Biomed. Inform. **96**, 103248 (2019)

37. Taylor, R.: Interpretation of the correlation coefficient: a basic review. J. Diagn. Med. Sonogr. **6**(1), 35–39 (1990)

38. Tiulpin, A., Thevenot, J., Rahtu, E., Lehenkari, P., Saarakkala, S.: Automatic knee osteoarthritis diagnosis from plain radiographs: a deep learning-based approach. Sci. Rep. **8**(1), 1–10 (2018)

39. Tschandl, P., et al.: Human-computer collaboration for skin cancer recognition. Nat. Med. **26**(8), 1229–1234 (2020)

40. Xie, S., Girshick, R., Dollár, P., Tu, Z., He, K.: Aggregated residual transformations for deep neural networks. In: Proceedings of the IEEE Conference on Computer Vision and Pattern Recognition, pp. 1492–1500 (2017)

41. Yang, S., Yin, B., Cao, W., Feng, C., Fan, G., He, S.: Diagnostic accuracy of deep learning in orthopaedic fractures: a systematic review and meta-analysis. Clin. Radiol. **75**(9), 713-e17 (2020)

42. Zhou, B., Khosla, A., Lapedriza, A., Oliva, A., Torralba, A.: Learning deep features for discriminative localization. In: Proceedings of the IEEE Conference on Computer Vision and Pattern Recognition (CVPR), June 2016

Effects of Fairness and Explanation on Trust in Ethical AI

Alessa Angerschmid[2], Kevin Theuermann[3], Andreas Holzinger[2,3,4,5],
Fang Chen[1], and Jianlong Zhou[1,2(✉)]

[1] Human-Centered AI Lab, University of Technology Sydney, Sydney, Australia
{fang.chen,jianlong.zhou}@uts.edu.au
[2] Human-Centered AI Lab, Medical University Graz, Graz, Austria
{alessa.angerschmid,andreas.holzinger}@human-centered.ai
[3] Graz University of Technology, Graz, Austria
kevin.theuermann@egiz.gv.at
[4] University of Natural Resources and Life Sciences Vienna, Vienna, Austria
[5] xAI Lab, Alberta Machine Intelligence Institute, Edmonton, Canada

Abstract. AI ethics has been a much discussed topic in recent years. Fairness and explainability are two important ethical principles for trustworthy AI. In this paper, the impact of AI explainability and fairness on user trust in AI-assisted decisions is investigated. For this purpose, a user study was conducted simulating AI-assisted decision making in a health insurance scenario. The study results demonstrated that fairness only affects user trust when the fairness level is low, with a low fairness level reducing user trust. However, adding explanations helped users increase their trust in AI-assisted decision making. The results show that the use of AI explanations and fairness statements in AI applications is complex: we need to consider not only the type of explanations, but also the level of fairness introduced. This is a strong motivation for further work.

Keywords: AI explanation · AI fairness · Trust · AI ethics

1 Introduction

Artificial Intelligence (AI) informed decision-making is claimed to lead to faster and better decision outcomes, and has been increasingly used in our society from the decision-making of daily lives such as recommending movies and books to making more critical decisions such as medical diagnoses, credit risk prediction, and shortlisting talents in recruitment. Among such AI-informed decision-making tasks, trust and perception of fairness have been found to be critical factors driving human behaviour in human-machine interactions [40,48]. Because of the black-box nature of AI models that make it hard for users to understand why a decision is made or how the data is processed for the decision-making [7,44,46], trustworthy AI has experienced a significant surge in interest from the research community to various application domains, especially in high

A. Holzinger et al. (Eds.): CD-MAKE 2022, LNCS 13480, pp. 51–67, 2022.
https://doi.org/10.1007/978-3-031-14463-9_4

stake domains which usually require testing and verification for reasonability by domain experts not only for safety but also for legal reasons [19,36,42,43]. Explanation and trust are common partners in everyday life, and extensive research has investigated the relations between AI explanations and trust from different perspectives ranging from philosophical to qualitative and quantitative dimensions [30]. For instance, Zhou et al. [45] showed that the explanation of influences of training data points on predictions significantly increased the user trust in predictions. Alam and Mueller [3] investigated the roles of explanations in AI-informed decision-making in medical diagnosis scenarios. The results show that visual and example-based explanations integrated with rationales had a significantly better impact on patient satisfaction and trust than no explanations, or with text-based rationales alone. The previous studies that empirically tested the importance of explanations to users in various fields consistently showed that explanations significantly increase user trust. Furthermore, with the advancement of AI explanation development, different explanation approaches such as local and global explanations, and feature importance-based and example-based explanations are proposed [44]. As a result, besides the explanation presentation styles such as visualisation and text [3], it is also critical to understand how different explanation approaches affect user trust in AI-informed decision-making. In addition, Edwards [10] stated that the main challenge for AI-informed decision-making is to know, whether an explanation that seems valid is accurate. This information is also needed to ensure transparency and accountability of the decision.

Besides, the data used to train machine learning models are often historical records or samples of events. They are usually not a precise description of events and conceal discrimination with sparse details, which are very difficult to identify. AI models are also imperfect abstractions of reality because of their statistical nature. All these lead to imminent imprecision and discrimination (bias) associated with AI. As a result, the investigation of fairness in AI has been becoming an indispensable component for responsible socio-technical AI systems in various decision-making tasks such as allocation of social benefits, hiring, and criminal justice [5,12]. And extensive research focuses on fairness definitions and unfairness quantification. Furthermore, human's perceived fairness (perception of fairness) plays an important role in AI-informed decision-making since AI is often used by humans and/or for human-related decision-making [35]. Duan et al. [9] argued that AI-informed decision-making can help users make better decisions. Furthermore, the authors propose that AI-informed decisions will be mostly accepted by humans, when used as a support tool. Considerable research on perceived fairness has evidenced its links to trust such as in management and organizations [25,32].

In addition, Dodge et al. [8] argued that AI explanations can also provide an effective interface for the human-in-the-loop, enabling people to identify and address fairness issues. They also demonstrated the need of providing different explanation types for different fairness issues. All these demonstrate the inter-connection relations between explanation and fairness in AI-informed decision-making. Despite the proliferation of investigations of effects of AI explanation on trust and perception of fairness, or effects of introduced fairness on trust and perception of fairness, it is critical to understand how AI explanation and introduced fairness concurrently affect user trust since AI explanation and fairness are common partners in AI-informed decision-making. Therefore, in this work, we aim to investigate the effects of both AI explanation and introduced fairness on user trust.

Our aim in this paper is to understand user trust under both different types of AI explanations and different levels of introduced fairness. In particular, two commonly used explanation approaches of example-based explanations and feature importance-based explanations are introduced into the AI-informed decision-making pipeline under different levels of introduced fairness. We aim to discover, whether AI explanations and introduced fairness with fairness statement benefit human's trust and if so, which explanation type or fairness level benefits more than others. A user study is designed by simulating AI-informed decision-making in health insurance through manipulating AI explanations and introduced fairness levels. Statistical analyses are performed to understand effects of AI explanations and introduced fairness on trust.

2   Related Work

2.1   AI Fairness and Trust

User trust in algorithmic decision-making has been investigated from different perspectives. Zhou et al. [41,47] argued that communicating user trust benefits the evaluation of effectiveness of machine learning approaches. Kizilcec [23] found that appropriate transparency of algorithms by explanation benefited the user trust. Other empirical studies found the effects of confidence score, model accuracy and users' experience of system performance on user trust [38,39,43].

Understanding relations between fairness and trust is nontrivial in the social interaction context such as marketing and services. Roy et al. [32] showed that perceptions of fair treatment on customers play a positive role in engendering trust in the banking context. Earle and Siegrist [11] found that the issue importance affected the relations between fairness and trust. They showed that procedural fairness did not affect trust when the issue importance was high, while procedural fairness had moderate effects on trust when issue importance was low. Nikbin et al. [28] showed that perceived service fairness had a significant effect on trust, and confirmed the mediating role of satisfaction and trust in the relationship between perceived service fairness and behavioural intention.

Kasinidou et al. [21] investigated the perception of fairness in algorithmic decision-making and found that people's perception of a system's decision as 'not fair' is affecting the participants' trust in the system. Shin's investigations [33,34] showed that perception of fairness had a positive effect on trust in an algorithmic decision-making system such as recommendations. Zhou et al. [48] got similar conclusions that introduced fairness is positively related to user trust in AI-informed decision-making, i.e. the high level of introduced fairness resulted in the high level of user trust.

These previous works motivate us to further investigate how multiple factors such as AI fairness and AI explanation together affect user trust in AI-informed decision-making.

2.2   AI Explanation and Trust

Explainability is indispensable to foster user trust in AI systems, particularly in sensible application domains. Holzinger et al. [17] introduced the concept of causability and demonstrated the importance of causability in AI explanations [18], [20]. Shin [34] used causability as an antecedent of explainability to examine their relations to trust, where causability gives the justification for what and how AI results should be explained to determine the relative importance of the properties of explainability. Shin argued that the inclusion of causability and explanations would help to increase trust and help users to assess the quality of explanations, e.g. with the Systems Causability Scale [15].

The influence of training data points on predictions is one of typical AI explanation approaches [24]. Zhou et al. [45] investigated the effects of influence on user trust and found that the presentation of influences of training data points significantly increased the user trust in predictions, but only for training data points with higher influence values under the high model performance condition. Papenmerer et al. [29] investigated the effects of model accuracy and explanation fidelity, and found that model accuracy is more important for user trust than explainability. When adding nonsensical explanations, explanations can potentially harm trust. Larasati et al. [26] investigated the effects of different styles of textual explanations on user trust in an AI medical support scenario. Four textual styles of explanations including contrastive, general, truthful, and thorough were investigated. It was found that contrastive and thorough explanations produced higher user trust scores compared to general explanation style, and truthful explanation showed no difference compared to the rest of the explanations. Wang et al. [37] compared different explanation types such as feature importance, feature contribution, nearest neighbour, and counterfactual explanation from three perspectives of improving people's understanding of the AI model, helping people recognize the model uncertainty, and supporting people's calibrated trust in the model. They highlighted the importance of selecting different AI explanation types in designing the most suitable AI methods for a specific decision-making task.

These findings confirmed the impact of explanation and its types on users trust in AI systems. In this paper, we investigate how different explanation types such as example-based and feature importance-based explanations affect user trust in AI-informed decision-making by considering the effects of AI fairness concurrently.

3   Method

3.1   Case Study

This research selected the health insurance decision-making as a case study for AI-informed decision-making. The decision of the monthly payment rate is a significant step in the health insurance decision-making process. It is often based on information about the age and lifestyle of applicants. For example, a 20-year old applicant, who does neither smoke nor drink and works out frequently, is less likely to require extensive medical care. Therefore, the insurance company most likely decides to put this applicant into the lower payment class with a lower monthly rate for insurance. The insurance will increase with the age of the applicant and pre-known illnesses or previous hospital admissions. AI is used to get faster results for these decisions while enhancing customer experience since AI allows the automatic calculation of key factors and guarantees an equal procedure for every applicant [22]. This decision-making process is simulated in the study by creating fake personas with different attributes and showing their prediction of a monthly insurance rate. The simulation determines the monthly rate based on the factors of age, gender, physical activities, as well as drinking and smoking habits.

The advisory organ of the EU on GDPR, Article 29 Working Party, added a guideline [4] with detailed descriptions and requirements for profiling and automated decision-making. They also state that transparency is a fundamental requirement for the GDPR. Two explanation approaches of example-based explanation and feature importance-based explanation with fairness conditions are introduced into the decision-making process to meet requirements for AI-informed decision-making by GDPR [2] and other EU regulations and guidelines [1,13].

3.2   Explanations

This study aims to understand how AI explanations affect user trust in decision-making. Two types of explanations are investigated in the experiment:

- Example-based explanation. Example-based explanation methods select particular instances of the dataset as similar or adverse examples to explain the behaviour of AI models. Examples are commonly used as effective explanations between humans for explaining complex concepts [31]. Example-based AI explanations have been used to help users gain intuition for AI that are otherwise difficult to explain through algorithms [6]. In this study, both similar and adverse examples are introduced into tasks to investigate user responses.
- Feature importance-based explanation. Feature importance is one of the most common AI explanations [43]. It is a measure of the individual contribution of a feature to AI outcomes. For example, a feature is "important" if changing its values increases the model error, as the model relied on the feature for the prediction. A feature is "unimportant" if changing its values leaves the model error unchanged. In this study, the importance of each feature on a specific AI prediction is presented to analyse user responses.

In addition, tasks without any specific explanations (called control condition in this study) are also introduced to see if the explanation is indeed helpful or provides a better understanding of the decision-making process.

3.3   Fairness

In this study, gender is used as a protected attribute in fairness investigations. Two levels of introduced fairness are used in the study:

– Low level of fairness. At this level, the decisions are completely biased to one gender. In this study, statements such as "male and female customers having a similar personal profile did receive a different insurance rate: male customers pay € 30 more than female customers." are used to show the least fairness of the AI system.
– High level of fairness. At this level, both males and females are fairly treated in the decision-making. In this study, statements such as "male and female having a similar personal profile were treated similarly" are used to show the most fairness of the AI system.

In addition, tasks without any fairness information (called control condition in this study) are also introduced to investigate the difference of user responses in decision-making with and without the fairness information.

3.4   Task Design

Table 1. Experiment task conditions.

		Fairness		
		Control	Low	High
Explanation	Control	T	T	T
	Example-based	T	T	T
	Feature importance	T	T	T

According to the application scenario as described above, we investigated the decisions made by participants under both explanation and fairness conditions. In each task, AI models automatically recommended a decision based on the use case. Participants were then asked to accept or reject this decision under the presentation of different explanation and fairness conditions (3 explanation conditions by 3 fairness conditions, see Table 1). Figure 1 shows an example of the use case statement, the decision recommended by AI models, as well as the presentation of fairness and explanation conditions. After the decision-making, different questions are asked to rate users' trust in AI models. All together, each participant conducts 9 tasks (3 explanation conditions × 3 fairness conditions = 9 tasks). The orders of tasks are randomised to avoid any bias introduced. In addition, 2 training tasks are conducted by each participant before formal tasks.

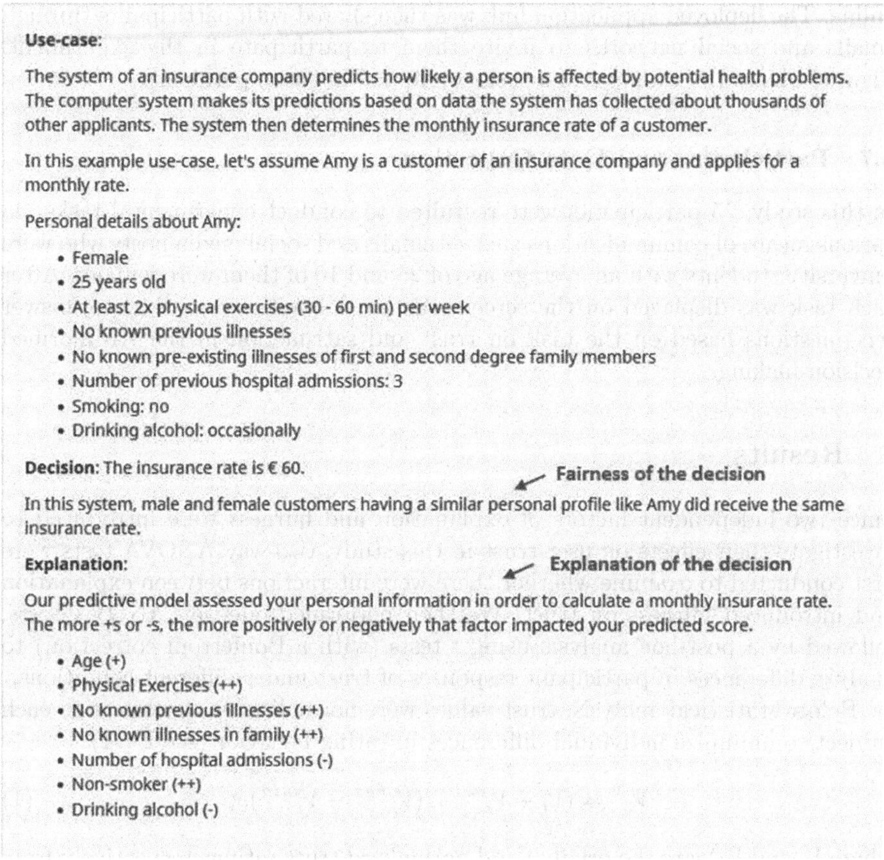

Use-case:

The system of an insurance company predicts how likely a person is affected by potential health problems. The computer system makes its predictions based on data the system has collected about thousands of other applicants. The system then determines the monthly insurance rate of a customer.

In this example use-case, let's assume Amy is a customer of an insurance company and applies for a monthly rate.

Personal details about Amy:

- Female
- 25 years old
- At least 2x physical exercises (30 - 60 min) per week
- No known previous illnesses
- No known pre-existing illnesses of first and second degree family members
- Number of previous hospital admissions: 3
- Smoking: no
- Drinking alcohol: occasionally

Decision: The insurance rate is € 60.   ◄─ **Fairness of the decision**

In this system, male and female customers having a similar personal profile like Amy did receive the same insurance rate.

Explanation: ◄─ **Explanation of the decision**

Our predictive model assessed your personal information in order to calculate a monthly insurance rate. The more +s or -s, the more positively or negatively that factor impacted your predicted score.

- Age (+)
- Physical Exercises (++)
- No known previous illnesses (++)
- No known illnesses in family (++)
- Number of hospital admissions (-)
- Non-smoker (++)
- Drinking alcohol (-)

Fig. 1. An example of the experiment

3.5 Trust Scales

In this study, trust is assessed with six items using self-report scales following approaches in [27]. The scale is on a 4-point Likert-type response scale ranging from 1 (strongly disagree), 2 (disagree), 3 (agree), to 4 (strongly agree).

- I believe the system is a competent performer.
- I trust the system.
- I have confidence in the advice given by the system.
- I can depend on the system.
- I can rely on the system to behave in consistent ways.
- I can rely on the system to do its best every time I take its advice.

3.6 Experiment Setup

Due to social distancing restrictions and lockdown policies during the COVID-19 pandemic, this experiment was implemented and deployed on the cloud server online. The deployed application link was then shared with participants through emails and social networks to invite them to participate in the experiment. Figure 1 shows the example of a task conducted in the experiment.

3.7 Participants and Data Collection

In this study, 25 participants were recruited to conduct experimental tasks via various means of communications such as emails and social media posts who were university students with an average age of 26 and 10 of them were females. After each task was displayed on the screen, the participants were asked to answer ten questions based on the task on trust and satisfaction in the AI-informed decision-making.

4 Results

Since two independent factors of explanation and fairness were introduced to investigate their effects on user trust in this study, two-way ANOVA tests were first conducted to examine whether there were interactions between explanation and introduced fairness on trust. We then performed one-way ANOVA tests, followed by a post-hoc analysis using t-tests (with a Bonferroni correction) to analyze differences in participant responses of trust under different conditions.

Before statistical analysis, trust values were normalised with respect to each subject to minimize individual differences in rating behavior (see Eq. 1):

$$V_i^N = (V_i - V_i^{min})/(V_i^{max} - V_i^{min}) \tag{1}$$

where V_i and V_i^N are the original and normalised trust ratings respectively from the participant i, V_i^{min} and V_i^{max} are the minimum and maximum of trust values respectively from the participant i in all of his/her tasks.

4.1 Effects of Fairness on Trust

Figure 2 shows normalised trust values over the introduced fairness conditions. A one-way ANOVA test was performed to compare the effect of introduced fairness on user trust. The one-way ANOVA test found that there were statistically significant differences in user trust among three introduced fairness conditions $F(2, 222) = 8.446, p < .000$. A further post-hoc comparison with t-tests (with a Bonferroni correction under a significance level set at $\alpha < .017$) was conducted to find pair-wise differences in user trust between three fairness conditions. The adjusted significance alpha level of .017 was calculated by dividing the original alpha of .05 by 3, based on the fact that we had three fairness conditions. It was found that participants had a statistically significant high level of trust under the high level of fairness compared to the low fairness condition ($t = 4.185, p < .000$). Moreover, it was found that participants also had a statistically significant higher level of trust under the control condition (no fairness information presented)

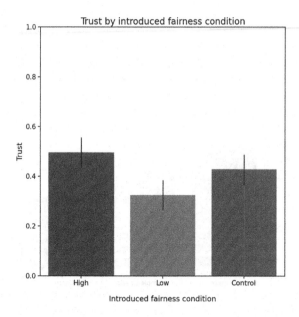

Fig. 2. User trust under introduced fairness conditions

than that under the low fairness condition ($t = 2.433, p < .016$). However, there was not a statistically significant difference found in user trust between the introduced high fairness condition and control condition ($t = 1.602, p < .111$).

These findings imply that the introduced fairness condition did affect user trust in AI-information decision-making only under the low fairness condition, where introduced fairness decreased user trust in AI-informed decision-making.

4.2 Effects of Explanation on Trust

Figure 3 shows normalised trust values over various explanation conditions. A one-way ANOVA test revealed statistically significant differences in user trust under different explanation types $F(2, 222) = 11.226, p < .000$. Then post-hoc tests with the aforementioned Bonferroni correction were conducted. It was found that participants had statistically significant lower level of trust under the control condition (no explanation presented) than that under the feature importance-based explanation ($t = 4.645, p < .000$) and example-based explanation ($t = 2.455, p < .015$) respectively. There was not a significant difference in user trust between feature importance-based explanation and example-based explanation ($t = 2.329, p < .021$).

The results showed that explanations did help users increase their trust significantly in AI-informed decision-making, but different explanation types did not show differences in affecting user trust.

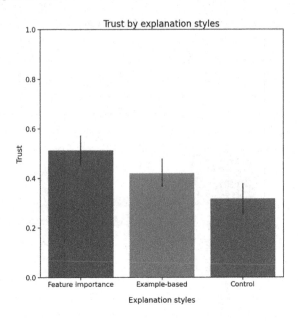

Fig. 3. User trust under explanation types

4.3   Effects of Fairness and Explanation on Trust

This subsection further analyses the effects of fairness on trust under different given explanation types, and the effects of explanation on trust under different given fairness levels.

Effects of Fairness on Trust Under Example-Based Explanations. Figure 4 shows normalised trust values over various fairness conditions under the example-based explanation condition. A one-way ANOVA test was conducted to compare the effect of introduced fairness on user trust under the example-based explanation. The test found a statistically significant difference in trust between introduced fairness levels, $F(2, 72) = 8.146, p < .001$. Further post-hoc t-tests (with Bonferroni correction) were then conducted to find differences in trust among different fairness levels. Participants showed a significant higher trust level under high introduced fairness than that under the low introduced fairness level ($t = 3.887, p < .000$). Moreover, user trust was significantly higher under the control condition (no fairness information presented) than that under the low introduced fairness level ($t = 3.266, p < .002$). However, there was not a significant difference in trust between the high introduced fairness and the control condition ($t = .436, p < .665$).

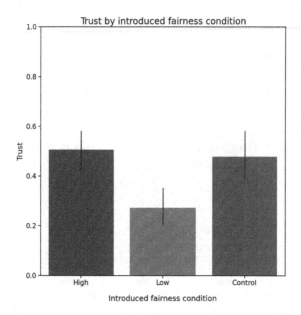

Fig. 4. Effects of fairness on user trust under the example-based explanation

The results showed that under the example-based explanation condition, the low level of fairness statement significantly decreased the user trust in decision-making but the high level of fairness statement did not affect user trust.

Effects of Fairness on Trust Under Feature Importance-Based Explanations. Figure 5 shows the normalized trust levels for introduced fairness levels under feature importance-based explanation. A one-way ANOVA test found no significant differences in trust in different introduced fairness levels under the feature importance-based explanation, $F(2, 72) = 2.353, p < .102$.

From the results, we can see that under the feature importance-based explanation condition, no fairness information seems to influence the user's trust.

Effects of Explanation on Trust Under Low Level Introduced Fairness. Figure 6 shows the normalized trust values with different explanation types under low-level introduced fairness. A one-way ANOVA test found statistical significant differences in trust among explanation types under the low level of introduced fairness, $F(2, 72) = 3.307, p < .042$. The further t-test found that participants showed no significant higher level of trust under feature importance-based explanation than that under the control condition (no explanation presented) $(t = 2.248, p < .046)$. Moreover, there were neither significant differences found in user trust between the control condition and example-based explanation $(t = .035, p < .972)$, nor between the two explanation types $(t = 2.296, p < .026)$.

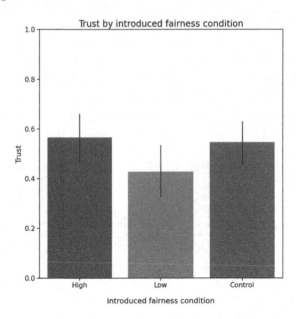

Fig. 5. Effects of fairness under feature importance-based explanations on user trust

Therefore, we can say that under the low level of introduced fairness, neither explanation type did significantly increase user's trust in the decision-making process.

Effects of Explanation on Trust Under High Level of Introduced Fairness. Figure 7 shows the normalised trust values in different explanation types under the high level of introduced fairness. A one-way ANOVA test revealed no statistical significant differences in user trust among explanation types under the high level of introduced fairness, $F(2, 72) = 2.369, p < .101$.

The explanation type under the high level of introduced fairness had no influence on user's trust.

5   Discussion

Explanation and fairness, along with robustness [14, 16], are among the indispensable components of AI-based decision making for trustworthy AI. AI-informed decision-making and automated aids have been becoming much popular with the advent of new AI-based intelligent applications. Therefore, we opted to study the effects of both AI explanations and fairness on human-AI trust in a specialised AI-informed decision-making scenario.

The study found that the fairness statement in the scenario did affect user trust in AI-information decision-making only under the low level of fairness condition, where the low-level fairness statement decreased user trust in AI-informed

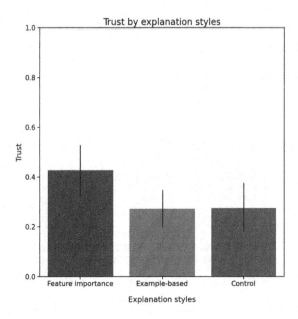

Fig. 6. Effects of explanation on user trust under low introduced fairness level

decision-making. However, the addition of explanations helped users increase their trust significantly in AI-informed decision-making, and different explanation types did not show differences in affecting user trust. We then drilled down into the effects on trust under specific conditions. From the explanation's perspective, it was found that under the example-based explanation condition, the low level of fairness statement significantly decreased the user trust in decision-making but the high level of fairness statement did not affect user trust. Nevertheless, the level of fairness under the feature importance-based explanation condition did not show any impact on the user trust. Furthermore, from the introduced fairness' perspective, it revealed that under the low level of introduced fairness, neither explanation type significantly increased user trust in decision-making. The high level of introduced fairness, on the other hand, showed no effects at all on user trust. It also implies that the introduced fairness levels did not affect user trust too much.

These findings suggest that the deployment of AI explanation and fairness statements in real-world applications is complex: we need to not only consider explanation types, but also levels of introduced fairness. In order to maximise user trust in AI-informed decision-making, the explanation types and the level of fairness statement can be adjusted in the user interface of intelligent applications.

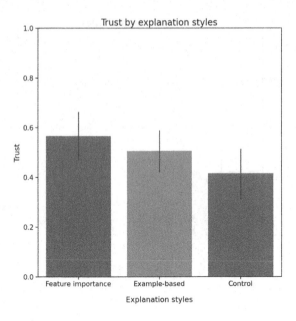

Fig. 7. Effects of explanation on trust under high introduced fairness level

6   Conclusion and Future Work

This paper investigated the effects of introduced fairness and explanation on user trust in AI-informed decision-making. A user study by simulating AI-informed decision-making through manipulating AI explanations and fairness levels found that the introduced fairness affected user trust in AI-informed decision-making only under the low level of fairness condition. It was also found that the AI explanations increased user trust in AI-informed decision-making, and different explanation types did not show differences in affecting user trust. The future work of this study will focus on the effects of explanation and introduced fairness on perception of fairness as well as accountability of AI.

Acknowledgements. This work does not raise any ethical issues. Parts of this work have been funded by the Austrian Science Fund (FWF), Project: P-32554 explainable Artificial Intelligence; and by the Australian UTS STEM-HASS Strategic Research Fund 2021.

References

1. European parliament resolution of 20 October 2020 with recommendations to the commission on a framework of ethical aspects of artificial intelligence, robotics and related technologies, 2020/2012(INL). https://eur-lex.europa.eu/legal-content/EN/TXT/?uri=CELEX:52020IP0275. Accessed 19 Jan 2022

2. Regulation (EU) 2016/679 of the European parliament and of the council of 27 April 2016 on the protection of natural persons with regard to the processing of personal data and on the free movement of such data, and repealing directive 95/46/EC (general data protection regulation) (2016). https://eur-lex.europa.eu/legal-content/EN/TXT/PDF/?uri=CELEX:02016R0679-20160504

3. Alam, L., Mueller, S.: Examining the effect of explanation on satisfaction and trust in AI diagnostic systems. BMC Med. Inform. Decis. Mak. **21**(1), 178 (2021). https://doi.org/10.1186/s12911-021-01542-6

4. Article 29 Working Party: Guidelines on automated individual decision-making and profiling for the purposes of regulation 2016/679. https://ec.europa.eu/newsroom/article29/items/612053/en. Accessed 19 Jan 2022

5. Berk, R., Heidari, H., Jabbari, S., Kearns, M., Roth, A.: Fairness in criminal justice risk assessments: the state of the art. Sociol. Methods Res. **50**, 0049124118782533 (2018)

6. Cai, C.J., Jongejan, J., Holbrook, J.: The effects of example-based explanations in a machine learning interface. In: Proceedings of the 24th International Conference on Intelligent User Interfaces, IUI 2019, pp. 258–262 (2019). https://doi.org/10.1145/3301275.3302289

7. Castelvecchi, D.: Can we open the black box of AI? Nature News **538**(7623), 20 (2016)

8. Dodge, J., Liao, Q.V., Zhang, Y., Bellamy, R.K.E., Dugan, C.: Explaining models: an empirical study of how explanations impact fairness judgment. In: Proceedings of the 24th International Conference on Intelligent User Interfaces, IUI 2019, pp. 275–285 (2019)

9. Duan, Y., Edwards, J.S., Dwivedi, Y.K.: Artificial intelligence for decision making in the era of big data - evolution, challenges and research agenda. Int. J. Inf. Manag. **48**, 63–71 (2019). https://doi.org/10.1016/j.ijinfomgt.2019.01.021

10. Dwivedi, Y.K., et al.: Artificial intelligence (AI): multidisciplinary perspectives on emerging challenges, opportunities, and agenda for research, practice and policy. Int. J. Inf. Manag. **57**, 101994 (2021). https://doi.org/10.1016/j.ijinfomgt.2019.08.002

11. Earle, T.C., Siegrist, M.: On the relation between trust and fairness in environmental risk management. Risk Anal. **28**(5), 1395–1414 (2008)

12. Feldman, M., Friedler, S.A., Moeller, J., Scheidegger, C., Venkatasubramanian, S.: Certifying and removing disparate impact. In: Proceedings of KDD 2015, pp. 259–268 (2015)

13. High-Level Export Group on Artificial Intelligence: Ethics guidelines for trustworthy AI. https://digital-strategy.ec.europa.eu/en/library/ethics-guidelines-trustworthy-ai. Accessed 19 Jan 2022

14. Holzinger, A.: The next frontier: AI we can really trust. In: Kamp, M. (ed.) ECML PKDD 2021. CCIS, vol. 1524, pp. 1–14. Springer, Cham (2021). https://doi.org/10.1007/978-3-030-93736-2_33

15. Holzinger, A., Carrington, A., Müller, H.: Measuring the quality of explanations: the system causability scale (SCS). KI - Künstliche Intelligenz **34**(2), 193–198 (2020). https://doi.org/10.1007/s13218-020-00636-z

16. Holzinger, A., et al.: Information fusion as an integrative cross-cutting enabler to achieve robust, explainable, and trustworthy medical artificial intelligence. Inf. Fusion **79**(3), 263–278 (2022). https://doi.org/10.1016/j.inffus.2021.10.007

17. Holzinger, A., Langs, G., Denk, H., Zatloukal, K., Müller, H.: Causability and explainability of artificial intelligence in medicine. Wiley Interdiscip. Rev. Data Mining Knowl. Discov. **9**(4), 1–13 (2019). https://doi.org/10.1002/widm.1312

18. Holzinger, A., Malle, B., Saranti, A., Pfeifer, B.: Towards multi-modal causability with graph neural networks enabling information fusion for explainable AI. Inf. Fusion **71**(7), 28–37 (2021). https://doi.org/10.1016/j.inffus.2021.01.008
19. Holzinger, K., Mak, K., Kieseberg, P., Holzinger, A.: Can we trust machine learning results? Artificial intelligence in safety-critical decision support. ERCIM News **112**(1), 42–43 (2018)
20. Hudec, M., Minarikova, E., Mesiar, R., Saranti, A., Holzinger, A.: Classification by ordinal sums of conjunctive and disjunctive functions for explainable AI and interpretable machine learning solutions. Knowl. Based Syst. **220**, 106916 (2021). https://doi.org/10.1016/j.knosys.2021.106916
21. Kasinidou, M., Kleanthous, S., Barlas, P., Otterbacher, J.: I agree with the decision, but they didn't deserve this: future developers' perception of fairness in algorithmic decisions. In: Proceedings of the 2021 ACM Conference on Fairness, Accountability, and Transparency, FAccT 2021, pp. 690–700 (2021). https://doi.org/10.1145/3442188.3445931
22. Kelley, K.H., Fontanetta, L.M., Heintzman, M., Pereira, N.: Artificial intelligence: implications for social inflation and insurance. Risk Manag. Insur. Rev. **21**(3), 373–387 (2018). https://doi.org/10.1111/rmir.12111
23. Kizilcec, R.F.: How much information? Effects of transparency on trust in an algorithmic interface. In: Proceedings of the 2016 CHI Conference on Human Factors in Computing Systems, CHI 2016, pp. 2390–2395. Association for Computing Machinery (2016). https://doi.org/10.1145/2858036.2858402
24. Koh, P.W., Liang, P.: Understanding black-box predictions via influence functions. In: Proceedings of ICML 2017, pp. 1885–1894, July 2017
25. Komodromos, M.: Employees' perceptions of trust, fairness, and the management of change in three private universities in Cyprus. J. Hum. Resour. Manag. Labor Stud. **2**(2), 35–54 (2014)
26. Larasati, R., Liddo, A.D., Motta, E.: The effect of explanation styles on user's trust. In: Proceedings of the Workshop on Explainable Smart Systems for Algorithmic Transparency in Emerging Technologies co-located with IUI 2020, pp. 1–6 (2020)
27. Merritt, S.M., Heimbaugh, H., LaChapell, J., Lee, D.: I trust it, but i don't know why: effects of implicit attitudes toward automation on trust in an automated system. Hum. Factors **55**(3), 520–534 (2013)
28. Nikbin, D., Ismail, I., Marimuthu, M., Abu-Jarad, I.: The effects of perceived service fairness on satisfaction, trust, and behavioural intentions. Singap. Manag. Rev. **33**(2), 58–73 (2011)
29. Papenmeier, A., Englebienne, G., Seifert, C.: How model accuracy and explanation fidelity influence user trust. In: IJCAI 2019 Workshop on Explainable Artificial Intelligence (xAI), pp. 1–7, August 2019
30. Pieters, W.: Explanation and trust: what to tell the user in security and AI? Ethics Inf. Technol. **13**(1), 53–64 (2011). https://doi.org/10.1007/s10676-010-9253-3
31. Renkl, A., Hilbert, T., Schworm, S.: Example-based learning in heuristic domains: a cognitive load theory account. Educ. Psychol. Rev. **21**, 67–78 (2009). https://doi.org/10.1007/s10648-008-9093-4
32. Roy, S.K., Devlin, J.F., Sekhon, H.: The impact of fairness on trustworthiness and trust in banking. J. Mark. Manag. **31**(9–10), 996–1017 (2015)
33. Shin, D.: User perceptions of algorithmic decisions in the personalized AI system: perceptual evaluation of fairness, accountability, transparency, and explainability. J. Broadcast. Electron. Media **64**(4), 541–565 (2020). https://doi.org/10.1080/08838151.2020.1843357

34. Shin, D.: The effects of explainability and causability on perception, trust, and acceptance: implications for explainable AI. Int. J. Hum. Comput. Stud. **146**, 102551 (2021). https://doi.org/10.1016/j.ijhcs.2020.102551
35. Starke, C., Baleis, J., Keller, B., Marcinkowski, F.: Fairness perceptions of algorithmic decision-making: a systematic review of the empirical literature (2021)
36. Stoeger, K., Schneeberger, D., Kieseberg, P., Holzinger, A.: Legal aspects of data cleansing in medical AI. Comput. Law Secur. Rev. **42**, 105587 (2021). https://doi.org/10.1016/j.clsr.2021.105587
37. Wang, X., Yin, M.: Are explanations helpful? A comparative study of the effects of explanations in AI-assisted decision-making. In: Proceedings of 26th International Conference on Intelligent User Interfaces, pp. 318–328. ACM (2021)
38. Yin, M., Vaughan, J.W., Wallach, H.: Does stated accuracy affect trust in machine learning algorithms? In: Proceedings of ICML2018 Workshop on Human Interpretability in Machine Learning (WHI 2018), pp. 1–2 (2018)
39. Zhang, Y., Liao, Q.V., Bellamy, R.K.E.: Effect of confidence and explanation on accuracy and trust calibration in AI-assisted decision making. In: Proceedings of the 2020 Conference on Fairness, Accountability, and Transparency, FAT* 2020, pp. 295–305 (2020)
40. Zhou, J., Arshad, S.Z., Luo, S., Chen, F.: Effects of uncertainty and cognitive load on user trust in predictive decision making. In: Bernhaupt, R., Dalvi, G., Joshi, A., Balkrishan, D.K., O'Neill, J., Winckler, M. (eds.) INTERACT 2017. LNCS, vol. 10516, pp. 23–39. Springer, Cham (2017). https://doi.org/10.1007/978-3-319-68059-0_2
41. Zhou, J., Bridon, C., Chen, F., Khawaji, A., Wang, Y.: Be informed and be involved: effects of uncertainty and correlation on user's confidence in decision making. In: Proceedings of the 33rd Annual ACM Conference Extended Abstracts on Human Factors in Computing Systems, CHI EA 2015, pp. 923–928. Association for Computing Machinery (2015). https://doi.org/10.1145/2702613.2732769
42. Zhou, J., Chen, F.: 2D transparency space—bring domain users and machine learning experts together. In: Zhou, J., Chen, F. (eds.) Human and Machine Learning. HIS, pp. 3–19. Springer, Cham (2018). https://doi.org/10.1007/978-3-319-90403-0_1
43. Zhou, J., Chen, F. (eds.): Human and Machine Learning: Visible, Explainable, Trustworthy and Transparent. HIS, Springer, Cham (2018). https://doi.org/10.1007/978-3-319-90403-0
44. Zhou, J., Gandomi, A.H., Chen, F., Holzinger, A.: Evaluating the quality of machine learning explanations: a survey on methods and metrics. Electronics **10**(5), 593 (2021)
45. Zhou, J., Hu, H., Li, Z., Yu, K., Chen, F.: Physiological indicators for user trust in machine learning with influence enhanced fact-checking. In: Machine Learning and Knowledge Extraction, pp. 94–113 (2019)
46. Zhou, J., Khawaja, M.A., Li, Z., Sun, J., Wang, Y., Chen, F.: Making machine learning useable by revealing internal states update - a transparent approach. Int. J. Comput. Sci. Eng. **13**(4), 378–389 (2016)
47. Zhou, J., et al.: Measurable decision making with GSR and pupillary analysis for intelligent user interface. ACM Trans. Comput. Hum. Interact. **21**(6), 1–23 (2015). https://doi.org/10.1145/2687924
48. Zhou, J., Verma, S., Mittal, M., Chen, F.: Understanding relations between perception of fairness and trust in algorithmic decision making. In: Proceedings of the International Conference on Behavioral and Social Computing (BESC 2021), pp. 1–5, October 2021

Towards Refined Classifications Driven by SHAP Explanations

Yusuf Arslan[1]([✉])[iD], Bertrand Lebichot[1][iD], Kevin Allix[1][iD], Lisa Veiber[1][iD],
Clément Lefebvre[2], Andrey Boytsov[2][iD], Anne Goujon[2], Tegawendé F. Bissyandé[1][iD],
and Jacques Klein[1][iD]

[1] SnT – University of Luxembourg, Esch-sur-Alzette, Luxembourg
yusuf.arslan@uni.lu
[2] BGL BNP Paribas, Luxembourg, Luxembourg

Abstract. Machine Learning (ML) models are inherently approximate; as a result, the predictions of an ML model can be wrong. In applications where errors can jeopardize a company's reputation, human experts often have to manually check the alarms raised by the ML models by hand, as wrong or delayed decisions can have a significant business impact. These experts often use interpretable ML tools for the verification of predictions. However, post-prediction verification is also costly. In this paper, we hypothesize that the outputs of interpretable ML tools, such as SHAP explanations, can be exploited by machine learning techniques to improve classifier performance. By doing so, the cost of the post-prediction analysis can be reduced. To confirm our intuition, we conduct several experiments where we use SHAP explanations directly as new features. In particular, by considering nine datasets, we first compare the performance of these "SHAP features" against traditional "base features" on binary classification tasks. Then, we add a second-step classifier relying on SHAP features, with the goal of reducing false-positive and false-negative results of typical classifiers. We show that SHAP explanations used as SHAP features can help to improve classification performance, especially for false-negative reduction.

Keywords: Interpretable machine learning · SHAP Explanations · Second-step classification

1 Introduction

Machine Learning (ML) is being massively explored to automate a variety of prediction and decision-making processes in various domains. However, the predictions of an ML model can be wrong since ML models are inherently approximate [28]. In the finance sector, for example, when an ML model predicts a transaction is suspicious, it raises an alarm, which can be a true-positive or a false-positive. Such predictions are automatically queued for further manual inspection by financial experts [9]. The existence of these false alarms increases the cost of post-prediction analysis, and wrong or delayed decisions can have a significant business impact [23].

© IFIP International Federation for Information Processing 2022
Published by Springer Nature Switzerland AG 2022
A. Holzinger et al. (Eds.): CD-MAKE 2022, LNCS 13480, pp. 68–81, 2022.
https://doi.org/10.1007/978-3-031-14463-9_5

Figure 1 summarizes the key steps of an ML pipeline where a domain expert must intervene to triage ML predictions. In such a setting, reducing the number of false-positive (or false-negative) predictions upstream is paramount in order to reduce the workload of financial experts and increase the customers' trust in companies. To do so, financial companies often rely on manual interventions by domain experts. It can decrease false-positive rates by several percentage points, but involving domain experts is also costly [35].

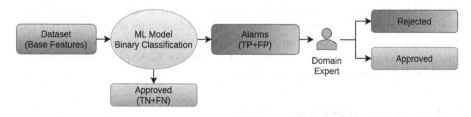

Fig. 1. Financial ML framework with human intervention (TP: true-positive, FP: false-positive, TN: true negative and FN: false negative)

Towards ensuring that domain experts are provided with relevant information for assessing model results, interpretable ML techniques and tools are increasingly leveraged. Among the state-of-the-art interpretable ML approaches, SHAP [20] is a popular technique that is widely used in the literature and by practitioners: its SHAP values, which are derived from SHAP explanations, are computed to evaluate the importance of the contributions of different features on the predictions, potentially enabling the identification of prediction errors as well as providing investigation directions.

Our hypothesis is that if SHAP can help humans to better understand a model, it could also help algorithms. Indeed, if humans are able to leverage information in SHAP explanations, such information may be automatically and systematically exploited in an automated setting. To confirm our intuition, we inspect SHAP values on binary classification tasks. This hypothesis is actually supported by recent works. For instance, [1, 36] show the usage of SHAP explanations by domain experts for the reasoning of case-based scenarios of frauds and anomalies.

To do the first step towards the automatization of the processing of the SHAP explanations, let us see SHAP as a transformation of the learning space. If n_f is the number of features and n_s is the number of samples for a given dataset, SHAP values can be seen as the result of a (nonlinear) transformation f of the learning space: $f : \mathbb{R}^{n_s \times n_f} \to \mathbb{R}^{n_s \times n_f}$. Indeed, each n_s sample will receive n_f SHAP values. SHAP, for each sample, provides a float per feature, reflecting its contribution to the prediction of that sample. The full set of SHAP features has the same size as the full set of base features. In the rest of this paper, the features obtained through SHAP explanations will be referred to as *SHAP features*, and the original features that were available for the classification will be referred to as *base features*.

The idea is to use this transformation to, hopefully, send the data to a **more separable space** (It seems to be the case for SHAP in practice, as shown in [2]). This idea is one of the cornerstones of SVM and is widely used in many domains [4,27,31]. According to these hypotheses, SHAP values may hold information, to be exploited, for improving the performance of ML classifiers.

This Paper. We present an empirical investigation of our hypothesis by focusing on seven publicly available binary classification datasets and two proprietary datasets from the financial domain. In particular, we study the added value of the information encoded in SHAP explanations compared to the features that were available for the classification. More precisely, we first compare the performance (in terms of accuracy) of a classifier where SHAP explanations are used as features in comparison with a "traditional" classifier relying on base features. We then investigate the feasibility of building a pipeline of cascaded classifiers where the second classifier leverages SHAP explanations to filter out incorrectly classified samples after the initial classifier. In the end, we show that this strategy indeed increases classifier performance.

2   Background and Related Work

The Cost of Being Wrong
Financial institutions are wary of the "cost of being wrong" [3]. This cost is two-fold. First, *bad* decisions–made by a human being or by an automatic system–carry severe risks of financial loss, direct or indirect. Furthermore, in a line of business where *Trust* is of prime importance, any loss in reputation, through scandals or mere negative hearsay, can quickly lead to substantial financial consequences. Second, trying to prevent automatic systems from making *wrong* decisions itself incurs significant costs in the form of increased workload for expensive experts, lack of flexibility arising from the delays needed to have automated decisions vetted by experts, and the massive extra cost of designing systems and processes that provably mitigate the risks of *bad* decisions.

Counter-intuitively, the fear of *bad* decisions may lead some actors to forgo approaches that could be more accurate but do not help analysts vet the decisions. In particular, Deep-Learning—despite its documented prowess—is sometimes deemed inappropriate [34] because it brings nothing to help justify the decisions and no explanations to archive for auditing purposes.

Overall, these costs and risks call for more precise techniques that enable and ease manual inspection and leave exploitable audit trails. Interpretable ML techniques can decrease these costs and risks. [37] uses SHAP explanations for case-based reasoning tasks and reports that the similarity of SHAP explanations is more helpful than the similarity of feature values for domain experts, though finding the most appropriate distance function that shows similarity for a specific dataset is not a fully resolved question. [18] evaluates SHAP and Local Interpretable Model-agnostic Explanations (LIME)[1] [25] to obtain useful information for domain experts and facilitate the FP reduction task. [18] suggests eliminating FPs by employing an ML filter that uses SHAP features instead of base features. According to their findings, the performance of the ML filter using SHAP features is better than the ML model using base features and thus can be leveraged [18].

---

[1] https://github.com/marcotcr/lime.

In the interpretable ML domain, little research has been conducted about the use of explanations to enhance model performance [13]. In the context of this study, we aim at inspecting the effect of SHAP values in a two-step classification pipeline. Two-step cascaded classification has been used for financial applications, as reported by various studies [5,7,14,33]. The idea behind two-step classification is to obtain SHAP features as SHAP values by using a first step classifier and then use a second step classifier with SHAP features to improve classification performance.

We inspect SHAP explanations, which are already investigated for various tasks, including but not limited to clustering [2], rule mining [18], case-based reasoning [37], and feature selection [15], from the classification point of view.

Shapley Values
Shapley Values [29], which guarantee a fair distribution of payout among players, derive from the cooperative game theory domain and have been quite influential in various domains for a long time [24,30]. Among attribution methods that aim at distributing the prediction scores of a model for the specific input to its base features, the Shapley values method is the one that satisfies the properties of symmetry, dummy, and additivity [21]. These three properties, namely, symmetry (interchangeable players should receive same the pay-offs), dummy (dummy players should receive nothing), and additivity (if the game is separated, so do the pay-offs), can be considered a definition of a fair payout. Recently, Shapley values have been investigated on the interpretable ML domain to solve the fairness issue of feature contribution values [20]. In this study, features are considered as players, and predictions are regarded as pay-offs. This implementation, which is SHAP[2] [20], shows how to distribute the payout fairly among the features. Besides, SHAP explanations are suitable for the needs of finance actors [10].

Overall, Shapley values can be seen as providing another *representation* of the original data [17] and hence might help an ML algorithm to learn patterns it would not have been able to infer from the raw base features of the datasets.

3   Empirical Datasets

We perform our empirical evaluation by relying on seven publicly available binary classification datasets and two proprietary binary classification datasets from our industrial partner, a major national bank.

The details of the datasets are as follows:

1. The *Adult* dataset[3], which is also known as "Census Income", contains 32 561 samples with 12 categorical and numerical features. The prediction task of the dataset is to find out whether a person makes more than $50K per year or not.
2. The *Bank Marketing* [22] dataset[4] contains marketing campaigns of a banking institution. It has 44 581 samples with 16 categorical and numerical features. The prediction task of the dataset is to identify whether a customer will subscribe to a term deposit or not.

---

[2] https://github.com/slundberg/shap.
[3] https://archive.ics.uci.edu/ml/datasets/adult.
[4] https://archive.ics.uci.edu/ml/datasets/Bank+Marketing.

3. The *Credit Card Fraud* dataset[5] contains credit card transactions. It has 284 807 samples with 30 numerical features. The prediction task is to identify whether a transaction is non-fraudulent or fraudulent.
4. The *Heloc* (Home equity line of credit) dataset[6] comes from an explainable machine learning challenge of the FICO company[7]. It contains anonymized Heloc applications of real homeowners. It has 10 459 samples with 23 numerical features. The prediction task of the dataset is to classify the risk performance of an applicant as *good* or *bad*. *Good* means that an applicant made payments within a three-month period in the past two years. *Bad* means that an applicant did not make payments at least one time in the past two years.
5. The *Lending Club* dataset[8] contains loans made through the Lending Club platform. It has 73 157 samples with 63 categorical and numerical features. The prediction task of the dataset is to identify whether a customer that is requesting a loan will be able to repay it or not.
6. The *Paysim* [19] dataset[9] is a financial mobile money simulator. It has 6 362 260 samples with 7 categorical and numerical features. The prediction task of the dataset is to identify whether a transaction is non-fraudulent or fraudulent.
7. The *ULB Fraud* [16] dataset[10] contains simulated transactions. It has 32 561 samples with 12 categorical and numerical features. The prediction task of the dataset is to identify fraudulent transactions.
8. *Proprietary-1*: The first proprietary dataset contains transaction records. It contains 29 200 samples with 10 categorical and numerical features. With this dataset, the goal is to classify a transaction as Type-A or Type-B (for confidentiality reasons, we cannot detail Type-A and Type-B). We will use *Proprietary-1* to name this dataset.
9. *Proprietary-2*: The second proprietary dataset contains financial requests. We will use *Proprietary-2* to name this dataset. It contains 389 451 samples with 87 categorical and numerical features. The prediction task is to classify financial requests as Type-A or Type-B (for confidentiality reasons, we cannot detail Type-A and Type-B). We will use *Proprietary-2* to name this dataset.

4   Experiment Setup

4.1   Research Questions

Our study takes form around the question of whether SHAP features (see Sect. 1), derived from SHAP explanations, could be useful to improve the classification performance or not. The intuition behind it is that just like SHAP can help humans, it may be able to help algorithms. More concretely, this study assesses whether the feature transformation induced by SHAP, i.e., the computation of the SHAP features, can be

---

[5] https://www.kaggle.com/mlg-ulb/creditcardfraud.

[6] https://aix360.readthedocs.io/en/latest/datasets.html.

[7] https://community.fico.com/s/explainable-machine-learning-challenge.

[8] https://www.kaggle.com/wordsforthewise/lending-club.

[9] https://www.kaggle.com/ealaxi/paysim1.

[10] https://github.com/Fraud-Detection-Handbook.

exploited by machine learning techniques. To that end, we answer two research questions.

RQ1: Can SHAP features outperform base features in a traditional one-step classification approach in terms of accuracy?

RQ2: If we compare a traditional (one-step) classification approach against a two-step classification approach where the second classifier uses either SHAP or base features, what is the best alternative in terms of accuracy? We will divide our answer in terms of false-positive reduction (RQ2.1) and false-negative reduction (RQ2.2). Through RQ2, we want to assess whether SHAP features used in a two-step approach could help domain experts quickly triage ML decisions, for instance, by reducing the number of false-positive decisions.

4.2 Experiment Process

Training step and Machine Learning Algorithms: The training step and the used algorithms are represented in Fig. 2-a.

- *Base Features, GBC, and MLP:* On each of our nine datasets, we first train two binary classifiers by using the base features that are proposed (cf. Sect. 3). For one classifier, we use Gradient Boosting Classifier (GBC). For the other one, we use Multi Layer Perceptron (MLP). There are two reasons for these choices: (1) GBC is the current state-of-the-art approach for tabular unbalanced classification problems, and MLP is better than tree-based approaches to capture additive structure [11], which is the case for SHAP explanations, (2) we tested various other classifiers (e.g., random forests (RF) and logistic regression (LR)) and GBC & MLP were the best on most of our datasets (results not reported here).
- *SHAP features:* The idea is to use the SHAP explanations as features to train a new classifier with these newly computed SHAP features. In practice, we compute the SHAP features by applying the SHAP explainer on the GBC classifier. Then, we use the obtained SHAP features as inputs of an MLP classifier. Note that we also consider RF, MLP, RL, and GBC, but the best results (not reported here) are achieved with MLP.

The training phase can be seen in Fig. 2-a.

Testing Steps: To answer our research questions, we implement three scenarios depicted in Fig. 2-b, Fig. 2-c, and Fig. 2-d respectively.

- One Step Binary Classification (Fig. 2-b): To answer RQ1, we compare our three classifiers - GBC with base features, MLP with base features, and MLP with SHAP features - in traditional binary classification tasks on the nine datasets and tasks described in Sect. 3. We compute the Precision-Recall curve and the ROC curve to compare the results for all decision thresholds.
- Two-Step Classification – Positive (Fig. 2-c): To answer RQ2.1, as a first step, we use a GBC classifier with base features. Then, as a second step, we apply two classifiers on the positively classified samples only: one by considering the base

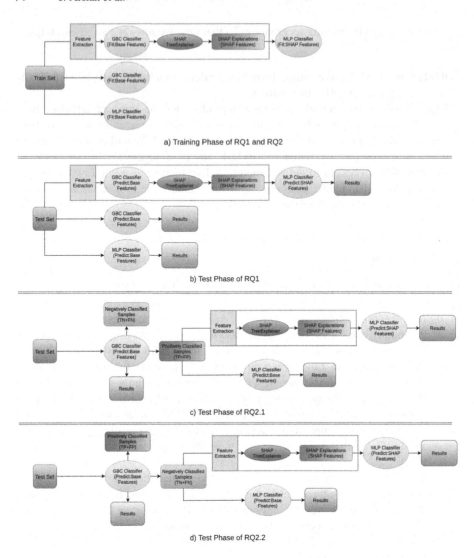

a) Training Phase of RQ1 and RQ2

b) Test Phase of RQ1

c) Test Phase of RQ2.1

d) Test Phase of RQ2.2

Fig. 2. Experiment process (TP: true-positive, FP: false-positive, TN: true negative and FN: false negative)

features and another one by considering the SHAP features. The number of positively classified samples depends on the decision threshold, which impacts the classification results. A second classification threshold is also considered for the second-step classifier. For each threshold, we test the values $[0.1, 0.2, 0.3, ..., 0.9]$ to identify the best results. We report the best F1 score and balanced classification rate (BCR) for these thresholds.

- Two-Step Classification – Negative (Fig. 2-d): To answer RQ2.2, we use a similar process as the one for RQ2.1, but we focus on negatively classified samples.

Finally, all our experiments are performed using 5-Fold cross-validation, and are repeated 5 times. The averaged results are then reported.

4.3 Evaluation Metrics

In this paper, we are using the following metrics and tests for evaluation of the results:

Receiver operating characteristic (ROC) curve: True-positive Rate (TPR) is the number of true-positives over the number of true-positives and false-negatives. It shows the performance of models in the prediction of the positive class when the actual outcome is positive. False-positive Rate (FPR) is the number of false-positives over the number of false-positives and true-negatives. It shows the number of positive classifications while the actual outcome is negative. ROC Curve is a visual representation of the trade-off between TPR and FPR [6].

Precision recall (PR) curve: Precision is the number of true-positives over the number of true-positives plus the number of false-positives. Recall is the number of true-positives over the number of true-positives plus the number of false negatives. PR curve shows the trade-off between precision and recall for different thresholds. PR metric evaluates output quality of a classifier. It is used especially in case of class imbalance. High precision implies a low false-positive rate and a high recall implies low false-negative rate. ROC curves are suitable for balanced datasets, whereas PR curves are suitable for imbalanced datasets [26]. We choose these two metrics (ROC and PR curve) since they show the performance of classification models for all thresholds.

F1 score: It is the (balanced) harmonic mean of precision and recall. A high value of F1 score means high classification performance [32].

Balanced classification rate (BCR): All classifiers aim at increasing the sensitivity without sacrificing the specificity [32]. BCR combines sensitivity (TPR) and specificity (1-FPR).

$$Balanced\,Classification\,Rate = \frac{\frac{True\,Positive}{(True\,Positive+False\,Negative)} + \frac{True\,Negative}{(True\,Negative+False\,Positive)}}{2}$$

5 Results and Discussion

Answers of RQ1: We use ROC Curves and PR Curves to answer this research question. The ROC Curves for each dataset can be seen in Fig. 3, and the PR Curves can be seen in Fig. 4.

According to ROC Curves (and the Area Under Curve - AUC), SHAP features obtain better results than base features for 6 out of 9 datasets.

Similar to ROC Curves, according to PR Curves, SHAP features obtain better results than base features for 5 out of 9 datasets.

One of the interesting findings in our experiments is that the MLP classifier with base features (orange line) is less successful than the MLP classifier with SHAP features (green line). This result can be explained by the fact that SHAP features are "well-separable" as indicated in [36]. More specifically, the data is better separated in the

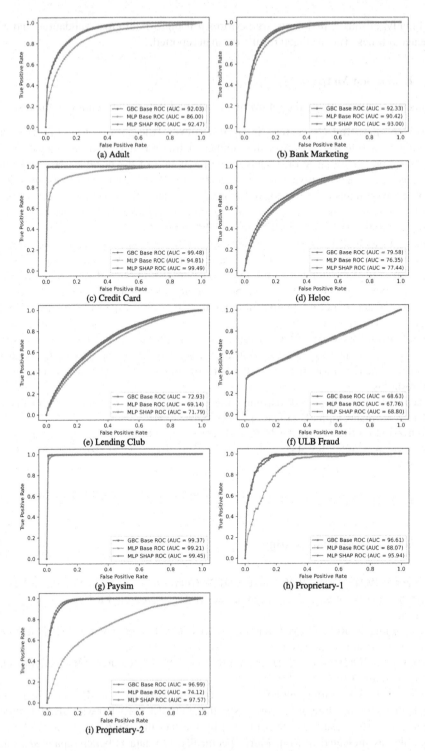

Fig. 3. ROC curves comparison of GBC with base features vs. MLP with base features vs. MLP with SHAP features (5-fold cross validation + repeated 5 times) (Color figure online)

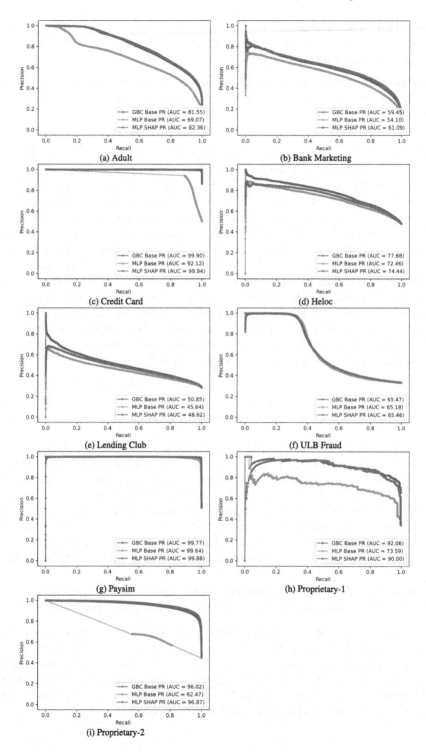

Fig. 4. PR curves comparison of GBC with base features vs. MLP with base features vs. MLP with SHAP features (5-fold cross validation + repeated 5 times) (Color figure online)

SHAP feature space, and an MLP classifier that uses SHAP features can work better in this space. Another reason could be related to the fact that more patterns come to the surface with SHAP features.

Answers of RQ2: We divided our analysis into two sub-research questions, RQ2.1 and RQ2.2, to report our analysis of positively and negatively classified samples, respectively.

RQ2.1 Answer: As described in Fig. 2-c), we compare a classical one-step GBC classifier with two two-step classifiers that focus on positively classified samples only. Both two-step classifiers rely on an MLP classifier, but one uses base features, and the other one uses SHAP features. A comparison of F1 and BCR scores can be seen in Table 1. The two-step (SHAP) obtains the best F1 score for 7 out of 9 datasets and the best BCR score for 6 out of 9 datasets. The one-step classifier obtains the best F1 score for 1 out of 9 datasets and the best BCR score for 2 out of 9 datasets. The two-step (Base) obtains the best F1 and BCR scores for only one out of 9 datasets. According to these findings, the two-step (SHAP) outperforms the other two classifiers on most of the datasets.

We rely on a Friedman/Nemenyi test [8] (with $\alpha = 0.1$) to confirm whether there is a statistically significant difference between the performance scores of the three classifiers. The test concludes that Two-step (SHAP) (The two-step classifier using SHAP features) is significantly better than Two-step (base) (The two-step classifier using base features). However, the test also concludes that there is no significant difference between the one-step classifier and the Two-step (SHAP).

RQ2.2 Answer: As described in Fig. 2-d), we follow an experimental process that is similar to the one used to answer RQ2.1, except that both two-step classifiers focus on negatively classified samples. A comparison of F1 and BCR scores can be seen in Table 2.

Two-step (SHAP) obtains the best F1 and BCR scores for 7 out of 9 datasets. The one-step classifier obtains the best F1 score for 1 out of 9 datasets and never obtains the best BCR score in any of the tested datasets. The two-step (Base) obtains the best F1 score for one out of 9 datasets and the best BCR score for two out of 9 datasets. According to these findings, the two-step (SHAP) obtains the best results overall.

We rely on a Friedman/Nemenyi test (with $\alpha = 0.1$) to confirm whether there is a statistically significant difference among the performance scores of the three classifiers. The test concludes that Two-step (SHAP) is superior to both Two-step (Base) and One-step.

Our findings show that a classifier with SHAP features can be applied to negatively or positively classified samples as a step to improve classification performance.

Comparing RQ2.1 and RQ2.2 Results: Two-step (SHAP) obtains better results on negatively classified samples than on positively classified samples. In all the datasets that we use, class distributions exhibit a slight to severe class imbalance, and positive samples are in minority class. Therefore, there is a relatively small amount of positively classified samples for some datasets. The higher number of negative samples, which means more data for training, can be the reason of better results.

Table 1. Best results of F1 and BCR for different thresholds on positively classified samples (5-fold cross validation + repeated 5 times).

F1 (positive)									
	Adult	Bank	Credit	Heloc	Lending	Paysim	ULB	Prop-1	Prop-2
One-step	71.40	60.04	99.19	72.56	52.93	98.97	52.44	**87.65**	90.38
Two-step (Base)	69.11	59.79	95.25	**72.59**	52.57	99.06	52.47	83.48	78.82
Two-step (SHAP)	**72.62**	**61.89**	**99.91**	72.47	**53.03**	**99.31**	**52.60**	87.63	**91.61**
BCR (positive)									
One-step	83.50	85.36	99.19	71.21	**66.49**	98.97	66.96	**92.38**	91.39
Two-step (Base)	81.49	83.85	95.51	**72.25**	66.28	99.06	67.35	87.98	83.59
Two-step (SHAP)	**83.97**	**85.77**	**99.91**	72.15	66.48	**99.31**	**67.00**	92.03	**92.56**

Table 2. Best results of F1 and BCR for different thresholds for negatively classified samples (5-fold cross validation + repeated 5 times).

F1 (negative)									
	Adult	Bank	Credit	Heloc	Lending	Paysim	ULB	Prop-1	Prop-2
One-step	71.40	60.04	99.19	**72.56**	52.93	98.97	51.95	87.65	90.38
Two-step (Base)	70.51	60.26	94.96	72.54	53.12	98.85	**52.14**	86.10	77.91
Two-step (SHAP)	**72.62**	**62.11**	**99.89**	72.55	**53.31**	**99.33**	52.13	**87.82**	**91.62**
BCR (negative)									
One-step	83.50	85.36	99.19	72.21	66.49	98.97	66.99	92.38	91.39
Two-step(Base)	83.13	85.37	94.05	**72.23**	66.54	98.84	**67.29**	91.69	76.56
Two-step(SHAP)	**83.97**	**86.41**	**99.89**	72.14	**66.59**	**99.33**	67.00	**92.55**	**92.56**

6 Conclusions

In this study, we leverage SHAP features to improve classification performance. Our experiments are performed on seven datasets from the literature and two datasets from our industrial partner. We start by showing that a classifier based on SHAP features can be as efficient as a classifier based on base features. We then show that a second-step classifier, based on the SHAP features, can easily be added to reduce both false-positives and false-negatives.

Our findings are important for several reasons. First, we detect that a classifier based on SHAP features is as powerful as a classifier based on base features. Second, our findings show that domain experts can infer from SHAP explanations comfortably, which is especially important when SHAP explanations offer better visualization. Third, the results reveal that it is possible to improve classification performance by the use of two-step classification.

As future work, we are planning to utilize SHAP explanations for detecting redundant samples in resampling strategies to tackle unbalanced datasets. Besides, it can be an interesting future work to use SHAP explanations in a positive-confidence classifier [12] in which SHAP values for each feature can be used instead of prediction probabilities.

References

1. Antwarg, L., Miller, R.M., Shapira, B., Rokach, L.: Explaining anomalies detected by autoencoders using Shapley Additive Explanations. Expert Syst. Appl. **186**, 115736 (2021)
2. Arslan, Y., et al.: On the suitability of SHAP explanations for refining classifications. In: Proceedings of the 14th International Conference on Agents and Artificial Intelligence (ICAART 2022) (2022)
3. Bank of England: Machine learning in UK financial services (2019). https://www.bankofengland.co.uk/-/media/boe/files/report/2019/machine-learning-in-uk-financial-services.pdf. Accessed Apr 2022
4. Becker, T.E., Robertson, M.M., Vandenberg, R.J.: Nonlinear transformations in organizational research: possible problems and potential solutions. Organ. Res. Methods **22**(4), 831–866 (2019)
5. Berger, C., Dohoon, K.: A two-step process for detecting fraud using ADW, oracle machine learning, APEX and oracle analytics cloud (2020). https://blogs.oracle.com/machinelearning/a-two-step-process-for-detecting-fraud-using-oracle-machine-learning. Accessed Apr 2022
6. Bradley, A.P.: The use of the area under the ROC curve in the evaluation of machine learning algorithms. Pattern Recogn. **30**(7), 1145–1159 (1997)
7. Darwish, S.M.: A bio-inspired credit card fraud detection model based on user behavior analysis suitable for business management in electronic banking. J. Ambient Intell. Human. Comput. **11**, 4873–48871 (2020). https://doi.org/10.1007/s12652-020-01759-9
8. Demšar, J.: Statistical comparisons of classifiers over multiple data sets. J. Mach. Learn. Res. **7**, 1–30 (2006)
9. Ghamizi, S., et al.: Search-based adversarial testing and improvement of constrained credit scoring systems. In: 28th ACM Joint Meeting on ESEC/FSE, pp. 1089–1100 (2020)
10. Misheva, B.H., Hirsa, A., Osterrieder, J., Kulkarni, O., Lin, S.F.: Explainable AI in credit risk management. Credit Risk Management, 1 March 2021
11. Hastie, T., Tibshirani, R., Friedman, J.: The Elements of Statistical Learning: Data Mining, Inference, and Prediction, vol. 2. Springer, New York (2009). https://doi.org/10.1007/978-0-387-84858-7
12. Ishida, T., Niu, G., Sugiyama, M.: Binary classification from positive-confidence data. In: Advances in Neural Information Processing Systems, vol. 31 (2018)
13. Jia, Y., Frank, E., Pfahringer, B., Bifet, A., Lim, N.: Studying and exploiting the relationship between model accuracy and explanation quality. In: Oliver, N., Pérez-Cruz, F., Kramer, S., Read, J., Lozano, J.A. (eds.) ECML PKDD 2021. LNCS (LNAI), vol. 12976, pp. 699–714. Springer, Cham (2021). https://doi.org/10.1007/978-3-030-86520-7_43
14. Khormuji, M.K., Bazrafkan, M., Sharifian, M., Mirabedini, S.J., Harounabadi, A.: Credit card fraud detection with a cascade artificial neural network and imperialist competitive algorithm. IJCA **96**(25), 1–9 (2014)
15. Komatsu, M., Takada, C., Neshi, C., Unoki, T., Shikida, M.: Feature extraction with SHAP value analysis for student performance evaluation in remote collaboration. In: 2020 15th International Joint Symposium on Artificial Intelligence and Natural Language Processing (iSAI-NLP), pp. 1–5 (2020)
16. Le Borgne, Y.A., Siblini, W., Lebichot, B., Bontempi, G.: Reproducible Machine Learning for Credit Card Fraud Detection - Practical Handbook. Université Libre de Bruxelles (2022)
17. Li, R., et al.: Machine learning-based interpretation and visualization of nonlinear interactions in prostate cancer survival. JCO Clin. Cancer Inform. **4**, 637–646 (2020)
18. Lin, C.F.: Application-grounded evaluation of predictive model explanation methods. Master's thesis, Eindhoven University of Technology (2018)

19. Lopez-Rojas, E., Elmir, A., Axelsson, S.: PaySim: a financial mobile money simulator for fraud detection. In: 28th European Modeling and Simulation Symposium, EMSS, Larnaca, pp. 249–255. Dime University of Genoa (2016)
20. Lundberg, S.M., Lee, S.I.: A unified approach to interpreting model predictions. In: Proceedings of the 31st International Conference on Neural Information Processing Systems, pp. 4768–4777 (2017)
21. Molnar, C.: Interpretable machine learning. Lulu.com (2020)
22. Moro, S., Cortez, P., Rita, P.: A data-driven approach to predict the success of bank telemarketing. Decis. Support Syst. **62**, 22–31 (2014)
23. Pascual, A., Marchini, K., Van Dyke, A.: Overcoming false positives: saving the sale and the customer relationship. White paper, Javelin strategy and research reports (2015). Accessed Apr 2022
24. Quigley, J., Walls, L.: Trading reliability targets within a supply chain using Shapley's value. Reliab. Eng. Syst. Saf. **92**(10), 1448–1457 (2007)
25. Ribeiro, M.T., Singh, S., Guestrin, C.: Why should I trust you?: Explaining the predictions of any classifier. In: ACM SIGKDD, pp. 1135–1144 (2016)
26. Saito, T., Rehmsmeier, M.: The precision-recall plot is more informative than the roc plot when evaluating binary classifiers on imbalanced datasets. PLoS ONE **10**(3), e0118432 (2015)
27. Shachar, N., et al.: The importance of nonlinear transformations use in medical data analysis. JMIR Med. Inform. **6**(2), e27 (2018)
28. Shalev-Shwartz, S., Ben-David, S.: Understanding Machine Learning: From Theory to Algorithms. Cambridge University Press, Cambridge (2014)
29. Shapley, L.S.: A value for n-person games. In: Contributions to the Theory of Games, vol. 2, no. 28, pp. 307–317 (1953)
30. Sheng, H., Shi, H., et al.: Research on cost allocation model of telecom infrastructure co-construction based on value Shapley algorithm. Int. J. Future Gener. Commun. Netw. **9**(7), 165–172 (2016)
31. Song, C., Liu, F., Huang, Y., Wang, L., Tan, T.: Auto-encoder based data clustering. In: Ruiz-Shulcloper, J., Sanniti di Baja, G. (eds.) CIARP 2013. LNCS, vol. 8258, pp. 117–124. Springer, Heidelberg (2013). https://doi.org/10.1007/978-3-642-41822-8_15
32. Tharwat, A.: Classification assessment methods. New Engl. J. Entrep. **17**(1), 168–192 (2020). https://www.emerald.com/insight/content/doi/10.1016/j.aci.2018.08.003/full/html
33. Thejas, G., Dheeshjith, S., Iyengar, S., Sunitha, N., Badrinath, P.: A hybrid and effective learning approach for click fraud detection. Mach. Learn. Appl. **3**, 100016 (2021)
34. Veiber, L., Allix, K., Arslan, Y., Bissyandé, T.F., Klein, J.: Challenges towards production-ready explainable machine learning. In: 2020 USENIX Conference on Operational Machine Learning (OpML 2020) (2020)
35. Wedge, R., Kanter, J.M., Veeramachaneni, K., Rubio, S.M., Perez, S.I.: Solving the false positives problem in fraud prediction using automated feature engineering. In: Brefeld, U., et al. (eds.) ECML PKDD 2018. LNCS (LNAI), vol. 11053, pp. 372–388. Springer, Cham (2019). https://doi.org/10.1007/978-3-030-10997-4_23
36. Weerts, H.J.: Interpretable machine learning as decision support for processing fraud alerts. Ph.D. thesis, Master's Thesis, Eindhoven University of Technology, 17 May 2019
37. Weerts, H.J., van Ipenburg, W., Pechenizkiy, M.: Case-based reasoning for assisting domain experts in processing fraud alerts of black-box machine learning models. In: KDD Workshop on Anomaly Detection in Finance (KDD-ADF 2019) (2019)

Global Interpretable Calibration Index, a New Metric to Estimate Machine Learning Models' Calibration

Federico Cabitza[1,2][✉], Andrea Campagner[1], and Lorenzo Famiglini[1]

[1] Dipartimento di Informatica, Sistemistica e Comunicazione,
University of Milano-Bicocca, viale Sarca 336, 20126 Milan, Italy
federico.cabitza@unimib.it
[2] IRCCS Istituto Ortopedico Galeazzi, Milan, Italy

Abstract. The concept of calibration is key in the development and validation of Machine Learning models, especially in sensitive contexts such as the medical one. However, existing calibration metrics can be difficult to interpret and are affected by theoretical limitations. In this paper, we present a new metric, called GICI (Global Interpretable Calibration Index), which is characterized by being local and defined only in terms of simple geometrical primitives, which makes it both simpler to interpret, and more general than other commonly used metrics, as it can be used also in recalibration procedures. Also, compared to traditional metrics, the GICI allows for a more comprehensive evaluation, as it provides a three-level information: a bin-level local estimate, a global one, and an estimate of the extent confidence scores are either over- or under-confident with respect to actual error rate. We also report the results from experiments aimed at testing the above statements and giving insights about the practical utility of this metric also to improve discriminative accuracy.

Keywords: Calibration · Re-calibration · Interpretability · Medical machine learning

1 Introduction

In very general terms, calibration regards the operation of comparing the actual output of a measurement instrument and the expected output, as well as adjusting the instrument (or its output) to make this difference as low as possible. In the Machine Learning (ML) domain, calibration is, intuitively speaking, the property by which the confidence scores (also called probability scores) associated with each prediction by a ML model can be interpreted, at least approximately, as the *frequentist probability* of being accurate [15]: that is, if a model predicts *any new single instance* to be associated with a class with a confidence score p,

© IFIP International Federation for Information Processing 2022
Published by Springer Nature Switzerland AG 2022
A. Holzinger et al. (Eds.): CD-MAKE 2022, LNCS 13480, pp. 82–99, 2022.
https://doi.org/10.1007/978-3-031-14463-9_6

then approximately $100 * p\%$ of the instances with the same confidence scores (should) belong to that class.

Thus, calibration is a fundamentally important property of any ML model [34], especially in high-stakes settings (e.g., in medicine) where accurate uncertainty estimation is paramount to adequately support human decision makers [17,24]; thus, increasing attention has been devoted to the development of methods and metrics to estimate the calibration of a model from finite sample data, as well as algorithms and techniques to improve calibration (i.e., make a model more calibrated, in the sense specified above).

One step backward: as mentioned in the beginning, the term *calibration* refers to two different, but related, concepts. First, calibration in the sense of the extent a model is *calibrated*, that is assessing whether a model satisfies the intuitive property mentioned above. Calibration in this former sense is usually assessed on a global level (that is, by considering the whole instance population or a sample of it) by metrics such as the Brier score [2] or the Expected Calibration Error [20]. In regard to the second meaning, calibration usually refers to the process of *making a model more calibrated*: given a model that has been deemed to be not sufficiently calibrated, higher calibration can be achieved by applying an algorithm or procedure that improves calibration on single instances.

Multiple approaches have been proposed to address also this latter sense of the term calibration, including Platt scaling [23], isotonic regression [21] and Venn predictors [35]. In what follows, for ease of expression and to avoid misunderstandings, we will mostly use the term *calibration* in regard to the former meaning mentioned above (i.e., the estimate), while we reserve the term *recalibration* for the second meaning.

Despite the two senses mentioned above being strictly related, research about them has been mostly disjoint [19]. The primary reason for this gap regards the different focus assumed in the two meanings. Indeed, *recalibration* methods focus on information about individual confidence scores: given a *single instance*, and the corresponding confidence score, the goal is to modify these latter in order to obtain a (more) accurate estimate of the model's probability of being correct *on that instance*. By contrast, calibration metrics provide information about the model's confidence scores only at a global level. While this allows calibration metrics to be reliably estimated, it impairs the ability to obtain information (w.r.t. calibration) about single instances and to act on it [19].

Arguably, however, local calibration is what matters most to the decision-makers ultimately employing and relying on the ML models [38]: what we care about is the *probability of the model being correct* on *this instance*, i.e., informing decision makers with a sort of instance-level *predictive value* or *precision*.

To bridge this gap, in this article we will propose a novel calibration metric, the Global Interpretable Calibration Index (GICI). Differently from existing proposals, the GICI is defined both on a local level and a global level (in the sense above). On a local level, the Local Interpretable Calibration Index (LICI) can be seen as the (normalized) deviation from perfect calibration relative to a (as small as desired) confidence score bin (and corresponding predicted instances);

on the other hand, on a global level the GICI can be defined in two different ways: either by a weighted average of the local LICI scores (thus providing a calibration metric akin to other existing proposals); or by separately considering the tendency of the model to under- or over-forecast.

Most relevantly, however, we show that our metric, since it is defined *also* on a local level, allows to also address the second meaning of calibration: this is done by means of a non-parametric recalibration algorithm that adjusts the confidence scores issued by the model, on the basis of sample statistics and information about the distribution of the GICI index. Finally, we will illustrate the soundness of our proposal on a selection of (medical) benchmark datasets, showing that it provides promising improvements in comparison to other existing recalibration methods. As we will notice in the end, better recalibration of a model also entails higher accuracy, if calibration improves in the neighborhood of the decision cut-off (usually .5). Thus, better ways to measure calibration means better models, *as well as* better ways to interpret them.

2   Background and Related Work

In what follows, let $S = \{(x_i, o_i)\}_{i=1}^N$ be a dataset consisting of N instances, where $x_i \in X$ is a representation of the instances (e.g., if $X = \mathbb{R}^n$, then instances are represented as n-dimensional vectors) and $o_i \in \{0, 1\}$ is the observed label for instance i. That is, we focus on the binary classification setting. A scoring classifier is a function $f : X \mapsto [0, 1]$. For each $x \in X$, the value $f(x)$ is the confidence score (for class 1) of f on instance x. A *k-binning* of S is a partition of S into k subsets $\{S_1, ..., S_k\}$ such that, if $i < j \in \{1, ..., k\}$, then $\max_{(x_i, o_i) \in S_i} f(x_i) \le \min_{(x_l, o_l) \in S_j} f(x_j)$.

A scoring classifier f is said to be *calibrated* [15,32] if $Pr(o = 1 | f(x) = p) = p$, where probability is taken with respect to the data generating process. Obviously, calibration cannot be directly assessed, since the data-generating process is usually unknown. Therefore, calibration is usually estimated from observed data in two ways: a qualitative manner and a quantitative one. For the former case, we refer to the use of calibration curves in reliability diagrams, which allow to qualitatively and visually estimate the distance of the model curve from the bisector, that is the extent the actual model output differs from the output of an ideal perfectly-calibrated model. This visual representation is called *calibration plot*, or *reliability diagram* [16], as shown in Fig. 1.

In regard to the quantitative approach, several methods have been devised to estimate the degree of calibration of an ML model; in what follows, we mention the main techniques that are commonly applied in studies reported in the ML specialist literature, starting from the most frequently used metric, the *Brier score*.

The *Brier Score* [2] is defined as in Eq. 1:

$$BS = \frac{1}{N} \sum_i^N (f(x_i) - o_i)^2 \tag{1}$$

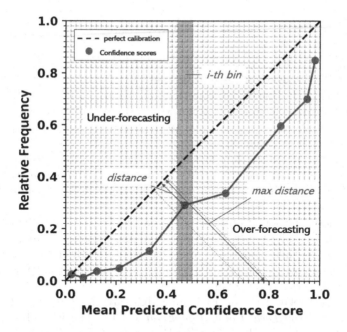

Fig. 1. An example of reliability diagram depicting the calibration curve of a classifier and the concepts mentioned in this contribution.

The Brier score has range in [0,1], where 0 denotes a perfectly calibrated model, while 1 denotes a totally un-calibrated one. Calibration is strongly non linear within this interval: for example, a maximally uncertain model (i.e., a model assigning uniformly equal confidence scores) would have a Brier score close to .25.

While the Brier's score is appealing from the theoretical point of view (in particular, it is a strictly proper scoring rule [14]), it also has limitations: for instance, as discussed in [1], the Brier score is greatly influenced by the distribution of the labels. This can have a negative impact on decision making, in all those high-stake settings, such as medicine, where the less represented classes are also the more relevant ones [36], since these classes regard high-impact, but infrequent, cases (such as serious complications or death) and therefore practitioners could favor sensitive models with over-forecast confidence scores [3]. However, this practical preference is not reflected in the Brier score [1].

Moreover, although consistently used in the specialist literature, still there is no clear consensus on how to interpret the Brier score, except for the common notion that scores above .25 are problematic [5].

Two other metrics that are sometimes used to express the concept of calibration error are the *Expected Calibration Error* (ECE) and the *Maximum Calibration Error* (MCE) [20]:

$$ECE = \sum_{i=1}^{N} P(i) \times |\hat{o}_i - e_i| \tag{2}$$

$$MCE = \max_{i=1}^{N}(|\hat{o}_i - e_i|) \tag{3}$$

where e_i is the average confidence score within *bin i* (i.e., $e_i = \frac{1}{|S_i|}\sum_{x \in S_i} f(x)$), $\hat{o}_i$ is the relative frequency of the positive class in bin i (i.e., $\hat{o}_i = \frac{1}{|S_i|}\sum_{x \in S_i} o_x$), and $P(i)$ is the proportion of instances that fall within S_i. The higher the ECE (or the MCE), the lower the model's calibration, so that the ideal value is 0, whereas the observed confidence score is equal to the relative frequency of the positive instance within the i-th bin. While the ECE and MCE are simpler to interpret than the Brier score, since they measure the linear deviation from calibration, they are less theoretically sound, as they are not *strictly proper scoring rules* [10], due to the fact that their definition grounds on that of the absolute deviation loss.

Furthermore, the definition of the ECE and MCE scores and the values that they can assume is affected by the adopted binning scheme, making their usage for comparing multiple models less immediate than the Brier score. Finally, the MCE provides an overly conservative estimate, as it considers only the maximum deviation from perfect calibration.

Two metrics related to the ECE and MCE are the *Integrated Calibration Index* (ICI) and the *Local Calibration Error* (LCE). The ICI [31] is obtained from the ECE by letting the size of the bins go to nil, and then estimating the frequency $\hat{o}_i$ by means of a locally weighted least square regression $r(x)$ using the Loess method [9]. Formally, letting $F(x) = |r(x) - f(x)|$, the ICI is defined as:

$$ICI = \int_0^1 F(x)\phi(x)dx \tag{4}$$

where $\phi(x)$ is the empirical density of the confidence scores. In addition to the problems related to the ECE, the estimation of the frequency $\hat{o}_i$ by means of the Loess regression is not robust when the data are not very dense or when the available sample is small [8,9].

One additional limitation concerning all of the metrics mentioned above (i.e., the Brier score, the ECE, MCE and ICI) is that, as mentioned in the Introduction, these scores provide an estimate of calibration only at the global level, that is with respect to a whole dataset, rather than on a local, bin-based or instance-based, level. By contrast, the recently proposed LCE [19] aims to measure calibration on a local level, that is relative to a a given new instance x, and the bin S_i to which x belongs. Formally, the LCE on instance x is defined as:

$$LCE_\gamma(x) = \frac{\sum_{x' \in S_i} |\mathbb{1}_{o_{x'}=f(x')^t} - f(x')|K_\gamma(x,x')}{\sum_{x' \in S_i} K_\gamma(x,x')} \tag{5}$$

$$MLCE_\gamma = \max_x LCE_\gamma(x) \tag{6}$$

where $K_\gamma(x,x') = g(\frac{x-x'}{\gamma})$, for $g : X^2 \mapsto \mathbb{R}$ and $\gamma \in \mathbb{R}$, is a kernel function, intuitively measuring the similarity between two instances. Thus, LCE measures the average calibration in a given bin, weighted by the similarity between each instance in the bin and the new instance: thus, MLCE is the maximum value of the LCE and it is then taken as a global measure of calibration.

It can be shown that, as $\gamma \to \infty$, $MLCE \to MCE$. Furthermore, being a local score, the LCE can also be used to *recalibrate* the confidence scores of a model [19]. Nonetheless, while the LCE and $MLCE$ allow for a more fine-grained definition of calibration compared to other existing metrics, they are still impacted by the same limitations as the ECE and MCE: in particular, the MLCE provides an overly conservative estimate of calibration, since it focuses only on the maximum observed deviation from perfect calibration.

Furthermore, compared to the ECE and MCE, the LCE and MLCE require two further hyper-parameters (other than the definition of the binning scheme), that is the value of γ and the kernel function g. Therefore, the robustness of LCE and MLCE is highly dependent on these parameters.

3 Definition of the GICI Index

As shown above, the metrics proposed so far in the specialist literature are affected by different limitations. The most relevant one, as already mentioned in the Introduction, regards the inability to provide information about calibration at the local level. Although LCE can provide this information and therefore it can be directly applicable for *recalibration*, this recent metric can be difficult to apply in real-world scenarios, as its application strongly depends on the underlying kernel function.

For these reasons, in this section, we introduce a novel metric we developed to address these shortcomings, the GICI. Similarly to the ECE and MCE, our metric only relies on geometric concepts, and this makes it simpler to interpret than the Brier score; moreover, it does not require additional parameters other than the selection of the proper binning scheme, making it easier to apply than the LCE. Finally, like the LCE our metric provides both a local and global estimation of the calibration of a classifier. As we will show in the next section, this latter property is particularly relevant, as it allows to define an instance-level recalibration method based on the estimation of a local/instance-based predictive value given the local distribution of the GICI index.

As before, let S be a dataset, $\mathcal{S}_k = \{S_i\}_{i=1}^k$ be a k-binning of S, and f a scoring classifier. The *calibration points* determined by $\mathcal{S}_k$ are defined as the sequence $\{(e_i, \hat{o}_i)\}_{i=1}^k$ where, as before, $e_i = \frac{1}{|S_i|} \sum_{x \in S_i} f(x)$ and $\hat{o}_i = \frac{1}{|S_i|} \sum_{x \in S_i} o_x$.

The GICI index is primarily defined at local level, that is at the level of the single bins (denoted as $LICI^i$, i.e. the local GICI index computed at the i-th bin). Intuitively, $LICI^i$ is defined as *the normalized distance between the calibration point $p_i = (e_i, \hat{o}_i)$ and the point p_i^* lying on the bisector closest to p_i*. In Fig. 1, the normalized distance is the ratio between the actual distance and the maximum distance from the bisector for that bin (it is worth of note that the maximum distance changes according to the bin and the corresponding distance between the x-axis of the reliability diagram and the bisector that passes through the predicted confidence score).

Since the bisector line is a 1-dimensional vector space with basis $b = \frac{1}{\sqrt{2}}(1,1)$, we can compute p_i^* as the orthogonal projection of p_i on the bisector line:

$$p_i^* = \langle b, p_i \rangle b,$$

where $|\cdot|_2$ is the Euclidean norm, and $\langle \cdot, \cdot \rangle$ is the standard inner product. Therefore, the distance between p_i and p_i^* is defined as:

$$d_i = |p_i - p_i^*|_2$$

Since the maximum value of d_i generally depends on p_i, we introduce a normalization step so that all d values are on the same scale. Given p_i, we define $\tilde{p}_i = (e_i, \tilde{o}_i)$ where

$$\tilde{o}_i = \begin{cases} 1 & e_i \leq 0.5, \\ 0 & otherwise \end{cases}$$

That is, $\tilde{p}_i$ is the point having the first component equal to e_i with maximum distance from the bisector line. We then define $\tilde{p}_i^* = \langle b, \tilde{p}_i^* \rangle b$, and d_i^{max} as

$$d_i^{max} = |\tilde{p}_i - \tilde{p}_i^*|_2$$

Then, the Local Interpretable Calibration Index (LICI) on bin i is defined as:

$$LICI^i = 1 - \frac{d_i}{d_i^{max}} \tag{7}$$

Intuitively, the local index measures the calibration within a bin by computing the distance of a calibration point (representing the average confidence score and average observed frequency) from the bisector line.

Starting from the local LICI index, we easily extend it to the whole k-binning $\mathcal{S}_k$ to obtain a global measure of calibration. First, we partition $\mathcal{S}_k$, into $\mathcal{S}_k^- = \{S_i : e_i > \hat{o}_i\}$ and $S_k^+ = \{S_i, e_i \leq \hat{o}_i\}$. Intuitively, $\mathcal{S}_k^-$ is the set of bins for which the corresponding calibration point lies below the bisector line (corresponding to over-forecasting), while $\mathcal{S}_k^+$ is the set of bins for which the corresponding calibration point lies above the bisector line (corresponding to under-forecasting). We then define two global versions of our index, namely the *Global Interpretable Calibration Index* (GICI) and the *Directional Interpretable Calibration Index* (DICI) indices:

$$GICI(\mathcal{S}_k) = \sum_{S_i \in \mathcal{S}_k} w_i \cdot LICI^i \tag{8}$$

$$DICI(\mathcal{S}_k) = GICI(\mathcal{S}_k^-) - GICI(\mathcal{S}_k^+) \tag{9}$$

where $w_i = \frac{|S_i|}{|S|}$. The ranges of $GICI$ and $DICI$ are, respectively, $[0,1]$ and $[-1,1]$.

Intuitively, the $GICI$ measures the total calibration of scoring classifier f on dataset S, relative to the k-binning $\mathcal{S}_k$, as the average of the local LICI indices on the k bins (weighted according to the number of data points observed in each bin). Hence, $GICI$ is the average normalized deviation from perfect calibration, as represented by the bisector line.

By contrast, the $DICI$ measures whether the scoring classifier tends to either over-estimate probabilities (i.e., over-forecasting), or under-estimate them (under-forecasting), on average. In particular, if $DICI < 0$ (i.e., $GICI(\mathcal{S}_k^-) < GICI(\mathcal{S}_k^+)$ then the model's confidence scores underestimate the observed frequencies; if $DICI > 0$ (i.e., $GICI(\mathcal{S}_k^-) > GICI(\mathcal{S}_k^+)$), then the model's confidence scores overestimate the observed frequencies) and if $DICI = 0$, then the model is on average balanced between over- and under-forecasting.

We note that the two above scores provide decision makers with complementary information: a perfectly calibrated model would be one for which $GICI = DICI = 0$, but $DICI = 0$ can hold even when a model is non-perfectly calibrated (e.g., this could happen when a model exhibits a typical symmetric sigmoid pattern of over-confidence on the high confidence scores and under-confidence on the low confidence scores).

As a final remark, we note that the GICI index is related to the ECE: indeed, it holds that $ECE = \sum_{\mathcal{S}_i \in \mathcal{S}_k} \sqrt{2} \cdot d_i^{max} \cdot w_i \cdot (1 - LICI^i)$. Thus, the ECE is the weighted sum of the unnormalized dual LICI scores, and the GICI score can be understood as a bin-wise normalized version of the ECE: being normalized, the GICI index facilitates interpretation in that local calibration scores are scaled so as to have uniform range across all bins. Indeed, the range of the bin-wise values of the ECE depends on the binning scheme, as well as on the distribution of data; their range, in general, is a proper subset of $[0, 1]$. This, in turn, implies that the ECE is obtained as the weighted average of values having a different scale; consequently, its interpretation may be more complex than in the case of GICI, for which the averaged values are always expressed in the same scale.

4 GICI-Based Recalibration

As mentioned in the previous sections, one of the advantages of the GICI index, compared to other existing proposals, is the fact that it is defined on a local level through the LICI index; this provides a twofold, connected advantage: the LICI index can be used to both define a computational method by which to correct the instance-level confidence scores, and therefore recalibrate the classifier on the basis of the LICI distribution and, consequently, to estimate the probability that the new instance x is positive (thus providing decision makers with a sort of instance-level predictive value). In this section, we illustrate the derivation of this recalibration method.

As in the previous sections, let S be a dataset and $\mathcal{S}_k$ a k-binning. Furthermore, let x a new instance, f the scoring classifier to be (re-)calibrated, and $c = f(x)$ be the confidence score of f on instance x. Intuitively, our aim is to estimate:

$$P(o_x = 1 | f(x) = c, LICI^i = p)$$

where $LICI^i = p$ is the local LICI score in the unique bin i s.t. $c_* = \min_{x' \in S_i} f(x') \leq f(x) \leq \max_{x' \in S_i} f(x') = c^*$. That is, given the model's confidence score and the corresponding local LICI index, we want to estimate the probability that the new instance x belongs to class 1. The proposed method to estimate this probability is given in what follows:

Theorem 1. *The probability $P(o_x = o, f(x) = c, LICI^i = p)$, for $o \in 0, 1$, can be estimated as:*

$$\frac{|S_i|}{|S|} \frac{\sum_{x' \in S_i} \mathbb{1}_{o_{x'} = o}}{|S_i|} \tilde{P} \tag{10}$$

where $\tilde{P}$ is the output of Algorithm 1.

Proof. By definition of conditional probability, the probability $P(o_x = 1 | f(x) = c, LICI^i = p)$ is equal to:

$$\frac{P(o_x = 1, f(x) = c, LICI^i = p)}{P(f(x) = c, LICI^i = p)},$$

where, by the law of total probability the denominator is equal to:

$$P(f(x) = c, LICI^i = p, o_x = 0)$$
$$+ P(f(x) = c, LICI^i = p, o_x = 1)$$

Therefore, we can focus on the estimation of $P(o_x = o, f(x) = c, LICI^i = p)$ to obtain an estimate of the quantity of interest. Letting $o \in \{0, 1\}$, then:

$$P(f(x) = c, LICI^i = p, o_x = o)$$
$$= P(LICI^i = p | o_x = o, f(x) = c)$$
$$* P(o_x = o | f(x) = c)$$
$$* P(f(x) = c)$$

By the law of the large numbers, and by assuming that as the size $|S|$ grows also the number of bins k similarly grows, then $P(o_x = o | f(x) = c), P(f(x) = c)$, can, respectively, be approximated as:

$$P(o_x = o | f(x) \in S_i) \sim \frac{\sum_{x' \in S_i} \mathbb{1}_{o_{x'} = o}}{|S_i|} \tag{11}$$

$$P(f(x) \in S_i) \sim \frac{|S_i|}{|S|} \tag{12}$$

As for $P(LICI^i = p | o_x = o, f(x) = c)$, by marginalization it can be expressed equivalently as the expectation:

$$\int P(LICI^i = p | o_x = o, f(x) = c, \mathbf{1}, \mathbf{c}) P(\mathbf{1}) P(\mathbf{c}) \tag{13}$$

where $\mathbf{1} \in \{0, 1\}^{|S_i|}$, $\mathbf{c} \in [c_*, c^*]^{|S_i|}$ are two $|S_i|$-dimensional random vectors, and the integration is taken with respect to the distributions of $\mathbf{1}$ and $\mathbf{c}$ in bin

S_i. Finally Eq. 13 can be approximated via a Monte Carlo estimate [26], as described by Algorithm 1. By the law of the large numbers, as $n \rightarrow \infty, \epsilon \rightarrow 0$, $\tilde{P}$ then approximates the desired probability.

Algorithm 1. Re-sampling algorithm to estimate $P(LICI^i = p | o_x = o,$ $f(x) = c)$

---

    procedure ESTIMATE_LICI_CONDITIONAL(S_i: bin, c: confidence score, o: label, n: num. of samples, ϵ: approximation parameter)

        $v_{true} \leftarrow LICI^i$

        $P \leftarrow 0$

        for $s = 1$ to n **do**

            $S_i' \leftarrow$ resample from S_i with replacement

            $S_i'.append((c, o))$

            Let v_{temp} be the value of $LICI$ on S_i'

            if $|v_{true} - v_{temp}| \leq \epsilon$ **then**

                $P \leftarrow P + 1$

            end if

        end for

        Output: $\frac{P}{n}$

    end procedure

Intuitively, Eq. (10) decomposes the desired instance-level probability in three terms. The first two terms are sample statistics, which estimate the probability of the outcome $o_x = o$ and of the confidence score $f(x) = c$ on a local level (that is, limited to bin S_i). By contrast, the third term is (an estimate of) the conditional distribution of the proposed metric $LICI^i$.

5 Comparative Experiments

In this section, we illustrate the applicability of our recalibration method by showing that, on a set of medical benchmark datasets chosen for their representativeness, the GICI index allows to improve the calibration of ML models, and to outperform other state-of-the-art recalibration methods.

5.1 Datasets

To illustrate the application of the proposed methods, we selected a collection of four, commonly used [25] medical benchmark datasets. We focused in particular on the healthcare domain as a paradigmatic example of setting where calibration is key for the development of reliable ML models [33].

    The four tabular datasets were selected based on four criteria: being openly available online; supporting binary classification tasks in medical context (as paradigmatic of settings where calibration matters); having sufficient cardinality and dimensionality; being affected by various degrees of imbalance of the target variable (see Table 1), i.e., imbalance either toward the positive class, or the negative class, or almost absent. In regard to this latter criterion, the choice was based on the previously mentioned impact of the distribution of labels on the metric most commonly used in the literature (i.e., the Brier score).

The selected datasets are: Breast Cancer Detection [37] (Breast), Pima Indians Diabetes [28] (Pima), Indian Liver Patient [27] (ILP), and Statlog (Heart) Data Set [13]. Table 1 shows the percentage of instances belonging to both classes, as well as the sample size and test size (%).

Table 1. Datasets used in the experiments.

Data set	Class **0**	Class **1**	# Obs	Test split
Breast	62.7 (%)	37.3 (%)	569	20 (%)
Pima	65.1 (%)	34.9 (%)	768	20 (%)
ILP	28.6 (%)	71.4 (%)	583	20 (%)
Heart	55.5 (%)	44.5 (%)	270	20 (%)

5.2   Experimental Setup

For each of the datasets, we decided to apply logistic regression as the classifier of choice. We decided to use this model as it is simple (in that no hyper-parameters has to be identified) and, nevertheless, is competitive in terms of performance to other more complex ML models [7,22], especially in imbalanced tasks [18]. Furthermore, logistic regression usually has good calibration compared to other common ML models [7], thus making it a more conservative baseline to compare recalibration methods (including the one proposed in this paper).

As a first illustrative experiment, we trained the baseline logistic regression model on the training set of each dataset, and then evaluated its calibration on the separate hold-out test sets, by means of different global calibration metrics, namely: the Brier score, ECE, GICI and DICI. This experiment had the aim to illustrate in a simple manner the application of our metrics.

Then, to evaluate the effectiveness of the proposed GICI recalibration method, we considered four different models: the standard logistic regression model (henceforth, baseline), the GICI recalibrated model, the sigmoid recalibrated (i.e., Platt scaled) model and the model recalibrated by means of isotonic regression. The baseline model was trained on the whole training set. In regard to sigmoid and isotonic regression, they were trained on the training set by means of cross-validation, as implemented in the scikit-learn library. In regard to the GICI recalibration, we performed a 5-fold cross-validation on the training set, in order to select the binning scheme (i.e., the number of bins: the search space was set within 3 to 20 quantiles) minimizing the Brier score across the fold (in particular, the number of bins was chosen based on the comparison of the Brier score 90% confidence intervals, as computed through the boostrap [12] method).

In all of the models, SMOTE over-sampling procedure [6] and standardization were applied as pre-processing steps on the training set. All methods were subsequently evaluated on the separate hold-out test set, to evaluate which model (if any) provided the best improvement in terms of calibration. In particular, comparison among the models was performed in terms of the Brier score since this is the most common metric and its definition is independent of the binning scheme. In addition to the point estimate, we also computed the 90% confidence intervals of the above mentioned Brier scores, and the average point estimate were compared for statistical significance using the non-parametric Friedman test (with the Conover post-hoc procedure) [11], declaring all p-values lower than .1 to be significant (at the 90% confidence level).

5.3 Results and Implications

The results of the illustrative experiment are reported in Table 2. The results of the experimental comparison, on the other hand, are reported in Table 3, in terms of average scores and 90% confidence intervals.

Table 2. Scores from different calibration metrics for baseline.

Dataset	GICI	DICI	ECE	Brier
Breast	.99	0	.02	.02
Pima	.86	.11	.10	.17
ILP	.75	−.25	.15	.20
Heart	.85	.02	.10	.12

Table 3. Calibration of the baseline and recalibrated models, in terms of the Brier score obtained on the hold-out test set. Square brackets denote the 90% confidence intervals. For each dataset, the best method (in terms of Brier score) is denoted in bold.

Dataset	Baseline	GICI	Sigmoid	Isotonic
Breast	.02 [.00, .04]	**.01 [.00, .03]**	.02 [.00, .04]	.03 [.00, .05]
Pima	.17 [.12, .22]	**.15 [.11, .20]**	.16 [.11, .21]	.16 [.11, .21]
ILP	.20 [.14, .25]	**.17 [.11, .23]**	**.17 [.12, .23]**	**.17 [.12, .23]**
Heart	.12 [.05, .19]	**.11 [.04, .18]**	.13 [.06, .20]	.12 [.05, .20]

As we see in Table 2, and as argued in the previous sections our metric is more interpretable than the Brier score and the ECE, e.g., showing that calibration for the ILP model is 25% far from perfection (but not worse), and also showing that the mis-calibration is towards under-forecasting (not the opposite). Also, the difference between the PIMA and Heart models' calibration, as shown to be

Table 4. P-values for the post-hoc pair-wise comparison among the considered methods. Significant p-values are denoted in bold.

	Baseline	GICI	Sigmoid	Isotonic
Baseline	-	**.049**	1	.698
GICI	-	-	**.049**	**.094**
Sigmoid	-	-	-	.698

wide by the Brier score (and yet undetected by the ECE score) is recognized by the GICI index as being small, and mainly due to over-forecasting. In regard to Table 3, the statistical analysis according to the Friedman test reported a p-value of .007, which was significant at the adopted confidence level. Therefore, we applied the post-hoc test, whose p-values are reported in Table 4.

As shown in Tables 3 and 4, the GICI recalibration method reported the greatest improvement in terms of calibration: in 3 datasets out of 4, the GICI recalibration reported the lowest Brier Score, and in the remaining dataset the GICI recalibration reported the same performance as the sigmoid scaling and isotonic regression. Furthermore, this improvement was significant at the 90% confidence level (and, with respect to the baseline model and sigmoid scaling, also at the 95% confidence level). Although these results were obtained on a small collection of benchmark datasets, and therefore cannot be generalized, they are promising since the datasets had different patterns of label balance, showing that the proposed calibration metrics (and the derived recalibration method) can be effective in improving calibration also in comparison with commonly adopted approaches.

A natural question that arises from these promising results is: "*so what?*" While being able to improve calibration can be by itself interesting, what we want to emphasize here is the potential connection between calibration and accuracy [29], at which we hinted in the Introduction. This connection arises from focusing on the impact of the GICI recalibration on the confidence scores in the neighborhood of the decision cut-off (that is, the threshold in the confidence score range that is used to determine whether an instance is classified as either positive or negative), which is usually set at the value .5. Let us consider an example whose confidence score falls in the neighborhood of the above threshold, e.g., .48. If the model was non-calibrated towards under-forecasting, the GICI recalibration would raise this value to, say, .51. This change in the confidence score would imply a change of classification from the negative to the positive label, and could thus affect the accuracy of the model. To better illustrate this point, we show the results of a simple experiment, in which we considered the effect of the GICI recalibration on the confidence scores within the range (.48 − .52) associated with the instances of the test set from the ILP dataset [27]. The results are reported in Table 5 and Fig. 2. We can make two observations; the first one is obvious: after applying the GICI recalibration method, the confidence scores became closer to the actual frequency of the positive class in the bin (i.e., 86%, 43% for the baseline model vs 71% for the GICI recalibrated one). Second, the GICI recalibration also improved the accuracy in the bin by 28 percent points, from 29% for the baseline model to 57% for the GICI recalibrated one, that is by a 30% relative improvement. This shows that the information about local calibration used in the definition of the GICI can be useful not only to improve calibration *per se* (as expected), but also to influence the accuracy of ML models.

Table 5. An example of GICI recalibration. CS denotes the confidence scores of the baseline model, while CS_{GICI} denotes the GICI recalibrated confidence scores. The accuracy of the baseline model in the bin was 29% (2 true positives, 4 false negatives and 1 false positive out of 7 instances), while the accuracy of the GICI recalibrated model was 57% (4 true positives and 1 false negative out of 7 instances).

CS	.51	.49	.51	.49	.49	.49	.51
CS_{GICI}	.68	.49	.68	.65	.49	.74	.51
True label	1	1	0	1	1	1	1

Concluding, and in support of the above argument, we note that the value of $DICI$ for the ILP dataset (see Table 2) is equal to $-.25$, which suggests that the baseline model suffers from under-forecasting, especially so around the decision threshold (see Fig. 2). Then, Table 5 highlights how the confidence scores in this neighborhood were adjusted towards a value higher than 0.5. This example shows how the GICI index can provide more comprehensive and interpretable information about a model's calibration.

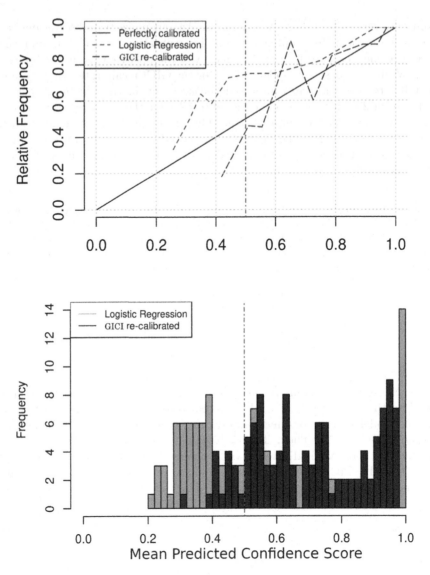

Fig. 2. Calibration curve and confidence score distribution, comparing the baseline and the GICI recalibrated models, on the ILP test set. The red dotted line denotes the .5 decision cut-off. (Color figure online)

6   Conclusions

In this paper we have presented a novel metric to assess the local-level degree of calibration of a classifying model, the GICI index. The property of being defined at local level allows to assess calibration in an easier to interpret manner, and at a much higher level of detail than other more common metrics, which nevertheless are only capable to provide an overall indication about calibration. Our index also allows to evaluate the calibration of a model at a global level, i.e. over the whole range of confidence scores that it produces, so as to allow an easy comparison between models; it also provides two other, even more important, indications: one indication, always defined at global level, regarding the tendency to make either over-forecasting or under-forecasting (i.e., what no metric commonly adopted allows to do); and also an estimate of the probability that the model is accurate regarding a specific prediction, that is a sort of *predictive value* or *precision* at the level of a single instance classified. This latter can be obtained by means of the GICI recalibration method, which we tested on a small but exemplificatory set of medical datasets with promising results about the capability to improve the calibration of a simple classifier over the best known and most widely used methods, i.e., Platt Scaling and Isotonic Regression.

In light of these results, we aim our future research toward a more comprehensive evaluation of the potential of the GICI index, for instance by comparing GICI-based recalibration performance with respect to different metrics (including ours), recalibration methods and binning schemes, by considering the sensitivity of the GICI index w.r.t. to the number of bins or the use of alternative binning procedures [30], and by embedding in its definition some way to take the similarity of the instance into account with respect to the predictions performed in the same confidence bin [4,19]. Similarly, we plan to investigate the potential of the local version of the GICI in XAI, to ascertain which parts of the input are impacting most on the bins where the local calibration error is higher. Moreover, from the human-AI interaction perspective, we will investigate whether providing decision makers with additional information about the classifier can actually improve human decision effectiveness (in terms of error rate), trust and satisfaction, in terms of higher transparency and explainability. Indeed, adopting the GICI index in such naturalistic settings will allow decision support systems to expose their global degree of calibration and whether they either tend to over- or under-forecast, as well as more information about their specific prediction, such as the instance-level calibration of the model (that is the extent it is calibrated in the neighborhood of the confidence score associated with the prediction), the local calibration of the model in the neighborhood of the decision cut-off (that is the confidence region where classification is performed, generally .5), and, as we hinted above, a sort of positive predictive value.

After all, we believe that what calibration is all about is mainly giving human decision makers reasons to trust computational decision aids more, especially for the very instance at hand. As we said above, better ways to measure calibration means better models, *as well as* better ways to interpret their output.

References

1. Assel, M., Sjoberg, D., Vickers, A.: The brier score does not evaluate the clinical utility of diagnostic tests or prediction models. Diagn. Progn. Res. **1**, 1–7 (2017)
2. Brier, G.W.: Verification of forecasts expressed in terms of probability. Mon. Weather Rev. **78**(1), 1–3 (1950)

3. Burt, T., Button, K., Thom, H., Noveck, R., Munafò, M.R.: The burden of the "false-negatives" in clinical development: analyses of current and alternative scenarios and corrective measures. Clin. Transl. Sci. **10**(6), 470–479 (2017)

4. Cabitza, F., Campagner, A., Sconfienza, L.M.: As if sand were stone. New concepts and metrics to probe the ground on which to build trustable AI. BMC Med. Inform. Decis. Making **20**(1), 1–21 (2020)

5. Cabitza, F., et al.: The importance of being external. Methodological insights for the external validation of machine learning models in medicine. Comput. Methods Programs Biomed. **208**, 106288 (2021)

6. Chawla, N.V., Bowyer, K.W., Hall, L.O., Kegelmeyer, W.P.: SMOTE: synthetic minority over-sampling technique. J. Artif. Intell. Res. **16**, 321–357 (2002)

7. Christodoulou, E., Ma, J., Collins, G.S., Steyerberg, E.W., Verbakel, J.Y., Van Calster, B.: A systematic review shows no performance benefit of machine learning over logistic regression for clinical prediction models. J. Clin. Epidemiol. **110**, 12–22 (2019)

8. Cleveland, W.S.: Robust locally weighted regression and smoothing scatterplots. J. Am. Stat. Assoc. **74**(368), 829–836 (1979)

9. Cleveland, W.S., Devlin, S.J.: Locally weighted regression: an approach to regression analysis by local fitting. J. Am. Stat. Assoc. **83**(403), 596–610 (1988)

10. DeGroot, M.H., Fienberg, S.E.: The comparison and evaluation of forecasters. J. Roy. Stat. Soc. Ser. D (Stat.) **32**(1–2), 12–22 (1983)

11. Demšar, J.: Statistical comparisons of classifiers over multiple data sets. J. Mach. Learn. Res. **7**, 1–30 (2006)

12. Efron, B.: Bootstrap methods: another look at the jackknife. Ann. Stat. **7**(1), 1–26 (1979)

13. Frank, A., Asuncion, A.: Statlog (heart) data set (2010). http://archive.ics.uci.edu/ml/datasets/Statlog+(Heart)

14. Gneiting, T., Raftery, A.E.: Strictly proper scoring rules, prediction, and estimation. J. Am. Stat. Assoc. **102**(477), 359–378 (2007)

15. Guo, C., Pleiss, G., Sun, Y., Weinberger, K.Q.: On calibration of modern neural networks. In: International Conference on Machine Learning, pp. 1321–1330. PMLR (2017)

16. Hartmann, H.C., Pagano, T.C., Sorooshian, S., Bales, R.: Confidence builders: evaluating seasonal climate forecasts from user perspectives. Bull. Am. Meteor. Soc. **83**(5), 683–698 (2002)

17. Kompa, B., Snoek, J., Beam, A.L.: Second opinion needed: communicating uncertainty in medical machine learning. NPJ Digit. Med. **4**(1), 1–6 (2021)

18. Luo, H., Pan, X., Wang, Q., Ye, S., Qian, Y.: Logistic regression and random forest for effective imbalanced classification. In: 2019 IEEE 43rd Annual Computer Software and Applications Conference (COMPSAC), vol. 1, pp. 916–917. IEEE (2019)

19. Luo, R., et al.: Localized calibration: metrics and recalibration. arXiv preprint arXiv:2102.10809 (2021)

20. Naeini, M.P., Cooper, G.F., Hauskrecht, M.: Obtaining well calibrated probabilities using Bayesian binning. In: Proceedings of the Twenty-Ninth AAAI Conference on Artificial Intelligence, AAAI 2015, pp. 2901–2907. AAAI Press (2015)

21. Niculescu-Mizil, A., Caruana, R.: Predicting good probabilities with supervised learning. In: Proceedings of the 22nd International Conference on Machine Learning, pp. 625–632 (2005)

22. Nusinovici, S., et al.: Logistic regression was as good as machine learning for predicting major chronic diseases. J. Clin. Epidemiol. **122**, 56–69 (2020)

23. Platt, J.: Probabilistic outputs for support vector machines and comparisons to regularized likelihood methods. Adv. Large Margin Classif. **10**(3), 61–74 (1999)
24. Raghu, M., et al.: Direct uncertainty prediction for medical second opinions. In: International Conference on Machine Learning, pp. 5281–5290. PMLR (2019)
25. Ramana, B.V., Boddu, R.S.K.: Performance comparison of classification algorithms on medical datasets. In: 2019 IEEE 9th Annual Computing and Communication Workshop and Conference (CCWC), pp. 0140–0145. IEEE (2019)
26. Robert, C., Casella, G.: Monte Carlo Statistical Methods, vol. 2. Springer, New York (2004). https://doi.org/10.1007/978-1-4757-4145-2
27. Rossi, R.A., Ahmed, N.K.: ILP, Indian liver patient dataset. In: AAAI (2015). https://networkrepository.com
28. Rossi, R.A., Ahmed, N.K.: Pima Indians diabets dataset. In: AAAI (2015). https://networkrepository.com
29. Sahoo, R., Zhao, S., Chen, A., Ermon, S.: Reliable decisions with threshold calibration. In: Advances in Neural Information Processing Systems, vol. 34 (2021)
30. Scargle, J.D.: Studies in astronomical time series analysis. V. Bayesian blocks, a new method to analyze structure in photon counting data. Astrophys. J. **504**(1), 405 (1998)
31. Steyerberg, E., et al.: Assessing the performance of prediction models a framework for traditional and novel measures. Epidemiology **21**, 128–38 (2010)
32. Vaicenavicius, J., Widmann, D., Andersson, C., Lindsten, F., Roll, J., Schön, T.: Evaluating model calibration in classification. In: The 22nd International Conference on Artificial Intelligence and Statistics, pp. 3459–3467. PMLR (2019)
33. Van Calster, B., McLernon, D.J., Van Smeden, M., Wynants, L., Steyerberg, E.W.: Calibration: the achilles heel of predictive analytics. BMC Med. **17**(1), 1–7 (2019)
34. Van Calster, B., Vickers, A.J.: Calibration of risk prediction models: impact on decision-analytic performance. Med. Decis. Making **35**(2), 162–169 (2015)
35. Vovk, V., Petej, I.: Venn-abers predictors. arXiv preprint arXiv:1211.0025 (2012)
36. Wallace, B.C., Dahabreh, I.J.: Class probability estimates are unreliable for imbalanced data (and how to fix them). In: 2012 IEEE 12th International Conference on Data Mining, pp. 695–704. IEEE (2012)
37. Wolbergs, W., et al.: Breast cancer wisconsin (diagnostic) data set. UCI Machine Learning Repository (1992). http://archive.ics.uci.edu/ml/
38. Zhao, S., Ma, T., Ermon, S.: Individual calibration with randomized forecasting. In: International Conference on Machine Learning, pp. 11387–11397. PMLR (2020)

The ROC Diagonal is Not Layperson's Chance: A New Baseline Shows the Useful Area

André M. Carrington[1](✉), Paul W. Fieguth[2], Franz Mayr[3], Nick D. James[4], Andreas Holzinger[5], John W. Pickering[6], and Richard I. Aviv[1]

[1] Department of Radiology, Radiation Oncology and Medical Physics, Faculty of Medicine, University of Ottawa and the Ottawa Hospital, Ottawa, Canada
{acarrington,raviv}@toh.ca

[2] Department of Systems Design Engineering, University of Waterloo, Waterloo, Canada
pfieguth@uwaterloo.ca

[3] Faculty of Engineering, Universidad ORT Uruguay, Montevideo, Uruguay
mayr@ort.edu.uy

[4] Software Solutions, Systems Integration and Architecture, The Ottawa Hospital, Ottawa, Canada
njames@toh.ca

[5] University of Natural Resources and Life Sciences Vienna, Vienna, Austria
andreas.holzinger@human-centered.ai

[6] Christchurch Heart Institute, Department of Medicine, University of Otago, Christchurch, New Zealand
john.pickering@otago.ac.nz

Abstract. In many areas of our daily lives (e.g., healthcare), the performance of a binary diagnostic test or classification model is often represented as a curve in a Receiver Operating Characteristic (ROC) plot and a quantity known as the area under the ROC curve (AUC or AUROC). In ROC plots, the main diagonal is often referred to as "chance" or the "random line". In general, however, this does not correspond to the layperson's concept of chance or randomness for binary outcomes. Rather, this represents a special case of layperson's chance, or the ROC curve for a classifier that has the same distribution of scores for the positive class and negative class. Where the ROC curve of a model deviates from the main diagonal, there is information. However, not all information is "useful information" compared to chance, including some areas and points above the diagonal. We define the binary chance baseline to identify areas and points in a ROC plot that are more useful than chance. In this paper, we explain this novel contribution about the state-of-art and provide examples that classify benchmark data.

Keywords: ROC · AUC · C-statistic · Chance · Explainable AI · Classification · Diagnostic tests

© IFIP International Federation for Information Processing 2022
Published by Springer Nature Switzerland AG 2022
A. Holzinger et al. (Eds.): CD-MAKE 2022, LNCS 13480, pp. 100–113, 2022.
https://doi.org/10.1007/978-3-031-14463-9_7

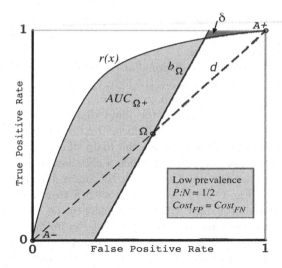

Fig. 1. Layperson's chance or 'binary chance', is a coin toss represented by the point $\Omega = (0.5, 0.5)$. Performance (cost-weighted accuracy) equal to Ω is depicted as the solid straight line b_Ω: the binary chance baseline. The ROC curve $r(x)$ has a useful area under the curve $AUC_{\Omega+}$ above and to the left of b_Ω—not the main diagonal d. There is negative utility in δ.

1 Introduction

In many fields which affect human life (e.g. health), the receiver operating characteristic (ROC) plot [3,19,23] depicts the performance of a binary diagnostic test or classification model as a ROC curve (Fig. 1, $r(x)$) which may be smooth (fitted) or staircase-like (empirical). Each point on the curve has a threshold value, often not displayed, where that threshold (e.g., $t = 0.7$) decides if a classifier's output (e.g., 0.6) is a positive outcome, if greater, or a negative outcome, if not.

In the ROC plot, a line drawn from the bottom left to the top right is called the main diagonal or chance diagonal (Fig. 1, dashed line) [12,27] because it is commonly said to represent chance [20,27], while others describe it as representing a random classifier [9].

The main diagonal is commonly used to interpret results in two ways. First, for the classifiers where a model's ROC curve is higher than the main diagonal, the model is said to be informative. Second, a model is thought to be better than chance where it is higher than the main diagonal—but we show that is **not** true for the most intuitive concept of chance for binary outcomes: a fair coin toss.

It is useful to compare any binary diagnostic test or classifier, to how a coin toss would decide the outcome for an instance or input, because if it performs worse, then we might as well use a coin toss. That is, we would compare a classifier against a black box classifier that had chance as a coin toss, as its internal mechanism.

A classifier starts to become useful when it performs better than a coin toss. However, let's posit the status quo, that a classifier starts to become useful at or just above the main diagonal.

The point (1, 1) on the main diagonal at the top-right of a ROC plot (Fig. 1, $A+$), is an all-positive classifier. It has the lowest possible threshold and predicts that all instances are positive. Predictions from $A+$ are mostly wrong for low prevalence data, yet $A+$ is also on the main diagonal which is also said to represent chance. Clearly, a coin toss performs better, being half right and half wrong, instead of mostly wrong. Hence the main diagonal does not represent chance as a coin toss, nor the point at which a classifier becomes useful.

Conversely, consider the point (0, 0) on the main diagonal at the bottom-left of a ROC plot (Fig. 1, $A-$). A classifier with a threshold at $A-$ classifies all inputs as negative. For data with low prevalence, i.e., few instances in the positive class, its predictions are mostly correct—more than the 50% correct predictions one obtains with a fair coin toss. Yet, $A-$ is on the main diagonal which is said to represent chance. Clearly, $A-$ performs better than chance as a coin toss—the layperson's concept of chance.

The main diagonal is commonly treated as a performance baseline, as in the examples above, but when it is used that way, we explain that what actually captures the user's intention and expectation is a line with performance equal to a fair coin toss.

This paper has four contributions, two that are clarifying and two that are novel. Our first contribution clarifies that the literature referring to the main diagonal as chance, is misleading, because it is not layperson's chance, nor any definition of chance found in a dictionary. Secondly, to evaluate and explain performance relative to layperson's chance (binary chance), we define a novel baseline. Iso-performance lines have not previously been applied for this purpose. Thirdly, we clarify that for realistic performance evaluation and explanation, one must express the prevalence and costs of error, which are either assumed implicitly or specified outright. These are relative costs incurred by the test subject or patient, not the health system—and these can be estimated without difficulty. Fourthly, we show the useful part of the area under the curve, as a novel contribution, demarcated by our novel baseline.

In the sections that follow we discuss: binary chance, the binary chance baseline, the main diagonal, the useful area under the ROC curve, counterbalancing effects on the slope of the binary baseline, examples in classification, related work and conclusions.

2   Binary Chance

We refer to "chance" in this manuscript as something which happens by chance, i.e., "luck" or "without any known cause or reason" [4]. Randomness is a synonym, and for binary outcomes, a fair coin toss produces a random outcome.

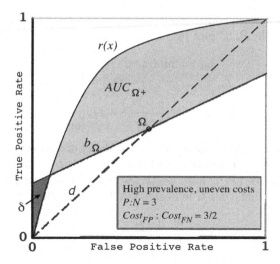

Fig. 2. In this example the binary chance baseline b_Ω has a low slope (lower than 1). This occurs for data with high prevalence and balanced costs, or when the class imbalance is greater than the cost imbalance, i.e., $3 > \frac{3}{2}$. For the ROC curve r(x), the useful area under the curve is $AUC_{\Omega+}$ (yellow) and negative utility is found in δ (red). (Color figure online)

We introduce the term "binary chance" for a fair coin toss. This is different from continuous chance as in the main diagonal (Sect. 4) and it is also distinguished from "a chance" of rain which implies a non-zero probability without fairness.

In this paper, we consider a coin toss, all-positive decisions and all-negative decisions to be classifiers with a hidden mechanism. The mechanism may consider the inputs or ignore them altogether when producing a corresponding output. Such classifiers are not good or useful by themselves—but they act as a baseline against which other classifiers are compared to determine if they are useful or not.

A fair coin toss (binary chance) is represented by the centre point $\Omega = (0.5, 0.5)$ [25] in a ROC plot (Fig. 2) because on average, the rate of heads (events) is 0.5 and the rate of tails (non-events) is 0.5.

To make comparison easy and visible between chance and points and areas in a ROC plot, we can draw a line through Ω that has performance equivalent to binary chance as established in the literature [16].

3   The Binary Chance Baseline

To find points with performance equivalent to binary chance, we draw an iso-performance line [9,10,16], denoted b_Ω through Ω (Fig. 2) and call it the binary chance baseline. It can represent cost-weighted accuracy, accuracy, or balanced accuracy—equivalent to chance.

We define the binary chance baseline $y = b_\Omega(x)$ or $x = b_\Omega^{-1}(y)$ for data $\mathcal{X}$ as a line with performance equal to chance $\Omega = (0.5, 0.5)$ with slope $m_\mathcal{X}$, except where the line is bottom-coded (no less than 0) and top-coded (no greater than 1) to stay within the ROC plot by $[\,\cdot\,]_{01}$, as follows:

$$[\,\cdot\,]_{01} := \min(\max(\,\cdot\,,0),1) \tag{1}$$

$$b_\Omega(x) := [m_\chi\,(x - \Omega_x) + \Omega_y]_{01} \tag{2}$$

$$b_\Omega^{-1}(y) := \left[\frac{(y - \Omega_y)}{m_\chi} + \Omega_x\right]_{01} \tag{3}$$

where the literature defines the slope m_χ (or skew) of iso-performance lines [10, 16, 21] as a function of: P positives and N negatives (or prevalence π); and costs $C_{(.)}$ of false positives (FP), false negatives (FN), treatment for true positives (TP) and non-treatment for true negatives (TN) as follows.

$$m_\chi := \frac{N}{P} \cdot \frac{C_{FP} - C_{TN}}{C_{FN} - C_{TP}} \tag{4}$$

$$= \frac{(1 - \pi)}{\pi} \cdot \frac{C_{FP} - C_{TN}}{C_{FN} - C_{TP}} \tag{5}$$

Notably, we are referring to the **relative cost** of a false negative versus a false positive, **incurred by a patient or subject**, such as reduced quality of life, risk of death, lost opportunity, or anxiety. We are **not** referring to system costs such as the test apparatus, labor, materials or administration.

We combine Eqs. (2) and (5), then (3) and (5) to obtain the explicit formula for the binary chance baseline, from a vertical and horizontal perspective, respectively:

$$b_\Omega(x) := \left[\frac{(1 - \pi)}{\pi} \cdot \frac{C_{FP} - C_{TN}}{C_{FN} - C_{TP}}(x - 0.5) + 0.5\right]_{01} \tag{6}$$

$$b_\Omega^{-1}(y) := \left[\frac{(y - 0.5)}{\frac{(1-\pi)}{\pi} \cdot \frac{C_{FP} - C_{TN}}{C_{FN} - C_{TP}}} + 0.5\right]_{01} \tag{7}$$

If we attempt to ignore costs, then we are *de facto* assuming equal costs which is worse than an estimate of costs. The estimate does not need to be perfect, just a better assumption. We usually know, for any given problem, if the cost of a false negative is worse than a false positive, or vice-versa, or about the same.

If we specify the prevalence and all of the cost parameters in the equation above (6) then the binary chance baseline represents **cost-weighted accuracy** or **average net benefit** equivalent to chance. This is the most realistic way to interpret performance relative to chance, and the baseline only aligns with the main diagonal in special cases when the class ratio and costs are balanced, together, or separately.

If we specify the prevalence but ignore costs, then the baseline represents **accuracy** equivalent to chance. When the prevalence is 50%, which is rarely the case, the baseline coincides with the main diagonal.

If we ignore prevalence and costs then the baseline represents **balanced accuracy** and coincides with the main diagonal. This approach and measure, while unrealistic, can be useful to view performance without the majority class dominating. To compare performance between different data sets (with different class imbalances) as an abstract sanity check or benchmark of sorts. For example, 70% AUC or 20% area under the curve and above the main diagonal, is generally considered to be a reasonable classifier, in an abstract sense. However, realistically, for some applications such as detecting melanoma or fraud, such performance may be inadequate.

Hence, there is a choice between two different goals: unrealistic theoretic performance that can be compared between data sets (e.g., AUC and the main diagonal), versus realistic performance measures for decision-making the include prevalence and costs (cost-weighted accuracy and average net benefit).

Historically, use of the ROC plot, AUC and the main diagonal have ignored prevalence and costs. However, artificial intelligence (AI) is being increasingly applied to situations that affect people in everyday life, which requires real-world explanations.

The binary chance baseline is meaningful for explanation because a model starts to become useful only when it performs better than chance—otherwise, a coin toss performs just as well. The main diagonal can be misleading if we use it for the wrong reasons, which leads into the next section: what the main diagonal does represent, if not layperson's chance.

4   The Main Diagonal

The literature on ROC plots sometimes refers to the ROC main diagonal as chance [20, 27]—a ROC curve produced by classification scores (or probabilities) drawn from *the same distribution* for events and non-events. The distribution does not need to be specified, as long as it is (almost everywhere[1]) the same. The classification scores (that underlie and precede the binary outcome) have an equal chance of being an event or non-event, hence we refer to this as "continuous chance" as distinguished from binary chance (layperson's chance).

When the distribution of scores is the same for events and non-events, the model is not informative, i.e., it has zero information to distinguish the two classes, resulting in a ROC curve along the main diagonal [13, Fig. 2]. The absence of information is demonstrated by divergence and distance measures which are zero in this case: the Jensen-Shannon (J-S) divergence [14, 15, 17] is zero, the Kullback-Leibler divergence [7] in either direction is zero, and the Hellinger distance [2] is zero.

The Hellinger distance fits the classification context nicely since it achieves a value of 1 when the two distributions are non-overlapping, or when two uni-model distributions are linearly separable—a situation where many classifiers

---

[1] Two probability density functions (PDF) are the same "almost everywhere" if they disagree on, at most, a set of isolated points (more formally, on a *set of measure zero*). This qualification is, admittedly, somewhat pedantic but necessary because any two such PDFs are effectively the same (and share the same cumulative distribution function). Changing a PDF at only individual points has no actual effect on the corresponding random variable it describes.

can perform perfectly with an AUC of 1 [13, Fig. 2]. That is, the [0, 1] range of the Hellinger distance correlates to the [0.5, 1] range of AUC achievable by most classifiers.

Points (and curves) on the main diagonal are said to be non-informative and of no predictive value [9, 18, 19]—however, that is typically not quite accurate. A model may not have any information along the diagonal, but if a threshold is chosen wisely based on prevalence as prior knowledge, then the resulting classifier may be useful at that threshold—informed by the threshold choice, not the model's own intelligence.

In all situations except those in which costs and prevalence exhibit a rare and unusual equilibrium, some points on the main diagonal if chosen with prior knowledge of prevalence, are more predictive than (binary) chance, as we have explained previously in the introduction for the points $A+$ and $A-$.

5   The Useful Area Under the ROC Curve

To explain performance in a ROC plot, we must consider both the binary chance baseline and the main diagonal. The area under the ROC curve and above the binary chance baseline (the yellow area denoted $AUC_{\Omega+}$ in Fig. 2), tells us where a model is useful, i.e., more useful than chance.

We define the useful area $AUC_{\Omega+}$ for an ROC curve $r(x)$ relative to binary chance Ω as follows, with a vertical and horizontal component, like the concordant partial AUC [3]. We use the notation $[\,\cdot\,]_+$ for a function or filter that only passes positive values (bottom codes to zero). For a partial (or whole) ROC curve in the range $\theta_{xy} = \{x \in [x_1, x_2],\ y \in [y_1, y_2]\}$:

$$[\cdot]_+ := \min(\,\cdot\,,\ 0) \tag{8}$$

$$AUC_{\Omega+}(\theta_{xy}) := \frac{1}{2}\int_{x_1}^{x_2} [r(x) - b_\Omega(x)]_+\ dx$$

$$+ \frac{1}{2}\int_{y_1}^{y_2} \left[\left(1 - r^{-1}(y)\right) - \left(1 - b_\Omega^{-1}(x)\right)\right]_+\ dy \tag{9}$$

In the special case of a whole ROC curve, $\theta_{01} = \{x, y \in [0, 1]\}$, the expression simplifies, because the horizontal and vertical areas are the same:

$$AUC_{\Omega+}(\theta_{01}) := \int_0^1 [r(x) - b_\Omega(x)]_+\ dx \tag{10}$$

Underneath the main diagonal and the binary chance baseline, the areas are both 0.5, but the location of those areas differ. Similarly, the areas between the curve and each baseline over the whole plot, are the same (Fig. 3), despite appearing to differ by δ (Figs. 1 and 2).

The difference is apparent in a range or region of interest (Fig. 4). Also, there are some points on a ROC curve that perform worse than chance—those which border δ (Fig. 2).

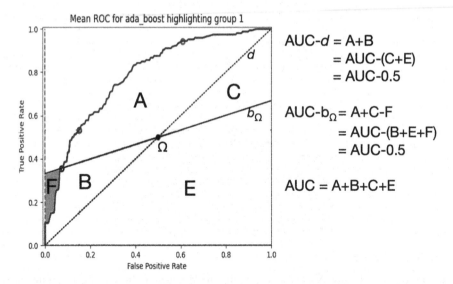

$$AUC\text{-}d = A+B$$
$$= AUC\text{-}(C+E)$$
$$= AUC\text{-}0.5$$

$$AUC\text{-}b_\Omega = A+C\text{-}F$$
$$= AUC\text{-}(B+E+F)$$
$$= AUC\text{-}0.5$$

$$AUC = A+B+C+E$$

Fig. 3. We illustrate that AUC_d, i.e., AUC minus the diagonal and AUC_Ω, i.e., AUC minus the binary chance baseline, are the same, using a ROC curve from Adaboost used to classify breast cancer data. While there is no difference over a whole ROC curve, there are differences for part of a ROC curve (in subsequent figures).

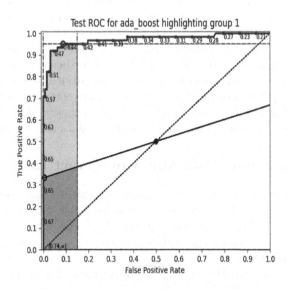

Fig. 4. A ROC plot for Adaboost applied to Wisconsin Breast Cancer recurrence data (size and texture). The vertical (sensitivity) aspect is highlighted for the region of interest $FPR = [0, 0.15]$, with yellow area better than binary chance. (Color figure online)

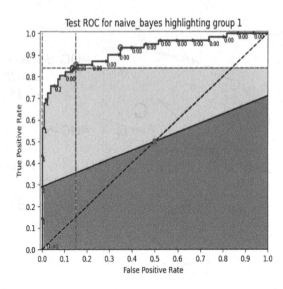

Fig. 5. A ROC plot for Naive Bayes applied to Wisconsin Breast Cancer recurrence data (size and texture). The horizontal (specificity) aspect is highlighted for the ROI $FPR = [0, 0.15]$.

The status quo approach to ROC plots, historically but unnecessarily, ignores prevalence and costs, limiting the use of ROC plots to abstract interpretations for initial model development, instead of interpreting a model in the context of real-life applications [11]. Halligan *et al.* [11] imply that this is an inherent limitation of ROC plots, but our work demonstrates that sometimes it is not. Others have also suggested additions to status quo ROC plots for better explanations [1].

The useful areas and points on ROC curves identified by the binary chance baseline in ROC plots, provide explanations that are complimentary to decision curve analysis. Furthermore, ROC plots can relate a variety of pre-test and post-test measures to each other, including predictive values and likelihood ratios.

6   Prevalence and Costs May Counteract Each Other

The previous section explains that ignoring prevalence and costs may cause errors in interpretation and choosing the best classifier. However, we may also incur errors by including prevalence while ignoring costs, or including costs while ignoring prevalence, because they often have counteracting effects in (2) as we explain in the following example.

Consider a medical condition, such as colon cancer. The cost C_{FN} of missing the disease in screening, a false negative, is much worse than a mistaken detection C_{FP}, a false positive, for which follow-up tests are conducted with some expense. The term $\frac{C_{FP} - C_{TN}}{C_{FN} - C_{TP}}$ causes a low slope, i.e., less than one.

However, as a condition with low prevalence, the term $\frac{N}{P}$ causes a high slope, i.e., it has an effect that counteracts the costs. These effects do not in general balance each other, and in some cases they act in the same direction. However, more often than not, the minority class tends to be the class of interest, and false negatives tend to be more costly, resulting in counteracting effects.

7 Examples in Classification

We conduct an experiment using the Wisconsin Diagnostic Breast Cancer (WDBC) data set [26] from the UCI data repository [8]. The Wisconsin data set was curated by clinicians for the prediction of whether or not breast cancer recurs by a time endpoint.

Our testing examines size, texture and shape features of nuclei sampled by thin needle aspiration. We omit shape, or size and shape in some experiments to show a range of different ROC curves and results for analysis with our new baseline.

For diagnostic tests (as opposed to screening) the SpPin rule [22] recommends a focus on high specificity: the left side of a ROC plot. Hence, we define the region of interest (ROI) as 85% specificity or above: FPR $= [0, 0.15]$ (Fig. 4). AUC in the ROI is computed from the combination of a vertical perspective (Fig. 4) and a horizontal perspective (Fig. 5) for the same ROC curve.

We interpret points (Table 1) in the ROC plot (Fig. 4) for the Adaboost algorithm. As discussed previously, the endpoints of the main diagonal (0, 0) and (1, 1) do **not** perform the same as chance (Table 1).

We compute **average net benefit** according to Metz [16, pp. 295], as the difference in the cost of using a test and not using it:

$$\overline{NB} = -(\overline{C}_{\text{use}} - \overline{C}_{\text{not\_use}}) \tag{11}$$

where cost of using the test is [16, pp. 295]:

$$\overline{C}_{\text{use}} = -(C_{FN} - C_{TP}) \cdot \pi \cdot TPR \; + (C_{FP} - C_{TN}) \cdot (1 - \pi) \cdot FPR \\ + C_{FN} \cdot \pi + C_{TN} \cdot (1 - \pi) + C_o \tag{12}$$

and the fixed cost, $\overline{C}_{\text{not\_use}}$ is set to zero for a diagnostic test. The overhead cost C_o is set to zero, since we are not interested in return on investment.

Cost-weighted accuracy is average net benefit normalized to the range [0, 1] or [0%, 100%]. Note that all points along the binary chance baseline, such as (0.5, 0.5) and the baseline's intersection with the ROC at (0, 0.33), have the same average net benefit and cost-weighted accuracy as chance.

In Fig. 4 about one third of the vertical area in the ROI performs no better than binary chance (Table 2). Thresholds in the gray region below the binary chance baseline perform worse than chance and should not be used, contrary to analysis with the main diagonal. The binary chance baseline has a gradual or low slope in this example (Fig. 4). Low prevalence (30%) with no other factors would cause a high slope, however the cost of false negatives are specified as five times worse than a false positive (a hypothetical cost), causing a low slope.

Table 1. Validation of expected cost-weighted accuracy at various points in the ROC plot for Wisconsin Diagnostic Breast Cancer classified with Adaboost (Fig. 4).

Description	ROC point	Accuracy	Average net benefit	Cost-weighted accuracy	Expectation for Avg NB, CW-Acc
A perfect classifier at a perfect threshold t	(0, 1)	100%	0	100%	Best possible value
The worst classifier at the worst threshold t	(1, 0)	0%	−2.49	0%	Worst possible value
Binary chance	(0.5, 0.5)	50%	−1.25	50%	Value for chance
All negative classifier A−, t = ∞	(0, 0)	67%	−1.86	25%	Worse than chance
All positive classifier A+, t = −∞	(1, 1)	33%	−0.63	75%	Better than chance
An optimal point on ROC curve, t = 0.45	(0.09, 0.95)	93%	−0.14	94%	Best value on ROC
An optimal point on ROC curve, in ROI_1, t = 0.45	(0.09, 0.95)	93%	−0.14	94%	Less than or equal to the best ROC value
Intersection of ROC curve and binary chance baseline	(0, 0.33)	75%	−1.25	50%	Equal to value for chance

Table 2. Results with useful areas highlighted in green cells and useful sensitivity and specificity highlighted in orange cells. *the same value as AUC_Ω

Description	AUC in a part	AUC in whole or normalized	Average Sens	Average Spec	Expectation
ROC	-	AUC =97.2%	97.2%	97.2%	Visually, AUC is nearly 100%
ROC above main diagonal	AUC_d =47.2%	-	47.2% above	47.2% above	Over the whole curve, $AUC_d = AUC - 0.5$
ROC above binary chance	AUC_Ω =47.2%	-	47.2% above	47.2% above	Over the whole curve, $AUC_\Omega = AUC - 0.5$, different from the 0.5 above $\|AUC_\Omega\| = \|AUC_d\|$ (Figure 3)
ROC in ROI_1 [0, 0.15]	AUC_1 =54.0%	$AUCn_1$ =98.1%	90.5%	99.3%	Visually, the ROI is not much better or worse than the whole curve (98.1% ≈ 97.2%)
ROC in ROI_1 above the main diagonal	AUC_{d1} =28.5%	-	83.0% above	46.8% above	
ROC in ROI_1 above the binary chance baseline	$AUC_{\Omega 1}$ =26.3%	-	54.8% above	46.7% above	In ROI_1 the useful area is smaller than what the diagonal identifies: 26.3% < 28.5%. Avg Sens above is smaller too.

8   Related Work

Zhou *et al.* [27] provide a classic work on the interpretation of ROC curves and plots including discussion of the main diagonal and chance. Numerous other sources [12, 20, 27] discuss and describe the main diagonal as chance or a random classifier [9], and the main diagonal has long been used as a point of comparison for performance measures in ROC plots.

Aside from binary chance and continuous chance as concepts we discuss in this paper, or "a chance" of rain, there is also the concept of chance agreement. Kappa [5] describes the amount of agreement beyond chance agreement, between any two models or people that produce scores. Kappa includes prevalence but does not include costs unless modified [6]. The unmodified version is more commonly known.

While Kappa and other priors may provide alternative baselines from which to judge utility, our paper focuses on the misunderstanding of the main diagonal as chance, and the clarification of the binary chance baseline as the layperson's concept of chance for binary outcomes. Other baselines may be investigated in other work.

Iso performance lines were introduced by Metz [16], and later used by others, such as Provost and Fawcett [9] for the purpose of identifying an optimal ROC point on the ROC curve. Flach [10] then investigated the geometry of ROC plots with iso performance lines. Subtil and Rabilloud [24] were the first, to apply iso performance lines as a baseline from which to measure performance. They examine performance equivalent to the all-negative and all-positive classifiers we discuss and denote as $A-$ and $A+$ respectively.

9   Conclusions and Future Work

ROC plots that label the main diagonal as chance are misleading for interpretation—whereas a label such as "no skill" is accurate. We newly illustrate the baseline that represents performance equal to layperson's chance, which is more intuitive and explainable. We showed that explanations based on the main diagonal, about several ROC points, are faulty, whereas the binary chance baseline is congruent with our expectations and computed values of cost-weighted accuracy.

While ROC plots were originally applied to disregard prevalence and costs to compare performance between different data sets (e.g., different radar scenarios with different prevalence)—that does not serve the need for realistic performance evaluation and explanation. We explained that prevalence and costs are always present—they are either assumed and implicit or they are explicit and more correct if estimated. We posit that the new methods in our paper will provide more insight into performance and ROC plots in past and present results.

Availability of Code

The measures and plots in this paper were created with the bayesianROC toolkit in Python, which can be installed at a command line, as follows. It requires the deepROC toolkit as well.

```
pip install bayesianroc
pip install deeproc
```

The associated links are:
https://pypi.org/project/bayesianroc/
https://github.com/DR3AM-Hub/BayesianROC
https://pypi.org/project/deeproc/
https://github.com/Big-Life-Lab/deepROC

Acknowledgements. Parts of this work has received funding by the Austrian Science Fund (FWF), Project: P-32554 "A reference model for explainable Artificial Intelligence in the medical domain".

Contributions. All authors contributed in writing this article. AC conceived the main ideas initially. In consultation with PF, JP, FM, and NJ various ideas were further developed and refined, with AH and RA providing guidance. Experiments were conducted and coded by AC and FM. All authors reviewed and provided edits to the article.

References

1. Althouse, A.D.: Statistical graphics in action: making better sense of the ROC curve. Int. J. Cardiol. **100**(215), 9–10 (2016)
2. Beran, R.: Minimum Hellinger distance estimates for parametric models. Ann. Stat. **5**(3), 445–463 (1977)
3. Carrington, A.M., et al.: A new concordant partial AUC and partial C statistic for imbalanced data in the evaluation of machine learning algorithms. BMC Med. Inform. Decis. Making **20**(1), 1–12 (2020)
4. Chance Noun: In the Cambridge Dictionary. Cambridge University Press. https://dictionary.cambridge.org/dictionary/english/chance
5. Cohen, J.: A coefficient of agreement for nominal scales. Educ. Psychol. Measur. **20**(1), 37–46 (1960)
6. Cohen, J.: Weighted kappa: nominal scale agreement provision for scaled disagreement or partial credit. Psychol. Bull. **70**(4), 213 (1968)
7. Cover, T.M., Thomas, J.A.: Elements of Information Theory. Wiley, Hoboken (2012)
8. Dua, D., Graff, C.: UCI machine learning repository (2017). http://archive.ics.uci.edu/ml
9. Fawcett, T.: An introduction to ROC analysis. Pattern Recogn. Lett. **27**(8), 861–874 (2006)
10. Flach, P.A.: The geometry of ROC space: understanding machine learning metrics through ROC isometrics. In: Proceedings of the Twentieth International Conference on Machine Learning (2003)

11. Halligan, S., Altman, D.G., Mallett, S.: Disadvantages of using the area under the receiver operating characteristic curve to assess imaging tests: a discussion and proposal for an alternative approach. Eur. Radiol. **25**(4), 932–939 (2015). https://doi.org/10.1007/s00330-014-3487-0

12. Hand, D.J.: Measuring classifier performance: a coherent alternative to the area under the ROC curve. Mach. Learn. **77**(1), 103–123 (2009). https://doi.org/10.1007/s10994-009-5119-5

13. Inácio, V., Rodríguez-Álvarez, M.X., Gayoso-Diz, P.: Statistical evaluation of medical tests. Ann. Rev. Stat. Appl. **8**, 41–67 (2021)

14. Lin, J.: Divergence measures based on the Shannon entropy. IEEE Trans. Inf. Theory **37**(1), 145–151 (1991). https://doi.org/10.1109/18.61115. http://ieeexplore.ieee.org/document/61115/

15. Menéndez, M., Pardo, J., Pardo, L., Pardo, M.: The Jensen-Shannon divergence. J. Franklin Inst. **334**(2), 307–318 (1997). Publisher: Elsevier

16. Metz, C.E.: Basic principles of ROC analysis. In: Seminars in Nuclear Medicine, vol. 8, pp. 283–298. Elsevier (1978)

17. Nielsen, F.: On a variational definition for the Jensen-Shannon symmetrization of distances based on the information radius. Entropy **23**(4), 464 (2021)

18. Obuchowski, N.A.: Receiver operating characteristic curves and their use in radiology. Radiology **229**(1), 3–8 (2003). https://doi.org/10.1148/radiol.2291010898

19. Obuchowski, N.A., Bullen, J.A.: Receiver operating characteristic (ROC) curves: review of methods with applications in diagnostic medicine. Phys. Med. Biol. **63**(7), 07TR01 (2018)

20. Powers, D.M.W.: Evaluation: from precision, recall and F-factor to ROC, informedness, markedness & correlation. Technical report, Flinders University, December 2007

21. Provost, F., Fawcett, T.: Robust classification for imprecise environments. Mach. Learn. **42**, 203–231 (2001). https://doi.org/10.1023/A:1007601015854

22. Sackett, D.L., Straus, S.: On some clinically useful measures of the accuracy of diagnostic tests. BMJ Evid.-Based Med. **3**(3), 68 (1998)

23. Streiner, D.L., Cairney, J.: What's under the ROC? An introduction to receiver operating characteristics curves. Can. J. Psychiatry **52**(2), 121–128 (2007)

24. Subtil, F., Rabilloud, M.: An enhancement of ROC curves made them clinically relevant for diagnostic-test comparison and optimal-threshold determination. J. Clin. Epidemiol. **68**(7), 752–759 (2015)

25. Van den Hout, W.B.: The area under an ROC curve with limited information. Med. Decis. Making **23**(2), 160–166 (2003). https://doi.org/10.1177/0272989X03251246

26. Wolberg, W.H., Mangasarian, O.L.: Multisurface method of pattern separation for medical diagnosis applied to breast cytology. Proc. Natl. Acad. Sci. **87**(23), 9193–9196 (1990)

27. Zhou, X.H., McClish, D.K., Obuchowski, N.A.: Statistical Methods in Diagnostic Medicine, vol. 569. Wiley, Hoboken (2002)

Debiasing MDI Feature Importance and SHAP Values in Tree Ensembles

Markus Loecher[(✉)] [iD]

Berlin School of Economics and Law, 10825 Berlin, Germany
`markus.loecher@hwr-berlin.de`

Abstract. We attempt to give a unifying view of the various recent attempts to (i) improve the interpretability of tree-based models and (ii) debias the default variable-importance measure in random forests, Gini importance. In particular, we demonstrate a common thread among the out-of-bag based bias correction methods and their connection to local explanation for trees. In addition, we point out a bias caused by the inclusion of inbag data in the newly developed SHAP values and suggest a remedy.

Keywords: Explainable AI · Gini impurity · SHAP values · Saabas value · Variable importance · Random forests

1 Variable Importance in Trees

Variable importance is not very well defined as a concept. Even for the case of a linear model with n observations, p variables and the standard $n >> p$ situation, there is no theoretically defined variable importance metric in the sense of a parametric quantity that a variable importance estimator should try to estimate [4]. Variable importance measures for random forests have been receiving increased attention in bioinformatics, for instance to select a subset of genetic markers relevant for the prediction of a certain disease. They also have been used as screening tools [3,13] in important applications highlighting the need for reliable and well-understood feature importance measures.

The default choice in most software implementations [9,16] of random forests [2] is the *mean decrease in impurity (MDI)*. The MDI of a feature is computed as a (weighted) mean of the individual trees' improvement in the splitting criterion produced by each variable. A substantial shortcoming of this default measure is its evaluation on the in-bag samples which can lead to severe overfitting [7]. It was also pointed out by [22] that *the variable importance measures of Breiman's original Random Forest method ... are not reliable in situations where potential predictor variables vary in their scale of measurement or their number of categories.*

There have been multiple attempts at correcting the well understood bias of the Gini impurity measure both as a split criterion as well as a contributor to importance scores, each one coming from a different perspective.

© IFIP International Federation for Information Processing 2022
Published by Springer Nature Switzerland AG 2022
A. Holzinger et al. (Eds.): CD-MAKE 2022, LNCS 13480, pp. 114–129, 2022.
https://doi.org/10.1007/978-3-031-14463-9_8

Strobl et al. [23] derive the exact distribution of the maximally selected Gini gain along with their resulting p-values by means of a combinatorial approach. Shi et al. [21] suggest a solution to the bias for the case of regression trees as well as binary classification trees [20] which is also based on p-values. Several authors [6,11] argue that the criterion for split variable and split point selection should be separated.

A different approach is to add so-called pseudo variables to a dataset, which are permuted versions of the original variables and can be used to correct for bias [19]. Recently, a modified version of the Gini importance called Actual Impurity Reduction (AIR) was proposed [14] that is faster than the original method proposed by Sandri and Zuccolotto with almost no overhead over the creation of the original RFs and available in the R package *ranger* [25].

2   Separating Inbag and Out-of-Bag (OOB) Samples

More recently, the idea to include OOB samples in order to compute a debiased version of the Gini importance yielded promising results [8,10,26]. Here, the original Gini impurity (for node m) for a categorical variable Y which can take D values $c_1, c_2, \ldots, c_D$ is defined as

$$G(m) = \sum_{d=1}^{D} \hat{p}_d(m) \cdot (1 - \hat{p}_d(m)), \text{ where } \hat{p}_d = \frac{1}{n_m} \sum_{i \in m} Y_i.$$

Loecher [10] proposed a *penalized Gini impurity* which combines inbag and out-of-bag samples. The main idea is to increase the impurity $G(m)$ by a penalty that is proportional to the difference $\Delta = (\hat{p}_{OOB} - \hat{p}_{inbag})^2$:

$$PG_{oob}^{\alpha,\lambda} = \alpha \cdot G_{oob} + (1 - \alpha) \cdot G_{in} + \lambda \cdot (\hat{p}_{oob} - \hat{p}_{in})^2 \tag{1}$$

In addition, Loecher [10] investigated replacing $G(m)$ by an unbiased estimator of the variance via the well known sample size correction.

$$\widehat{G}(m) = \frac{N}{N-1} \cdot G(m) \tag{2}$$

In this paper we focus on the following three special cases, $[\alpha = 1, \lambda = 2]$, $[\alpha = 0.5, \lambda = 1]$ as well as $[\alpha = 1, \lambda = 0]$:

$$PG_{oob}^{(1,2)} = \sum_{d=1}^{D} \hat{p}_{d,oob} \cdot (1 - \hat{p}_{d,oob}) + 2(\hat{p}_{d,oob} - \hat{p}_{d,in})^2$$

$$PG_{oob}^{(0.5,1)} = \frac{1}{2} \cdot \sum_{d=1}^{D} \hat{p}_{d,oob} \cdot (1 - \hat{p}_{d,oob}) + \hat{p}_{d,in} \cdot (1 - \hat{p}_{d,in}) + (\hat{p}_{d,oob} - \hat{p}_{d,in})^2$$

$$\widehat{PG}_{oob}^{(1,0)} = \frac{N}{N-1} \cdot \sum_{d=1}^{D} \hat{p}_{d,oob} \cdot (1 - \hat{p}_{d,oob})$$

Our main contributions are to

- unify other debiasing attempts into the penalized Gini framework; in particular, we show that $PG_{oob}^{(1,2)}$ is equivalent to the *MDI-oob* measure defined in [8] and therefore has close connections to the *conditional feature contributions* (CFCs) defined in [18]. Furthermore, $PG_{oob}^{(0.5,1)}$ is equivalent to the *unbiased split-improvement* measure defined in [26].
- demonstrate that similarly to MDI, both the CFCs as well as the related *SHapley Additive exPlanation* (SHAP) values defined in [12] are susceptible to "overfitting" to the training data.
- illustrate early results on two novel methods to debias SHAP values.

We refer the reader to [10] for a proof that $\widehat{PG}_{oob}^{(1,0)}$ and $PG_{oob}^{(0.5,1)}$ are unbiased estimators of feature importance in the case of non-informative variables.

3   Conditional Feature Contributions (CFCs)

The conventional wisdom of estimating the impact of a feature in tree based models is to measure the **node-wise reduction of a loss function**, such as the variance of the output Y, and compute a weighted average of all nodes over all trees for that feature. By its definition, such a *mean decrease in impurity* (MDI) serves only as a global measure and is typically not used to explain a *per-observation, local impact*. Saabas [18] proposed the novel idea of explaining a prediction by following the decision path and attributing changes in the expected output of the model to each feature along the path. Figure 1 illustrates the main idea of decomposing each prediction through the sequence of regions that correspond to each node in the tree. Each decision either adds or subtracts from the value given in the parent node and can be attributed to the feature at the node. So, each individual prediction can be defined as the global mean plus the sum of the K feature contributions:

$$f_{pred}(x_i) = \bar{Y} + \sum_{k=1}^{K} f_{T,k}(x_i) \tag{3}$$

where $f_{T,k}(x_i)$ is the contribution from the k-th feature (for tree T), written as a sum over all the inner nodes t such that $v(t) = k$ [8][1]:

$$f_{T,k}(x_i) = \sum_{t \in I(T):v(t)=k} \left[\mu_n \left(t^{left} \right) \mathbb{1} \left(x_i \in R_{t_{left}} \right) + \mu_n \left(t^{right} \right) \mathbb{1} \left(x_i \in R_{t_{right}} \right) - \mu_n(t) \mathbb{1} \left(x \in R_t \right) \right]$$

$$\tag{4}$$

where $v(t)$ is the feature chosen for the split at node t.

A "local" feature importance score can be obtained by summing Eq. (4) over all trees. Adding these local explanations over all data points yields a "global" importance score:

---

[1] Appendix A.1 contains expanded definitions and more thorough notation.

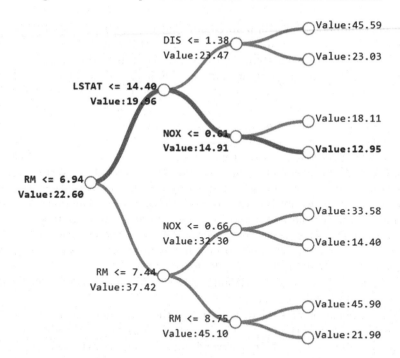

Fig. 1. Taken from [17]: Depicted is a regression decision tree to predict housing prices. The tree has conditions on each internal node and a value associated with each leaf (i.e. the value to be predicted). But additionally, the value at each internal node i.e. the mean of the response variables in that region, is shown. The red path depicts the decomposition of an example prediction $Y = \mathbf{12.95} \approx \mathbf{22.60}(\bar{Y}) - \mathbf{2.64}$(loss from RM) $-$ $\mathbf{5.04}$(loss from LSTAT) $- \mathbf{1.96}$(loss from NOX) (Color figure online)

$$Imp_{global}(k) = \sum_{i=1}^{N} |Imp_{local}(k, x_i)| = \sum_{i=1}^{N} \frac{1}{n_T} \sum_{T} |f_{T,k}(x_i)| \tag{5}$$

In the light of wanting to explain the predictions from tree based machine learning models, the "Saabas algorithm" is extremely appealing, because

- The positive and negative contributions from nodes convey directional information unlike the strictly positive purity gains.
- By combining many local explanations we can represent global structure while retaining local faithfulness to the original model.
- The expected value of every node in the tree can be estimated efficiently by averaging the model output over all the training samples that pass through that node.
- The algorithm has been implemented and is easily accessible in a python [18] and R [24] library.

However, Lundberg et al. [12] pointed out that it is strongly biased to alter the impact of features based on their distance from the root of a tree. This causes Saabas values to be inconsistent, which means one can modify a model to make a feature clearly more important, and yet the Saabas value attributed to that feature will decrease. As a solution, the authors developed an algorithm ("Tree-Explainer") that computes local explanations based on exact Shapley values in polynomial time[2]. This provides local explanations with theoretical guarantees of local accuracy and consistency. One should not forget though that the same idea of adding *conditional feature contributions* lies at the heart of *TreeExplainer*.

In this section, we call attention to another source of bias which is the result of using the same (inbag) data to (i) greedily split the nodes during the growth of the tree and (ii) computing the node-wise changes in prediction. We use the well known Titanic data set to illustrate the perils of putting too much faith into importance scores which are based entirely on training data - not on OOB samples - and make no attempt to discount node splits in deep trees that are spurious and will not survive in a validation set.

In the following model[3] we include *passengerID* as a feature along with the more reasonable *Age*, *Sex* and *Pclass*.

Figure 2 below depicts the distribution of the individual, "local" feature contributions, preserving their sign. The large variations for the variables with high cardinality (*Age, passengerID*) are worrisome. We know that the impact of *Age* was modest and that *passengerID* has no impact on survival but when we sum the absolute values, both features receive sizeable importance scores, as shown in Fig. 3. This troubling result is robust to random shuffling of the ID. Sect. A.2 will point out a close analogy between the well known MDI score and the more recent measure based on the conditional feature contributions.

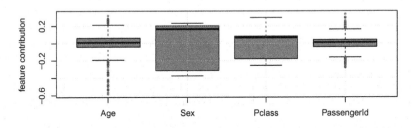

Fig. 2. Conditional feature contributions (TreeInterpreter) for the Titanic data.

---

[2] A python library is available: https://github.com/slundberg/shap.

[3] In all random forest simulations, we choose $mtry = 2, ntrees = 100$ and exclude rows with missing *Age*.

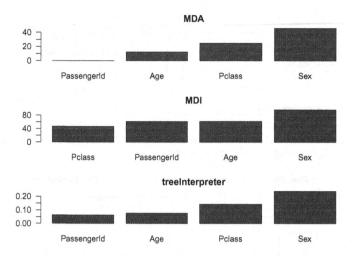

Fig. 3. Permutation importance (MDA, top panel) versus Mean decrease impurity (MDI, middle panel) versus conditional feature contributions (TreeInterpreter) for the Titanic data. The permutation based importance (MDA) is not fooled by the irrelevant ID feature. This is maybe not unexpected as the IDs should bear no predictive power for the out-of-bag samples.

4 SHAP Values

Lundberg et al. [12] introduce a new local feature attribution method for trees based on **SHapley Additive exPlanation** (SHAP) values which fall in the class of *additive feature attribution methods*. The authors point to results from game theory implying that Shapley values are the only way to satisfy three important properties: *local accuracy, consistency, and missingness*. In this section we show that even (SHAP) values suffer from (i) a strong dependence on feature cardinality, and (ii) assign non zero importance scores to uninformative features. We begin by extending the Titanic example from the previous section and find that *TreeExplainer* also assigns a non zero value of feature importance to *passengerID*, as shown in Fig. 4, which is due to mixing inbag and out-of-bag data for the evaluation. Simply separating the inbag from the OOB SHAP values is not a remedy as shown in the left graph of Fig. 5. However, inspired by Eq. (6) in [8], we compute weighted SHAP values by multiplying with y_i before averaging. Why multiply by Y ? The global importance scores are averages of the absolute values of the SHAP values, so they reflect merely variation; regardless whether that variation reflects the truth at all. So for e.g. passengerID the model will still produce widely varying SHAP values even on a testset (or oob) -leading to inflated importance - but we would want to "penalize" the wrong direction! (Not possible on the training data as the model was fit in order to optimize the agreement with Y_{train}). The right graph of Fig. 5 illustrates the elimination of the spurious contributions due to *passengerID* for the OOB SHAP values. Further support for the claims above is given by the following two kinds of simulations.

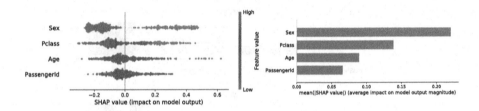

Fig. 4. Local SHapley Additive exPlanation (SHAP) values (left panel) vs. global SHAP scores (right panel) for the Titanic data.

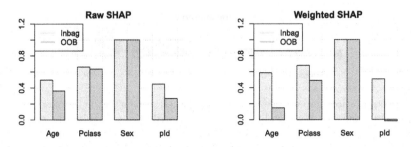

Fig. 5. Left graph: "raw" SHAP values for the Titanic data, separately computed for inbag and OOB. Right graph: weighted SHAP values are multiplied by y_i before averaging which eliminates the spurious contributions due to *passengerID* for OOB. Note that we scaled the SHAP values to their respective maxima for easier comparison. (*pId* is short for *passengerID*)

4.1   Null/Power Simulations

We replicate the simulation design used by [22] where a binary response variable Y is predicted from a set of 5 predictor variables that vary in their scale of measurement and number of categories. The first predictor variable X_1 is continuous, while the other predictor variables $X_2, \ldots, X_5$ are multinomial with $2, 4, 10, 20$ categories, respectively. The sample size for all simulation studies was set to n = 120. In the first *null case* all predictor variables and the response are sampled independently. We would hope that a reasonable variable importance measure would not prefer any one predictor variable over any other. In the second simulation study, the so-called *power case*, the distribution of the response is a binomial process with probabilities that depend on the value of x_2, namely $P(y = 1|X_2 = 1) = 0.35, P(y = 1|X_2 = 2) = 0.65$.

As is evident in the two leftmost panels of Fig. 6, both the Gini importance (MDI) and the SHAP values show a strong preference for variables with many categories and the continuous variable. This bias is of course well-known for MDI but maybe unexpected for the SHAP scores which clearly violate the *missingness* property. Encouragingly, both $PG_{OOB}^{(0.5,1)}$ and AIR [14] yield low scores for all predictors. The notable differences in the variance of the distributions for predictor variables with different scale of measurement or number of cate-

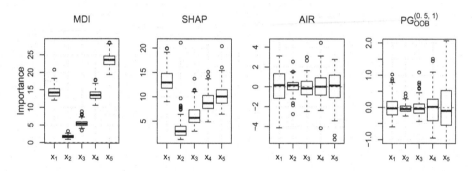

Fig. 6. Results of the null case, where none of the predictor variables are informative but vary greatly in their number of possible split points: the first predictor variable X_1 is continuous, while the other predictor variables $X_2, \ldots, X_5$ are multinomial with $2, 4, 10, 20$ categories, respectively.

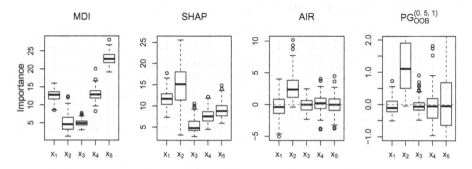

Fig. 7. Results of the power study, where only X_2 is informative. Other simulation details as in Fig. 6.

gories are unfortunate but to be expected. (The larger the numbers of categories in a multinomial variable, the fewer the numbers of observations per category and the larger therefore the variability of the measured qualities of the splits performed using the multinomial variable) The results from the power study are summarised in Fig. 7. MDI and SHAP again show a strong bias towards variables with many categories and the continuous variable. At the chosen signal-to-noise ratio MDI fails entirely to identify the relevant predictor variable. In fact, the mean value for the relevant variable X_2 is lowest and only slightly higher than in the null case. We mention in passing that Adler et al. [1] observe a very similar bias for SHAP scores in gradient boosting trees.

Both $PG_{OOB}^{(0.5,1)}$ and AIR clearly succeed in identifying X_2 as the most relevant feature. The large fluctuations of the importance scores for X_4 and especially X_5 are bound to yield moderate "false positive" rates and incorrect rankings in single trials. The signal-to-noise separation for the SHAP values is moderate but can be greatly improved by multiplying with y_i before averaging (in analogy to Fig. 5 and Sect. A.2), as shown in Fig. 8.

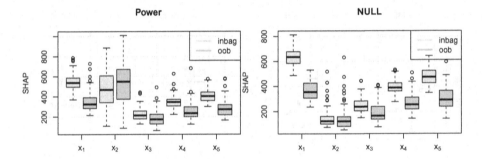

Fig. 8. Weighted SHAP values, as explained in the text. Left graph: power study, where only X_2 is informative. Right graph: null case, where none of the predictor variables is informative. Other simulation details as in Figs. 6, 7

4.2   Noisy Feature Identification

For a more systematic comparison of the 4 proposed penalized Gini scores, we closely follow the simulations outlined in [8] involving discrete features with different number of distinct values, which poses a critical challenge for MDI. The data has 1000 samples with 50 features. All features are discrete, with the jth feature containing $j + 1$ distinct values $0, 1, \ldots, j$. We randomly select a set S of 5 features from the first ten as relevant features. The remaining features are noisy features. All samples are i.i.d. and all features are independent. We generate the outcomes using the following rule:

$$P(Y = 1|X) = \text{Logistic}(\frac{2}{5} \sum_{j \in S} x_j / j - 1)$$

Treating the noisy features as label 0 and the relevant features as label 1, we can evaluate a feature importance measure in terms of its area under the receiver operating characteristic curve (AUC). We grow 100 deep trees (minimum leaf size equals 1, $m_{try} = 3$), repeat the whole process 100 times and report the average AUC scores for each method in Table 1. For this simulated setting, $\widehat{PG}_{oob}^{(0.5,1)}$ achieves the best AUC score under all cases, most likely because of the separation of the signal from noise mentioned above. We notice that the AUC score for the OOB-only $\widehat{PG}_{oob}^{(1,0)}$ is competitive to the permutation importance, SHAP and the AIR score.

Table 1. Average AUC scores for noisy feature identification. MDA = permutation importance, MDI = (default) Gini impurity. The $\widehat{PG}_{oob}$ scores apply the variance bias correction $n/(n-1)$. The SHAP_{in}, SHAP_{oob} scores are based upon separating the inbag from the oob data.

$\widehat{PG}_{oob}^{(1,0)}$	$PG_{oob}^{(1,0)}$	$\widehat{PG}_{oob}^{(0.5,1)}$	$PG_{oob}^{(0.5,1)}$	SHAP	SHAP_{in}	SHAP_{oob}	AIR	MDA	MDI
0.66	0.28	0.92	0.78	0.66	0.56	0.73	0.68	0.65	0.10

4.3   Shrunk SHAP

Extensive simulations suggest that the correlation between inbag and oob SHAP values can serve as an indicator for the degree of overfitting. The correlation between inbag and oob SHAP values is high for informative and nearly zero for uninformative features, which motivates the following smoothing method. For each feature

1. Fit a linear model $SHAP_{i,oob} = \beta_i \cdot SHAP_{i,inbag} + u_i$
2. Use the estimates $\widehat{SHAP}_{i,oob} = \hat{\beta}_i \cdot SHAP_{i,inbag}$ as local explanations instead of either $SHAP_{i,oob}$ or $SHAP_{i,inbag}$,

We can think of the predictions as a "regularized" or "smoothed" version of the original local SHAP values. The "shrinkage" depends on the correlation between inbag and oob. Figure 9 shows how effective this approach is for the simulated data introduced in Sect. 4.1. These are highly promising results. Notice that by definition of importance as the "variance" or MAD (mean absolute deviation) of the SHAP scores, we will always deal with positive values which makes it difficult to define an unbiased estimator with expected value of zero.

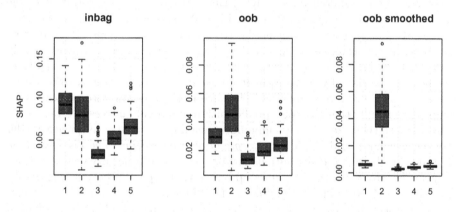

Fig. 9. Power study, where only X_2 is informative. The right most panel shows the smoothed SHAP values - as explained in the text. Other simulation details as in Fig. 7

5   Discussion

Random forests and gradient boosted trees are among the most popular and powerful [15] non-linear predictive models used in a wide variety of fields. Lundberg et al. [12] demonstrate that tree-based models can be more accurate than neural networks and even more interpretable than linear models. In the comprehensive overview of variable importance in regression models, Groemping [5] distinguishes between methods based on (i) variance decomposition and (ii) standardized coefficient sizes, which is somewhat analogous to the difference between

(i) MDI and (ii) CFCs. The latter measure the directional impact of $x_{k,i}$ on the outcome y_i, whereas MDI based scores measure a kind of *partial* R_k^2 (if one stretched the analogy to linear models). Li et al. [8] ingeniously illustrate the connection between these seemingly fundamentally different methods via Eq. (6). And the brilliant extension of CFCs to their Shapley equivalents by [12] bears affinity to the game-theory-based metrics LMG and PMVD ([5] and references therein), which are based on averaging the sequential R_k^2 over all orderings of regressors.

In this paper we have (i) connected the proposals to reduce the well known bias in MDI by mixing inbag and oob data [8,26] to a common framework [10], and (ii) pointed out that similar ideas would benefit/debias the *conditional feature contributions* (CFCs) [18] as well as the related SHAP values [12]. Applying a simple and well-known finite sample correction to the node impurity computation also results in an unbiased feature importance measure and marked improvements in discerning signal from noise. In addition, we proposed a novel debiasing scheme which is based on inbag-oob correlations yielding smoothed or "shrunk" SHAP scores that appear to nearly eliminate the observed cardinality bias of non-informative features.

While the main findings are applicable to any tree based method, they are most relevant to random forests (RFs) since (i) oob data are readily available and (ii) RFs typically grow deep trees. Li et al. [8] showed a strong dependence of the MDI bias on the depth of the tree: splits in nodes closer to the roots are much more stable and supported by larger sample sizes and hence hardly susceptible to bias. RFs "get away" with the individual overfitting of deep trees to the training data by **averaging** many (hundreds) of separately grown deep trees and often achieve a favourable balance in the bias-variance tradeoff. One reason is certainly that the noisy predictions from individual trees "average out", which is not the case for the summing/averaging of the strictly positive MDI leading to what could be called *interpretational overfitting*. The big advantage of conditional feature contributions is that positive and negative contributions can cancel across trees making it less prone to that type of overfitting. However, we have provided evidence that both the CFCs as well as the SHAP values are still susceptible to "overfitting" to the training data and can benefit from evaluation on oob data.

A   Appendix

A.1   Background and Notations

Definitions needed to understand Eq. (4). (The following paragraph closely follows the definitions in [8].)

Random Forests (RFs) are an ensemble of classification and regression trees, where each tree T defines a mapping from the feature space to the response. Trees are constructed independently of one another on a bootstrapped or subsampled data set $\mathcal{D}^{(T)}$ of the original data $\mathcal{D}$. Any node t in a tree T represents a subset (usually a hyper-rectangle) R_t of the feature space. A split of the node t is a pair (k, z) which divides the hyper-rectangle R_t into two hyper-rectangles $R_t \cap \mathbb{1}(X_k \leq z)$ and $R_t \cap \mathbb{1}(X_k > z)$ corresponding to the left child t left and right child t right of node t, respectively. For a node t in a tree T, $N_n(t) = \left|\{i \in \mathcal{D}^{(T)} : \mathbf{x}_i \in R_t\}\right|$ denotes the number of samples falling into R_t and

$$\mu_n(t) := \frac{1}{N_n(t)} \sum_{i:\mathbf{x}_i \in R_t} y_i$$

We define the set of inner nodes of a tree T as $I(T)$.

A.2 Debiasing MDI via OOB Samples

In this section we give a short version of the proof that $PG_{oob}^{(1,2)}$ is equivalent to the *MDI-oob* measure defined in [8]. For clarity we assume binary classification; Appendix A.2 contains an expanded version of the proof including the multi-class case. As elegantly demonstrated by [8], the MDI of feature k in a tree T can be written as

$$MDI = \frac{1}{|\mathcal{D}^{(T)}|} \sum_{i \in \mathcal{D}^{(T)}} f_{T,k}(x_i) \cdot y_i \tag{6}$$

where $\mathcal{D}^{(T)}$ is the bootstrapped or subsampled data set of the original data $\mathcal{D}$. Since $\sum_{i \in \mathcal{D}^{(T)}} f_{T,k}(x_i) = 0$, we can view MDI essentially as the sample covariance between $f_{T,k}(x_i)$ and y_i on the bootstrapped dataset $\mathcal{D}^{(T)}$.

MDI-oob is based on the usual variance reduction per node as shown in Eq. (34) (proof of Proposition (1)), but with a "variance" defined as the mean squared deviations of y_{oob} from the inbag mean μ_{in}:

$$\Delta_I(t) = \frac{1}{N_n(t)} \cdot \sum_{i \in D(T)} (y_{i,oob} - \mu_{n,in})^2 \mathbb{1}(x_i \in R_t) - \ldots$$

We can, of course, rewrite the variance as

$$\frac{1}{N_n(t)} \cdot \sum_{i \in D(T)} (y_{i,oob} - \mu_{n,in})^2 = \frac{1}{N_n(t)} \cdot \sum_{i \in D(T)} (y_{i,oob} - \mu_{n,oob})^2 + (\mu_{n,in} - \mu_{n,oob})^2 \tag{7}$$

$$= p_{oob} \cdot (1 - p_{oob}) + (p_{in} - p_{oob})^2 \tag{8}$$

where the last equality is for Bernoulli y_i, in which case the means $\mu_{in/oob}$ become proportions $p_{in/oob}$ and the first sum is equal to the binomial variance $p_{oob} \cdot (1 - p_{oob})$. The final expression is effectively equal to $PG_{oob}^{(1,2)}$.

Lastly, we now show that $PG_{oob}^{(0.5,1)}$ is equivalent to the *unbiased split-improvement* measure defined in [26]. For the binary classificaton case, we can rewrite $PG_{oob}^{(0.5,1)}$ as follows:

$$PG_{oob}^{(0.5,1)} = \frac{1}{2} \cdot \sum_{d=1}^{D} \hat{p}_{d,oob} \cdot (1 - \hat{p}_{d,oob}) + \hat{p}_{d,in} \cdot (1 - \hat{p}_{d,in}) + (\hat{p}_{d,oob} - \hat{p}_{d,in})^2$$

$$\text{(9)}$$

$$= \hat{p}_{oob} \cdot (1 - \hat{p}_{oob}) + \hat{p}_{in} \cdot (1 - \hat{p}_{in}) + (\hat{p}_{oob} - \hat{p}_{in})^2 \tag{10}$$

$$= \hat{p}_{oob} - \hat{p}_{oob}^2 + \hat{p}_{in} - \hat{p}_{in}^2 + \hat{p}_{oob}^2 - 2\hat{p}_{oob} \cdot \hat{p}_{in} + \hat{p}_{in}^2 \tag{11}$$

$$= \hat{p}_{oob} + \hat{p}_{in} - 2\hat{p}_{oob} \cdot \hat{p}_{in} \tag{12}$$

A.3   Variance Reduction View

Here, we provide a full version of the proof sketched in Sect. A.2 which leans heavily on the proof of Proposition (1) in [8].

We consider the usual variance reduction per node but with a "variance" defined as the mean squared deviations of y_{oob} from the inbag mean μ_{in}:

$$\Delta_{\mathcal{I}}(t) = \frac{1}{N_n(t)} \sum_{i \in \mathcal{D}^{(T)}} [y_{i,oob} - \mu_{n,in}(t)]^2 \mathbb{1}(\mathbf{x}_i \in R_t)$$

$$- \left[y_{i,oob} - \mu_{n,in}\left(t^{\text{left}}\right)\right]^2 \mathbb{1}(\mathbf{x}_i \in R_{t\text{left}}) - \left[y_{i,oob} - \mu_{n,in}\left(t^{\text{right}}\right)\right]^2 \mathbb{1}(\mathbf{x}_i \in R_{\text{right}})$$

$$\text{(13)}$$

$$= \frac{1}{N_n(t)} \sum_{i \in \mathcal{D}^{(T)}} \left([y_{i,oob} - \mu_{n,oob}(t)]^2 + [\mu_{n,in}(t) - \mu_{n,oob}(t)]^2\right) \mathbb{1}(\mathbf{x}_i \in R_t)$$

$$- \left(\left[y_{i,oob} - \mu_{n,oob}(t^{\text{left}})\right]^2 + \left[\mu_{n,in}(t^{\text{left}}) - \mu_{n,oob}(t^{\text{left}})\right]^2\right) \mathbb{1}(\mathbf{x}_i \in R_{t\text{left}})$$

$$- \left(\left[y_{i,oob} - \mu_{n,oob}(t^{\text{right}})\right]^2 + \left[\mu_{n,in}(t^{\text{right}}) - \mu_{n,oob}(t^{\text{right}})\right]^2\right) \mathbb{1}(\mathbf{x}_i \in R_{\text{right}})$$

$$\text{(14)}$$

$$= \frac{1}{N_n(t)} \underbrace{\sum_{i \in \mathcal{D}^{(T)}} \left\{[y_{i,oob} - \mu_{n,oob}(t)]^2 \mathbb{1}(\mathbf{x}_i \in R_t)\right\}}_{N_n(t) \cdot p_{oob}(t) \cdot (1 - p_{oob}(t))} + \underbrace{[\mu_{n,in}(t) - \mu_{n,oob}(t)]^2}_{[p_{oob}(t) - p_{in}(t)]^2}$$

$$- \frac{1}{N_n(t)} \underbrace{\sum_{i \in \mathcal{D}^{(T)}} \left\{\left[y_{i,oob} - \mu_{n,oob}(t^{\text{left}})\right]^2 \mathbb{1}(\mathbf{x}_i \in R_{t\text{left}})\right\}}_{N_n(t^{\text{left}}) \cdot p_{oob}(t^{\text{left}}) \cdot (1 - p_{oob}(t^{\text{left}}))} + \underbrace{\left[\mu_{n,in}(t^{\text{left}}) - \mu_{n,oob}(t^{\text{left}})\right]^2}_{[p_{oob}(t^{\text{left}}) - p_{in}(t^{\text{left}})]^2}$$

$$- \frac{1}{N_n(t)} \underbrace{\sum_{i \in \mathcal{D}^{(T)}} \left\{\left[y_{i,oob} - \mu_{n,oob}(t^{\text{right}})\right]^2 \mathbb{1}(\mathbf{x}_i \in R_{\text{right}})\right\}}_{N_n(t^{\text{right}}) \cdot p_{oob}(t^{\text{right}}) \cdot (1 - p_{oob}(t^{\text{right}}))} + \underbrace{\left[\mu_{n,in}(t^{\text{right}}) - \mu_{n,oob}(t^{\text{right}})\right]^2}_{[p_{oob}(t^{\text{right}}) - p_{in}(t^{\text{right}})]^2}$$

where the last equality is for Bernoulli y_i, in which case the means $\mu_{in/oob}$ become proportions $p_{in/oob}$ and we replace the squared deviations with the binomial variance $p_{oob} \cdot (1 - p_{oob})$. The final expression is then

$$\Delta_{\mathcal{I}}(t) = p_{oob}(t) \cdot (1 - p_{oob}(t)) + [p_{oob}(t) - p_{in}(t)]^2$$
$$- \frac{N_n(t^{\text{left}})}{N_n(t)} \left(p_{oob}(t^{\text{left}}) \cdot \left(1 - p_{oob}(t^{\text{left}})\right) + \left[p_{oob}(t^{\text{left}}) - p_{in}(t^{\text{left}})\right]^2 \right)$$
$$- \frac{N_n(t^{\text{right}})}{N_n(t)} \left(p_{oob}(t^{\text{right}}) \cdot \left(1 - p_{oob}(t^{\text{right}})\right) + \left[p_{oob}(t^{\text{right}}) - p_{in}(t^{\text{right}})\right]^2 \right)$$
$$(15)$$

which, of course, is exactly the impurity reduction due to $PG_{oob}^{(1,2)}$:

$$\Delta_{\mathcal{I}}(t) = PG_{oob}^{(1,2)}(t) - \frac{N_n(t^{\text{left}})}{N_n(t)} PG_{oob}^{(1,2)}(t^{\text{left}}) - \frac{N_n(t^{\text{right}})}{N_n(t)} PG_{oob}^{(1,2)}(t^{\text{right}}) \quad (16)$$

Another, somewhat surprising view of MDI is given by Eqs. (6) and (4), which for binary classification reads as:

$$MDI = \frac{1}{|\mathcal{D}^{(T)}|} \sum_{t \in I(T):v(t)=k} \sum_{i \in \mathcal{D}^{(T)}} \left[\mu_n\left(t^{left}\right) \mathbb{1}\left(x_i \in R_{t^{left}}\right) \right.$$
$$\left. + \mu_n\left(t^{right}\right) \mathbb{1}\left(x_i \in R_{t^{right}}\right) - \mu_n(t) \mathbb{1}\left(x \in R_t\right) \right] \cdot y_i$$
$$= \frac{1}{|\mathcal{D}^{(T)}|} \sum_{t \in I(T):v(t)=k} -p_{in}(t)^2 + \frac{N_n(t^{\text{left}})}{N_n(t)} p_{in}(t^{\text{left}})^2 + \frac{N_n(t^{\text{right}})}{N_n(t)} p_{in}(t^{\text{right}})^2$$
$$(17)$$

and for the oob version:

$$MDI_{oob} = -p_{in}(t) \cdot p_{oob}(t) + \frac{N_n(t^{\text{left}})}{N_n(t)} p_{in}(t^{\text{left}}) \cdot p_{oob}(t^{\text{left}}) + \frac{N_n(t^{\text{right}})}{N_n(t)} p_{in}(t^{\text{right}}) \cdot p_{oob}(t^{\text{right}})$$
$$(18)$$

A.4 $E(\widehat{\Delta PG}_{oob}^{(0)}) = 0$

The decrease in impurity (ΔG) for a parent node m is the weighted difference between the Gini importance[4] $G(m) = \hat{p}_m(1 - \hat{p}_m)$ and those of its left and right children:

$$\Delta G(m) = G(m) - [N_{m_l} G(m_l) - N_{m_r} G(m_r)] / N_m$$

We assume that the node m splits on an **uninformative** variable X_j, i.e. X_j and Y are independent.

We will use the short notation $\sigma_{m,.}^2 \equiv p_{m,.}(1 - p_{m,.})$ for . either equal to oob or in and rely on the following facts and notation:

1. $E[\hat{p}_{m,oob}] = p_{m,oob}$ is the "population" proportion of the class label in the OOB test data (of node m).
2. $E[\hat{p}_{m,in}] = p_{m,in}$ is the "population" proportion of the class label in the inbag test data (of node m).

---

[4] For easier notation we have (i) left the multiplier 2 and (ii) omitted an index for the class membership.

3. $E[\hat{p}_{m,oob}] = E[\hat{p}_{m_l,oob}] = E[\hat{p}_{m_r,oob}] = p_{m,oob}$
4. $E[\hat{p}_{m,oob}^2] = var(\hat{p}_{m,oob}) + E[\hat{p}_{m,oob}]^2 = \sigma_{m,oob}^2/N_m + p_{m,oob}^2$

$$\Rightarrow E[G_{oob}(m)] = E[\hat{p}_{m,oob}] - E[\hat{p}_{m,oob}^2] = \sigma_{m,oob}^2 \cdot \left(1 - \frac{1}{N_m}\right)$$

$$\Rightarrow E[\widehat{G}_{oob}(m)] = \sigma_{m,oob}^2$$

5. $E[\hat{p}_{m,oob} \cdot \hat{p}_{m,in}] = E[\hat{p}_{m,oob}] \cdot E[\hat{p}_{m,in}] = p_{m,oob} \cdot p_{m,in}$

Equalities 3 and 5 hold because of the independence of the inbag and out-of-bag data as well as the independence of X_j and Y.

We now show that $\mathbf{E(\Delta PG_{oob}^{(0)}) \neq 0}$ We use the shorter notation $G_{oob} = PG_{oob}^{(0)}$:

$$E[\Delta G_{oob}(m)] = E[G_{oob}(m)] - \frac{N_{m_l}}{N_m}E[G_{oob}(m_l)] - \frac{N_{m_r}}{N_m}E[G_{oob}(m_r)]$$

$$= \sigma_{m,oob}^2 \cdot \left[1 - \frac{1}{N_m} - \frac{N_{m_l}}{N_m}\left(1 - \frac{1}{N_{m_l}}\right) - \frac{N_{m_r}}{N_m}\left(1 - \frac{1}{N_{m_r}}\right)\right]$$

$$= \sigma_{m,oob}^2 \cdot \left[1 - \frac{1}{N_m} - \frac{N_{m_l} + N_{m_r}}{N_m} + \frac{2}{N_m}\right] = \frac{\sigma_{m,oob}^2}{N_m}$$

We see that there is a bias if we used only OOB data, which becomes more pronounced for nodes with smaller sample sizes. This is relevant because visualizations of random forests show that the splitting on uninformative variables happens most frequently for "deeper" nodes.

The above bias is due to the well known bias in variance estimation, which can be eliminated with the bias correction, as outlined in the main text. We now show that the bias for this modified Gini impurity is zero for OOB data. As before, $\widehat{G}_{oob} = \widehat{PG}_{oob}^{(0)}$:

$$E[\Delta\widehat{PG}_{oob}(m)] = E[\widehat{G}_{oob}(m)] - \frac{N_{m_l}}{N_m}E[\widehat{G}_{oob}(m_l)] - \frac{N_{m_r}}{N_m}E[\widehat{G}_{oob}(m_r)]$$

$$= \sigma_{m,oob}^2 \cdot \left[1 - \frac{N_{m_l} + N_{m_r}}{N_m}\right] = 0$$

References

1. Adler, A.I., Painsky, A.: Feature importance in gradient boosting trees with cross-validation feature selection. Entropy **24**(5), 687 (2022)
2. Breiman, L.: Random forests. Mach. Learn. **45**, 5–32 (2001). https://doi.org/10.1023/A:1010933404324
3. Díaz-Uriarte, R., De Andres, S.A.: Gene selection and classification of microarray data using random forest. BMC Bioinformatics **7**(1), 3 (2006)
4. Grömping, U.: Variable importance assessment in regression: linear regression versus random forest. Am. Stat. **63**(4), 308–319 (2009)
5. Grömping, U.: Variable importance in regression models. Wiley Interdiscip. Rev. Comput. Stat. **7**(2), 137–152 (2015)
6. Hothorn, T., Hornik, K., Zeileis, A.: Unbiased recursive partitioning: a conditional inference framework. J. Comput. Graph. Stat. **15**(3), 651–674 (2006)

7. Kim, H., Loh, W.Y.: Classification trees with unbiased multiway splits. J. Am. Stat. Assoc. **96**(454), 589–604 (2001)
8. Li, X., Wang, Y., Basu, S., Kumbier, K., Yu, B.: A debiased mdi feature importance measure for random forests. In: Wallach, H., Larochelle, H., Beygelzimer, A., d Alché-Buc, F., Fox, E., Garnett, R. (eds.) Advances in Neural Information Processing Systems 32, pp. 8049–8059 (2019)
9. Liaw, A., Wiener, M.: Classification and regression by randomForest. R News **2**(3), 18–22 (2002). https://CRAN.R-project.org/doc/Rnews/
10. Loecher, M.: Unbiased variable importance for random forests. Commun. Stat. Theory Methods **51**, 1–13 (2020)
11. Loh, W.Y., Shih, Y.S.: Split selection methods for classification trees. Stat. Sin. **7**, 815–840 (1997)
12. Lundberg, S.M., et al.: From local explanations to global understanding with explainable AI for trees. Nat. Mach. Intell. **2**(1), 56–67 (2020)
13. Menze, B.H., Kelm, B.M., Masuch, R., Himmelreich, U., Bachert, P., Petrich, W., Hamprecht, F.A.: A comparison of random forest and its Gini importance with standard chemometric methods for the feature selection and classification of spectral data. BMC Bioinformatics **10**(1), 213 (2009)
14. Nembrini, S., König, I.R., Wright, M.N.: The revival of the Gini importance? Bioinformatics **34**(21), 3711–3718 (2018)
15. Olson, R.S., Cava, W.L., Mustahsan, Z., Varik, A., Moore, J.H.: Data-driven advice for applying machine learning to bioinformatics problems. In: Pacific Symposium on Biocomputing 2018: Proceedings of the Pacific Symposium, pp. 192–203. World Scientific (2018)
16. Pedregosa, F., et al.: Scikit-learn: machine learning in python. J. Mach. Learn. Res. **12**, 2825–2830 (2011)
17. Saabas, A.: Interpreting random forests (2019). http://blog.datadive.net/interpreting-random-forests/
18. Saabas, A.: Treeinterpreter library (2019). https://github.com/andosa/treeinterpreter
19. Sandri, M., Zuccolotto, P.: A bias correction algorithm for the Gini variable importance measure in classification trees. J. Comput. Graph. Stat. **17**(3), 611–628 (2008)
20. Shih, Y.S.: A note on split selection bias in classification trees. Comput. Stat. Data Anal. **45**(3), 457–466 (2004)
21. Shih, Y.S., Tsai, H.W.: Variable selection bias in regression trees with constant fits. Comput. Stat. Data Anal. **45**(3), 595–607 (2004)
22. Strobl, C., Boulesteix, A.L., Zeileis, A., Hothorn, T.: Bias in random forest variable importance measures: illustrations, sources and a solution. BMC Bioinformatics **8**, 1–21 (2007). https://doi.org/10.1186/1471-2105-8-25
23. Strobl, C., Boulesteix, A.L., Augustin, T.: Unbiased split selection for classification trees based on the Gini index. Comput. Stat. Data Anal. **52**(1), 483–501 (2007)
24. Sun, Q.: tree. interpreter: Random Forest Prediction Decomposition and Feature Importance Measure (2020). https://CRAN.R-project.org/package=tree.interpreter. R package version 0.1.1
25. Wright, M.N., Ziegler, A.: ranger: a fast implementation of random forests for high dimensional data in C++ and R. J. Stat. Softw. **77**(1), 1–17 (2017). https://doi.org/10.18637/jss.v077.i01
26. Zhou, Z., Hooker, G.: Unbiased measurement of feature importance in tree-based methods. ACM Trans. Knowl. Discov. Data (TKDD) **15**(2), 1–21 (2021)

The Influence of User Diversity on Motives and Barriers when Using Health Apps - A Conjoint Investigation of the Intention-Behavior Gap

Eva Rössler[1], Patrick Halbach[1], Laura Burbach[1], Martina Ziefle[1], and André Calero Valdez[2]([✉])

[1] Human-Computer Interaction Center, RWTH Aachen University,
Campus Boulevard 57, 52076 Aachen, Germany
`eva.roessler@rwth-aachen.de`, {`halbach,burbach,ziefle`}`@comm.rwth-aachen.de`
[2] Institute for Multimedia and Interactive Systems, University of Lübeck,
Ratzeburger Allee 160, 23562 Lübeck, Germany
`calerovaldez@imis.uni-luebeck.de`

Abstract. Currently, there is a major health problem in our society, which partially is the result of an insufficient level of physical activity. Despite existing intentions, people sometimes fail to turn them into action and engage in physical activity. This intention-behavior gap provides a framework for the topic under study. Fitness apps offer a way to assist and support people in implementing physical activity in their daily routine. Therefore, this paper investigates the influence of user diversity on motives and barriers to fitness app use. For this purpose, a choice-based conjoint study was conducted in which 186 subjects were asked to repeatedly choose their favorite between three fictitious constellations of fitness apps. The apps were configured based on selected attributes. Differences in decision-making between men and women, exercisers and non-exercisers, as well as influences of certain personality dimensions and motivational types have been found. The results provide important clues that may help to customize fitness apps to specific user groups and for further research.

Keywords: Fitness apps · Health apps · Privacy · Group recommendation · User modelling · Human-computer interaction

1 Introduction

Well-balanced nutrition and physical activity (PA) are essential for a healthy lifestyle. Even though this is general knowledge, people tend to neglect these aspects [19]. Especially insufficient levels of physical activity and resulting consequences represent a huge risk in our society today. According to the World Health Organization (WHO), the PA levels of many adults and adolescents do

© IFIP International Federation for Information Processing 2022
Published by Springer Nature Switzerland AG 2022
A. Holzinger et al. (Eds.): CD-MAKE 2022, LNCS 13480, pp. 130–149, 2022.
https://doi.org/10.1007/978-3-031-14463-9_9

not meet the organization's recommendations for a healthy lifestyle [19]. Since the beginning of the COVID-19 pandemic, the topic has become even more important due to the increase in working in the home office and the decrease of opportunities to participate in sports clubs and activities in groups.

Even though many people have intentions to exercise, they sometimes fail to turn them into action [8]. This dissonance between intention and action is also called the intention-behavior-gap. It provides a framework for the topic under study and will be integrated in the theoretical part of this study. To support people in exercising, there is a broad offer of health-apps today. Especially in the given situation of the pandemic, apps like this provide an opportunity to exercise without the risk of infecting.

This study examines, which aspects should be integrated with a fitness-app for supporting fitness-activities and motivation. The study especially focuses on the aspects of user diversity. For this purpose, a choice-based conjoint study was conducted in which 186 subjects were asked to repeatedly choose their favorite out of three fictitious constellations of fitness apps. Specifically, our study examines how gender, age, and sports activity affect decision-making. In addition, the influence of personality dimensions and types of motivation towards exercising will be investigated. Our findings point out how fitness-app use can influence the intention behavior-gap in the context of PA and provide important clues for tailoring fitness apps to specific user groups.

2   Related Work

2.1   Relevance of Physical Activity and Health Apps

Although physical activity is essential for a healthy lifestyle, some people tend to neglect exercising. According to WHO, the PA level of 1 out of 4 adults and 3 out of 4 adolescents did not meet the organization's recommendations for a healthy lifestyle in the year 2018 [19]. This physical inactivity is the fourth most common risk factor of worldwide mortality at 5.5% [18] and "[...] is estimated to cause around 21–25% of breast and colon cancer burden, 27% of diabetes and about 30% of ischaemic heart disease burden" [18].

During the COVID-19 pandemic, the amount of PA decreased even more in some cases. In a review, Stockwell et al. bring together 66 studies to find out to what extent PA and sedentary behavior have changed during lockdown. Most studies demonstrate an overall decrease in PA and an increase in sedentary behavior [16]. It is noteworthy that most of the studies measured the level of PA by subjective assessments [16], but since many activities were severely restricted during the pandemic, it can be assumed that PA actually decreased.

These findings are alarming since PA seems to be even more important in a pandemic. Chastin et al. reviewed 55 studies and found that habitual PA can influence the immune system in a positive way [2]. "[...] higher levels of habitual physical activity are associated with a 31% lower prospective risk of infectious disease and 37% lower risk of infectious disease-related mortality" [2]. In addition, individuals who exercised an average of three times for 60 min over a period of about 20 weeks prior to vaccination show significantly higher levels of antibodies, compared to others who did not exercise [2].

Therefore, it is more important than ever to increase people's motivation for PA. Digital health apps can be one part of the solution. The number of such apps has risen rapidly in the past few years. While at the beginning of 2015 there were around 37,000 health and fitness apps in Apple's App Store, the number had already risen to over 82,000 by the third quarter of 2020 [15]. Among these are different types of health apps, e.g. for diagnosis, meditation, or medication intake. The number of health app users has also increased in recent years. While there were 9.7 million users in Germany in 2017, there are already 13.9 million in 2020. According to the forecast, the number of users could increase to 18.3 million by 2024 [14]. Again, not only apps to support sports activities, but also nutrition apps were taken into account. Both statistics show an increasing interest in digital health offerings. Due to the large number of these tools, the question arises of how an app should be designed to support users in achieving their goals and promoting their health. This task is essential because people often cannot bring themselves to engage in physical activity despite having the intention to do so.

Rhodes and De Bruijn were able to prove this discrepancy between the intention and the implementation regarding PA. They found this by reviewing 10 studies that examined the intention and implementation of PA in a total of 3899 individuals of different genders. The time between measurements of intention and measurements of PA ranged from 2 weeks to 6 months [8]. Rhodes and De Bruijn obtained the following results: 21% of the subjects did not intend PA and thus did not carry it out, whereas 2% of the subjects were active despite the lack of intention. Subjects who did not implement PA despite intention represent 36% of the sample, intenders who followed their plans 42%. Thus, only 54% of the intenders were able to implement their intended behavior [8].

2.2   Understanding the Process of Health Behavior Change

To get a better understanding of the aspects of changing health behaviors, the *Health Action Process Approach* (HAPA) by Ralf Schwarzer is a helpful tool. Schwarzer addresses the intention-behavior gap by identifying two phases in the process. "The model suggests a distinction between (a) pre-intentional motivation processes that lead to a behavioral intention, and (b) post-intentional volition processes that lead to the actual health behavior" [12]. In the motivation phase, risk perception, outcome expectancies, and action self-efficacy influence the formation of an intention. When this intention is formed, coping self-efficacy and recovery self-efficacy influence action initiation and maintenance in the volition phase. Here, action plans and coping plans also have an impact. Schwarzer defines action plans as When-Where-How plans, that can help individuals to implement an intended behavior. Coping plans, on the other hand, are used when the first-choice plan cannot be carried out for some reason. They represent alternatives for different situations that may emerge and should be designed beforehand [13]. This understanding of the different factors involved in health behavior change can be helpful in designing an app to optimally support this process.

2.3 Previous Findings on the Impact of Gender and Age on Health App Use

There is already some evidence of the effect of demographic features on fitness app use. In two studies, Klenk et al. [6] investigate the motives of German users of the app Runtastic, which is a mobile app for tracking sports activities. The aim of the studies was to identify differences between, e.g., men and women. Regarding gender, findings showed that for women, having fun and achieving their goals plays a greater role, while men are more inclined to share their results with others and use social functions [6]. Accordingly, differences based on gender regarding the desired app functions are also to be expected in the study conducted here. In terms of age, Cho et al. found that younger people use health apps more frequently than older people [3]. Klenk et al. found that the willingness to share results in the app Runtastic is rising with increasing age. However, other studies obtained opposite results on this issue [6]. Although the studies examined fitness app use from another angle than we will, it can be assumed that a difference between age groups is also evident here.

2.4 Scales Used in the Questionnaire

To examine various aspects of user diversity, a short version of the Big Five Inventory and the Sports Motivation Scale were integrated in the questionnaire. Both scales are briefly explained below.

Short Version of the Big Five Inventory. To capture personality traits of the individual subjects, the German short version of the Big Five Inventory (BFI-K) [7] was integrated into the questionnaire. It was chosen because it reliably represents personality traits while taking only about two minutes to complete. The scale allows to examine possible correlations between certain personality traits and motives in the use of fitness apps during the evaluation. The five personality traits measured by the BFI-K are extroversion, agreeableness, conscientiousness, neuroticism and openness to experience, which are examined by 21 items in total. An example item for conscientiousness is "I make plans and execute them". There is already evidence on how personality affects physical activity. In a meta-analysis, Wilson and Dishman examine correlations between the named personality dimensions and PA. For this purpose, they consult 64 studies with a total of 88,400 participants [17]. They find that people with higher levels of extroversion, conscientiousness, or openness are more likely to engage in PA than people who show fewer of these personality traits. They find the opposite results for neuroticism: People with a higher level of neuroticism are less likely to engage in PA. No significant correlation could be found between agreeableness and PA [17].

Sports Motivation Scale. For this study, the German version of the Sports motivation scale (SMS28) by Burtscher et al. [1] was integrated into the questionnaire. The SMS28 is used to examine seven types of motivation, distinguishing between extrinsic motivation, intrinsic motivation and amotivation. *Intrinsic motivation (IM)* includes IM toward knowledge, IM toward accomplishments and IM toward stimulation. Extrinsic motivation (EM) includes external regulation, introjected regulation and identified regulation. Each type of motivation type contains four items, which refer to the question "Why do you practice your sport?". An example item for IM toward knowledge is "Because of the good feeling of knowing more about the sport I practice". The items for amotivation were not used in the questionnaire, because they are not relevant for examining the research questions.

2.5   Underlying Research Questions

In this study, our general question will be:

What are the motives and barriers while using fitness apps?

Concerning previous findings, our question will be divided in different parts. Since there were already age- and gender-related differences regarding PA and fitness app use found, we ask:

1. *How do age and gender influence the motives and barriers?*

Based on Schwarzer's HAPA, the question also arises whether the use of fitness apps is influenced by whether a person has already been able to realize the intention to be physically active. Therefore, the following questions are investigated:

2. *Is there a difference between physically active and inactive individuals?*
3. *How does the source of motivation in physically active individuals influence the use of fitness apps?*

Since Wilson and Dishman were able to find correlations between certain personality traits and physical activity, this aspect is also examined here. The following question arises:

4. *How do personality traits affect motives and barriers?*

Following is a description of the methodology used to investigate the mentioned research questions, before the findings are reported and discussed.

3   Method

To find out what motives and barriers occur in the use of fitness apps, a quantitative study was conducted in the form of an online questionnaire. It was created using the Sawtooth Lighthouse software. The main part of the questionnaire was a choice-based conjoint study, in which the subjects were asked to repeatedly choose their individual favorite out of three fictitious app configurations.

3.1   Sample

The study was conducted in Germany, mostly North Rhine-Westphalia, so the results are based on a German perspective. The questionnaire was sent privately to people in the area around Aachen, and was shared in location-based Facebook groups. To ensure that the sample was not limited to a small area, people from small towns in the greater area of Aachen as well as from large cities such as Cologne and Berlin were contacted in this way. A total of 405 records were collected, of which 195 were complete. Of the participants with incomplete data sets, most dropped out on the first page ($N = 70$) or on one of the first four CBC tasks ($N = 87$). In addition, nine more data sets had to be excluded due to insufficient processing time. After data cleansing, 186 records remained that were qualified for analysis. The sample consists of 74% female participants and 26% male participants. One person identifies as non-binary. The age ranges from 16 to 78 years, with a mean age of 42.2 years ($SD = 14.3$). While 62% of the respondents stated that they exercise, 38% negated that.

3.2   The Questionnaire

In the first part of the questionnaire, participants were briefly introduced into the topic of the study. After a few questions regarding demographic data, the German short version of the Big Five Inventory was integrated to assess personality traits of the respondents. This scale consists of 21 statements, each of which subjects were asked to rate on a six-point Likert scale (from 1 = strongly disagree up to 6 = strongly agree). Since the questionnaire was already very time-consuming due to the conjoint study, the short version was chosen to avoid increasing the length even more while still preserving a reliable reflection of the full Big Five Inventory [7]. Extroversion, agreeableness, conscientiousness and neuroticism each contain four items. Openness to experience contains five items, because it shows less intern consistency compared to the other traits [7]. Further, respondents were asked about their exercising habits and current fitness app usage. Physically active participants then evaluated their motivation towards sports by rating statements of the German version of the sports motivation scale. The underlying question was "Why do you practice your sport?" and the statements should be rated on a five-point Likert scale (from 1 = does not apply at all, up to 5 = applies exactly). The items on amotivation were not included, as the reasons why a respondent is not motivated are irrelevant for the investigation. Subsequently, the decision scenarios of the choice-based conjoint study followed, which will be explained further in the following chapter.

   The full German questionnaire and additional material is available at the Open Science Framework at https://osf.io/xq43m/.

3.3   Description of the Choice-Based Conjoint Study and Decision Scenarios

Main part of the study was the choice-based conjoint study (CBC). In this type of study, participants are repeatedly confronted with several concepts out of which they are asked to choose their individual favorite. Compared to a ranking, this allows to identify relative importances of the attributes and relative part worth utilities. In addition, the decision tasks are more realistic than a ranking and reflect real purchase decisions, which is why the subjects should typically find it easy to empathize with the situations [9].

The apps were configured based on the attributes *data, optimization* and *search for training partners* (Table 1).

Table 1. Attributes and levels of the CBC.

Attribute	Levels				
Data	Activity data	Calendar data		Health data	
Optimization	Achievement progress	Motivation		Planning	
Search for training partners	Alone	in pairs		in a group	
		with friends	with App-Matching	with friends	with App-Matching

Data. We distinguish between three types of data that the fitness app can use. The first is *activity data*, which can be used to record steps, kilometers or the duration of activity. This enables users to view historical training data and evaluate the progress already made. If an app accesses *calendar data*, it can recognize when the user has free time slots and create corresponding training suggestions. For example, the app can recommend more intense training when there is little available time and longer activities when there is a lot of free time. When *health data* is used, weight, resting pulse or relevant pre-existing conditions, such as high blood pressure, are used to individualize the app and to support the search for suitable training partners. An app can access multiple data simultaneously. Activity data is required in each configuration.

Optimization. This attribute is also categorized into three aspects. This factor is used to optimize the training suggestions of the app in order to support the user most effectively. The aspect *training progress* optimizes the training suggestions for the most effective activities possible on basis of the available data. This helps the user to increase their performance and achieve training goals. With the *motivation* feature, an app suggests activities that increase motivation to exercise. For example, small challenges or visualizations can be integrated, which can motivate and encourage the user to reach their goals. In case of the aspect *planning*, the app ensures that the suggestions fit the user's schedule. For example, light workouts can be integrated on heavily scheduled days. A condition for this function is access to calendar data. The CBC was designed to support only one of these functions at a time.

Search for Training Partners. This attribute is divided into five levels. A distinction is made between whether an app supports training *alone*, *in pairs* or *in a group*. When training with others, there is a further distinction between training with known or unknown users. In case of training *with friends*, the app supports connecting with known users. These can be friends or other users with whom training has been done before. *App matching* supports training with unknown users. For this purpose, the app takes into account all data available, to find the optimal training partners or groups for the user. Criteria such as age, gender and search radius can be specified in advance. As with optimization, the CBC was also designed to support only one of these functions at a time (e.g., in a group of friends).

The participants were asked to repeatedly choose their individual favorite out of three fictitious app configurations (see Fig. 1).

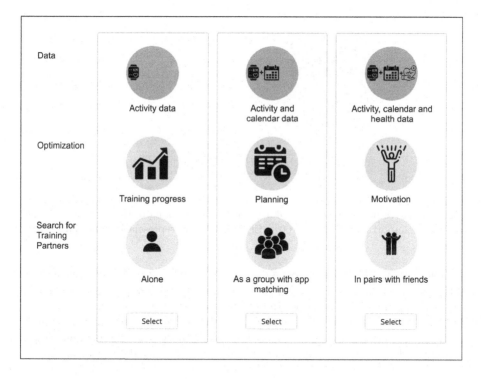

Fig. 1. Example of a CBC decision task

To examine the collected data in terms of the research questions, analyses were carried out with the programs Sawtooth Lighthouse and Jamovi after data cleansing. This included a latent class analysis (LC) and a hierarchical Bayes analysis (HB). The HB analysis is suitable because it can calculate estimates of the part-worth utilities even if the respondents make few decisions [10]. LC analysis is useful for additionally calculating which different types of decision-makers are represented in the sample. The findings are particularly valuable if the identified groups can be distinguished from each other on the basis of additional characteristics [11].

4   Results

4.1   Hierarchical Bayes Analysis

The Hierarchical Bayes analysis (HB) was conducted first because it provides an overview of decision-making for the overall sample. The results show that for the entire sample, training partner search has by far the largest effect on decision making. The attributes data and optimization show an equally large effect (see Fig. 2).

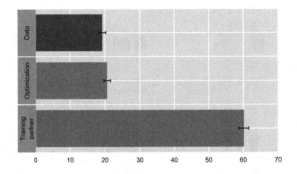

Fig. 2. Relative importance of attributes in decision making of the total sample. The sum of the importance scores is 100%. Error indicators represent the standard error.

The part worth utilities show a clear preference for training alone or in pairs with friends. In contrast, training in a group is clearly less preferred, especially with the app matching function. In optimization, there is a tendency to training progress. Subjects benefit significantly less from optimization in the sense of planning. In terms of data, the subjects benefit most from the combination of all data and least from an app that only uses the activity data (see Fig. 3).

4.2   Latent Class Analysis

Latent class analysis (LC) was used to identify different types of decision-makers. It turns out that at least two and at most five groups can be distinguished from each other. First, we will briefly explain how the number of groups was selected before they are defined in more detail. Various information criteria can be used to determine the number of groups. "An information criterion can be defined (roughly) as the distance between the model at hand and the real model." [4].

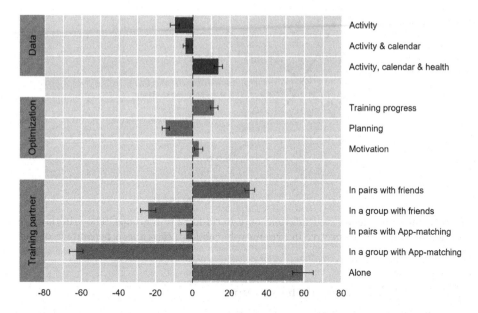

Fig. 3. Part-worth utilities of attribute levels in decision making of the total sample. The part-worth utilities are summed to zero for each attribute. Error indicators represent the standard error.

Consistent Akaike Information Criterion (CAIC), Percent Certainty (Pct Cert), and Relative Chi-Square are used here. CAIC is one of the most common information criteria when deciding on the number of segments. Smaller values are preferred [10]. Percent Certainty "indicates how much better the solution is compared to an 'ideal' solution than the null solution". Relative Chi-Square is based on Chi-Square, which indicates whether a solution fits significantly better than the null solution.

Now, based on the largest curvature in the individual graphs, it could be decided which number of segments is most suitable (see Fig. 4). Here, however, no clear decision can be made. Percent Certainty speaks for 3 groups, Relative Chi-Square for 2. CAIC, on the other hand, cannot be clearly interpreted. Therefore, group sizes and differences in group decision making are additionally considered. When distinguishing three groups, the group sizes equal about one third of the sample and they can be clearly distinguished from each other based on the relative importance of the attributes. Four groups are no longer as clearly distinguishable from one another in terms of content. Therefore, three groups are differentiated in the following.

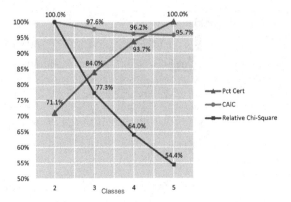

Fig. 4. Values of the selected information criteria for 2 to 5 groups. The criteria were normalized with the respective maximum value and are shown as a percentage of this value.

4.3 Decision-Making of the Groups

At the level of attribute importance, the training partner search is most important for decision-making for all three groups. Group 2 stands out in particular, as the other attributes have almost no influence on the decision-making process. Groups 1 and 3 make more balanced decisions in this respect (see Table 2). These different ratios show up clearly in the network diagrams in Fig. 5. For subjects in the first group, data usage is the second most relevant factor for decision-making; for group 3, optimization takes this place.

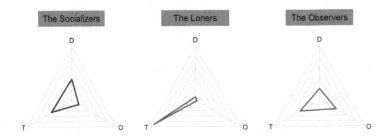

Fig. 5. Web charts for relative attribute importance in the LC groups. D = Data, O = Optimization, T = Training partner search.

To determine which specific characteristics influence the groups' decision-making, the relative part worth utilities must be considered (see Fig. 6). Based on these, the groups' decisions are now defined in more detail.

The Socializers. The decision-making of the first group is most influenced by the search for a training partner. They prefer training with friends, both in pairs

Table 2. Relative importance of attributes in LC-groups.

	Group 1 ($N = 62$)	Group 2 ($N = 67$)	Group 3 ($N = 57$)
Data	39.06	5.33	19.97
Optimization	15.76	3.39	36.51
Training partner	45.18	91.28	43.52

and in a group. In contrast, the use of an app matching function is clearly less preferred. In terms of data usage, they benefit most from the combination of activity, calendar and health data. At the optimization level, only minor differences in relevance are evident. Here, the subjects benefit most from an app that focuses on increasing motivation.

The Loners. For subjects in the second group, the search for a training partner is by far the most relevant for decision-making. It is most important for them to perform their workout alone. They benefit least from an app that supports training in groups, both in pairs and via app matching. At the data level, they are most likely to benefit from the combination of all data. With regard to optimization, they benefit most from the training progress function. For both attributes, however, the differences in relevance are very small.

The Observers. For subjects in this group, the highest value is optimization in terms of training progress. As the Loners, they also prefer training alone. In second place, they benefit most from training in pairs via app matching and with friends. On the data level, they benefit most from the usage of only data activity.

4.4 Gender, Age and Sport-Activity in the Groups

Next, we will examine whether the groups differ in terms of gender and age and whether there is a significant difference in the number of active or inactive participants between the groups. In terms of gender, the socializers and observers hardly differ from each other: socializers comprises 31% men, observers around 35%. For loners, on the other hand, the proportion of men is only about 13%. The number of sport-active individuals increases from socializers to observers. Whereas for socializers about half of the subjects do sports, observers has about 70% active in sports (see Fig. 7).

There is a significant difference in age between the socializers ($M = 40.4$, $SD = 13.6$) and the loners ($M = 44.4$, $SD = 14.5$). Observers ($M = 41.6$, $SD = 14.6$) lie with the average age between the other groups and shows no significant difference to them. Over the entire age range from 16 to 78 years, the distribution of age is similarly balanced in all three groups.

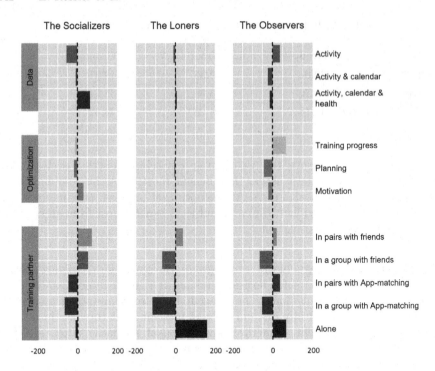

Fig. 6. Part-worth utilities of attribute levels in the three groups. The part-worth utilities are summed to zero for each attribute.

4.5 Further Influence of Gender, Age and Sport-Activity

To further examine whether gender, age and sport-activity influence the decision-making of the sample, they are also examined independently of the groups found in the LC-analysis. For this purpose, correlations and independent T-tests were calculated with the individual values of the HB-analysis. For age, there is a negative correlation with optimization (r (184) $= -.19$, $p < .05$) and a positive correlation with training partner search (r (184) $= .16$, $p < .05$). No correlations were found between individual part-worth utilities and age.

Gender and sport-activity were examined with independent T-tests. To examine the influence of gender, the binary person was filtered out, since this statement was only made once. Significant correlations with gender exist for optimization (t (183) $= -3.81$, $p < .001$) and training partner search (t (183) $= 3.66$, $p < .001$). Optimization influences men's decision making ($M = 26.5$, $SD = 14.5$) more than women's ($M = 18.6$, $SD = 11.5$). In contrast, training partner search is more relevant to women's decision making ($M = 63.2$, $SD = 18.6$) than to men's ($M = 51.8$, $SD = 18.8$). In addition, there is a significant correlation between sport-activity and optimization, (t (183) $= 2.05$, $p < .05$). This influences the sport-active respondents ($M = 22.1$, $SD = 13.5$) in their decision more than those who do not participate in sports ($M = 18.2$, $SD = 10.8$).

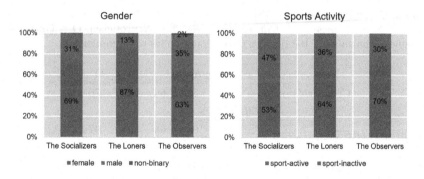

Fig. 7. Distribution of gender and sport-active subjects in the LC groups.

We also examined how the frequency of practicing sports per week affects decision-making. In this regard, no correlations could be found with the relative importance of the attributes. However, at the level of the part-worth utilities, there is a positive correlation with training progress (r (114) = .24, p < .05) and a negative correlation with motivation (r (114) = −.28, p < .01). No significant correlations were found with other levels of attributes.

4.6   Influence of Personality and Motivation Sources on Decision-Making

To further investigate the influence of user diversity, dimensions of BFI-K and SMS28 were also examined with regard to the relative importance of the attributes. For this purpose, correlations were calculated in each case. Some of the variables of BFI-K first had to be recoded in order to subsequently combine them into scales. Prior to this, Cronbach's α was calculated in each case in order to test the reliability of the variables. Next, correlations with Pearson's r were calculated between the personality dimensions and the relative importance of the attributes. The results show no correlation between a personality trait and the importance of an attribute for the sample (see Table 3).

The same procedure was applied to the dimensions of SMS28 in order to also investigate how the motivational source of sport-active participants affects the decision-making. Here, linear correlations were found between identification and optimization (r (114) = .27, p < .01), and between identification and training partner search (r (114) = −.28, p < .01). No correlations were found for other types of motivation (see Table 4).

In addition, the dimensions of BFI-K and SMS28 were examined for correlations with the relative part-worth utilities. The most significant result for the personality dimensions is a correlation between agreeableness and training alone. Individuals with a higher level of agreeableness are less likely to choose this training option (r (184) = −.22, p < .01). They prefer training with a friend (r (184) = .17, p < .05).

Table 3. Correlations between BFI-K dimensions and CBC attributes. E = Extroversion, A = Agreeableness, C = Conscientiousness, N = Neuroticism, O = Openness to experience. The value of the degrees of freedom applies to all dimensions.

		E	A	C	N	O
Cronbach's α		.85	.59	.66	.79	.70
	df	184	-	-	-	-
Data	r	.06	.09	.08	.08	.07
	p	.40	.20	.27	.31	.38
Optimization	r	.01	−.04	.06	−.03	.01
	p	.87	.56	.44	.66	.94
Training partner	r	−.05	−.04	−.10	−.03	−.05
	p	.48	.60	.19	.66	.50

Further, there are significant correlations between some of the SMS28 dimensions and the part-worth utilities of training partner search. For both intrinsic motivation toward knowledge and accomplishment, it appears that subjects with higher levels of each type of motivation are more likely to choose training with a group of friends (r (114) = .20 / .21, $p < .05$) and less likely to choose training in pairs using an app matching function (r (114) = −.20 / −.21, $p < .05$). There are also significant correlations with identification. There is a negative correlation to training in pairs using app matching (r (114) = −.20, $p < .05$) and a positive correlation to training in a group using app matching (r (114) = .19, $p < .05$). Strong significant correlations also exist positively to training in a group of friends (r (114) = .34, $p < .001$), and negatively to training alone (r (114) = −.31, $p < .001$). Further correlations can be found in Table 5.

5   Discussion

In the following, the results are discussed on the basis of the previously identified research questions.

1. *How do age and gender influence the motives and barriers?*

The LC groups cannot be clearly distinguished from one another with regard to gender and age. Although there is a significant age difference between the socializers ($M = 40.4$) and the loners ($M = 44.4$), this result does not indicate a clear difference in terms of the life stages of the participants. It is therefore questionable to what extent the groups actually differ from each other on the basis of age. Independently of the LC groups, however, a gender-dependent difference could be found concerning the importance of the attributes. Here it appears that women's decisions are more influenced by training partner search, while men make their decisions more dependent on optimization. Overall, these results do not necessarily correspond to the expectations derived from other studies [3,6]

Table 4. Correlations between SMS28 dimensions and CBC attributes. **p < .01; K = IM to knowledge, A = IM to accomplishment, S = IM to stimulation, Id = Identification, In = Introjection, E = External regulation. The value of the degrees of freedom applies to all dimensions.

		K	A	S	Id	In	E
Cronbach's α		.82	.87	.82	.77	.71	.78
	df	114	-	-	-	-	-
Data	r	.02	−.10	.05	.14	−.09	.04
	p	.80	.28	.62	.15	.33	.66
Optimization	r	.08	.16	.11	.27	.08	.16
	p	.40	.08	.25	.003**	.39	.09
Training partner	r	−.07	−.03	−.11	−.28	.01	−.14
	p	.45	.72	.26	.002**	.89	.15

regarding these attributes. In particular, the expectation that significant differences in age would be evident cannot be confirmed here.

2. *Is there a difference between physically active and inactive individuals?*

It was found that sports activity affects how important optimization is for decision making. In addition, among the participants who are active in sports, the question of how often they do sports is also significant. The more often subjects exercise per week, the more likely they are to choose optimization in terms of training progress and the less likely they are to choose motivation. These results could be explained by the assumption that people who exercise several times a week already have a high level of motivation and thus no longer need a corresponding motivational function. Since the motivation hurdle has already been overcome, the person is then more interested in observing the own performance.

3. *How does the source of motivation in physically active individuals influence the use of fitness apps?*

Some significant correlations were found in investigations with the dimensions of the SMS28. Here, the results at the level of training partner search stand out in particular. Subjects with a high degree of intrinsic motivation for knowledge and performance more often choose training in a group with friends and less often training in pairs via an app matching function. This could be explained by the assumption that people with a higher expectation of performance have already built up an environment with other athletes and thus do not need a function to search for other possible partners. Most correlations were found for the motivation type identification. Here a strong positive correlation to training in a group of friends and a strong negative correlation to training alone is found. There are also opposite results with regard to app matching: here the group is

Table 5. Correlations between BFI-K/SMS28 and part-worth utilities. *p < .05, **p < .01, ***p < .001; Table shows significant correlations only. A = Agreeableness, O = Openness to experience, K = IM toward knowledge, A = IM toward accomplishment, Id = Identification, E = External regulation.

		BFI-K		SMS28			
		A	O	K	A	Id	E
	df	184	-	114	-	-	-
Activity data	r	-	-	-	-	-	.20
	p	-	-	-	-	-	.03*
Motivation	r	-	−.16	-	-	-	-
	p	-	.04*	-	-	-	-
Planning	r	-	-	-	-	-	−.20
	p	-	-	-	-	-	.03*
In pairs - friends	r	-	-	-	-	-	-
	p	-	-	-	-	-	-
In a group - friends	r	.17	-	.20	.21	.34	-
	p	.02*	-	.03*	.03*	<.001***	-
In pairs - App Matching	r	-	-	−.20	−.21	−.20	-
	p	-	-	.03*	.03*	.03*	-
In a group - App Matching	r	-	-	-	-	.19	-
	p	-	-	-	-	.04*	-
Alone	r	−.22	-	-	-	−.31	-
	p	.003**	-	-	-	<.001***	-

chosen more frequently, while training in pairs is chosen less frequently. This can be attributed to the fact that the items of the SMS28 examine, at the level of identification, the extent to which sport is an opportunity for the respondent to maintain contact with other people.

4. *How do personality traits affect motives and barriers?*

It can be noted that people with a higher level of agreeableness would rather train in a group with friends and are less likely to train alone. This result was expected, as people with a high level of agreeableness get along well with other people and, accordingly, probably prefer training alone less. However, this does not explain why there are no other correlations, for example, to exercising with another friend. In addition, individuals with a higher degree of openness to new experiences were less likely to choose to optimize app use in terms of motivation. One possible explanation would be that these individuals have a lower motivational barrier due to their openness. Overall, with respect to the results of Wilson and Dishman [17], even clearer correlations were expected, but nevertheless the assumption that personality influences the use of fitness apps can be supported.

5.1   Referring to the *Health Action Process Approach*

In order to optimize the adaptation of fitness apps to users, it is not only important to recognize different user groups, but also to understand how the process
from an intention to an action proceeds. Here, the HAPA model [12,13] can
provide assistance, as it shows the different components it takes to change a
health behavior. With this understanding, an app could be designed to intervene in the different phases and support the user accordingly. A person who does
not exercise regularly might need features at the beginning of the process that
reinforce their intentions, while later features that encourage the maintenance
of the implemented behavior would help. Someone who already does a lot of
exercise, on the other hand, may only need the last. In addition, functions that
support the creation of action and coping plans could be integrated to increase
the likelihood of implementing physical activity. Since, as Rhodes & De Bruijn
showed, many people have at least the intention of doing sports [8], a suitable
fitness app could intervene at this point and support the conversion of intention
into action.

5.2   Limitations of the Study

In general, a choice-based conjoint study is very useful for investigating the
research questions addressed here. In contrast to a ranking, this type of study
makes it possible to determine the relative importance of attributes and attribute
levels for the overall sample, subgroups or individual subjects. However, this
study does not examine rejected attributes. Since subjects are forced to choose
an attribute in each decision situation, it is possible that attributes are selected
that would actually be rejected. On the other hand, a feature can have a strong
negative value, while not being rejected. In order to be able to make statements
about the rejection of attributes and attribute characteristics, combining the
CBC with an additional study would be appropriate. In addition, it should
be noted here that the search for participants can be difficult due to the size
of the study. A large part of the subjects dropped out during one of the first
decision tasks. This could be related to the complexity of the tasks and the high
effort required to complete them. Although the tasks reflect real-life purchase
decisions, the subjects have to process a lot of information each time before they
can make a decision [9]. Therefore, in order to obtain a larger sample, a different
concept for participant acquisition might need to be established. Also, in some
aspects our sample was not balanced. Regarding gender our sample contained
a lot more female than male respondents. A more balanced sample could show
more significant differences in decision-making between men and women. Since
the majority of the subjects in this study were acquired via Facebook groups, it
can be assumed that they have a certain willingness to share data regardless of
age, as well as a similar level of app knowledge. A sample with a wider spread
in this regard could therefore demonstrate further differences.

6   Conclusion

The aim of this study was to find out which functions of fitness apps could motivate or discourage potential users. Differences in the decisions between men and women, sport-active and sport-inactive respondents, as well as influences of certain personality dimensions and motivation types could be found. The results offer a clue for adapting fitness apps to different user groups. It is a big task to change health behavior in our society. However, it is necessary to increase the level of physical activity in order to maintain a healthy lifestyle. Digital offerings such as fitness can be a powerful tool to support people in their health behavior and promote the level of physical activity as recommended by the WHO. Considering the large number of these offerings on the market and the differences in peoples needs, fitness apps have to address users individually to ensure long-term results.

Furthermore, our results can contribute to human-centered design of software applications, that are based on artificial intelligence (AI). In the area of physical activity, AI can be useful for responding even more individually to potential users of fitness apps and taking even greater account of the user's current state. For example, training suggestions could always be adapted to the current training progress, while additional information like the current weather could be considered. To develop products that focus on the individual user's needs, abilities, and preferences, creating personas can be a helpful tool, as Holzinger et al. showed recently [5]. Our study provides information that could be used to create personas for AI in the field of PA, like fitness app preferences, personality traits or motivation sources. To address the different types of users even more individually, further research would be required on how the different types of decision-makers can be distinguished from one another. It could be helpful, to select a broader sample, in which the gender distribution is more balanced. People with less technology or app experience could have other needs regarding fitness apps and therefore should also be in focus in further research.

Since this study only theoretically addresses the intention-behavior gap, further research is also needed to determine how a fitness app can actually influence the implementation of an intention into an action. Therefore, intentions for physical activity could first be recorded in order to subsequently test which functions of fitness apps support the implementation of the intentions and which inhibit them. Here, the use of different prototypes of fitness apps would be appropriate in order to compare the effects of different functions with each other.

References

1. Burtscher, J., Furtner, M., Sachse, P., Burtscher, M.: Validation of a German version of the sport motivation scale (SMS28) and motivation analysis in competitive mountain runners. Percept. Mot. Skills **112**(3), 807–820 (2011)
2. Chastin, S.F., et al.: Effects of regular physical activity on the immune system, vaccination and risk of community-acquired infectious disease in the general population: systematic review and meta-analysis. Sports Med. **51**(8), 1673–1686 (2021)

3. Cho, J., Park, D., Lee, H.E., et al.: Cognitive factors of using health apps: systematic analysis of relationships among health consciousness, health information orientation, eHealth literacy, and health app use efficacy. J. Med. Internet Res. **16**(5), e3283 (2014)
4. Hoijtink, H.: Confirmatory latent class analysis: model selection using Bayes factors and (pseudo) likelihood ratio statistics. Multivar. Behav. Res. **36**(4), 563–588 (2001)
5. Holzinger, A., Kargl, M., Kipperer, B., Regitnig, P., Plass, M., Müller, H.: Personas for artificial intelligence (AI) an open source toolbox. IEEE Access **10**, 23732–23747 (2022)
6. Klenk, S., Reifegerste, D., Renatus, R.: Gender differences in gratifications from fitness app use and implications for health interventions. Mobile Media Commun. **5**(2), 178–193 (2017)
7. Rammstedt, B., John, O.P.: Kurzversion des big five inventory (BFI-K). Diagnostica **51**(4), 195–206 (2005)
8. Rhodes, R.E., de Bruijn, G.J.: How big is the physical activity intention-behaviour gap? A meta-analysis using the action control framework. Br. J. Health. Psychol. **18**(2), 296–309 (2013)
9. Sawtooth Software Inc: CBC Technical Paper (2017) (2022). https://sawtoothsoftware.com/resources/technical-papers/cbc-technical-paper
10. Sawtooth Software Inc: CBC/HB Technical Paper (2009) (2022). https://sawtoothsoftware.com/resources/technical-papers/cbc-hb-technical-paper
11. Sawtooth Software Inc: Latent class technical paper (2021) (2022). https://sawtoothsoftware.com/resources/technical-papers/latent-class-technical-paper
12. Schwarzer, R.: Modeling health behavior change: how to predict and modify the adoption and maintenance of health behaviors. Appl. Psychol. **57**(1), 1–29 (2008)
13. Schwarzer, R.: Health action process approach (HAPA) as a theoretical framework to understand behavior change. Actualidades en Psicología **30**(121), 119–130 (2016)
14. Statista: Digital-Health - Nutzerentwicklung bei Wearables und Fitness-Apps bis 2024 (2022). https://de.statista.com/statistik/daten/studie/1046996/umfrage/marktentwicklung-von-wearables-und-fitness-apps-in-deutschland/
15. Statista: mHealth-Apps - Anzahl im Apple App-Store 2020 (2022). https://de.statista.com/statistik/daten/studie/1191205/umfrage/anzahl-der-bei-apple-verfuegbaren-mhealth-apps/
16. Stockwell, S., et al.: Changes in physical activity and sedentary behaviours from before to during the COVID-19 pandemic lockdown: a systematic review. BMJ Open Sport Exerc. Med. **7**(1), e000960 (2021)
17. Wilson, K.E., Dishman, R.K.: Personality and physical activity: a systematic review and meta-analysis. Pers. Individ. Differ. **72**, 230–242 (2015)
18. World Health Organization: Global health risks: mortality and burden of disease attributable to selected major risks. World Health Organization (2009)
19. World Health Organization: Global action plan on physical activity 2018–2030: more active people for a healthier world. World Health Organization (2019)

Identifying Fraud Rings Using Domain Aware Weighted Community Detection

Shaik Masihullah, Meghana Negi$^{(\boxtimes)}$, Jose Matthew,
and Jairaj Sathyanarayana

Swiggy, Bangalore, Karnataka, India
{shaik.masihullah1,meghana.negi,jose.matthew,jairaj.s}@swiggy.in

Abstract. With the increase in online platforms, the surface area of malicious activities has increased manifold. Bad actors abuse policies and services like claims, coupons, payouts, etc., to gain material benefits. These fraudsters often work collusively (rings), and it is difficult to identify underlying relationships between them when analyzing individual actors. Fraud rings identification can be modeled as a community detection problem on graphs where nodes are the actors, and the edges represent common attributes between them. However, the challenge lies in incorporating the attributes' domain-informed importance and hierarchy in coming up with edge weights. Treating all edge types as equal (and binary) can be fairly naive; we show that using domain knowledge considerably outperforms other methods. For community detection itself, while the weight information is expected to be learned automatically in deep learning-based methods like Graph Neural Networks (GNN), it is explicitly provided in traditional methods. In this paper, we propose a scalable and extensible end-to-end framework based on domain-aware weighted community detection to detect fraud rings. We first convert a multi-edge weighted graph into a homogeneous weighted graph and perform domain-aware edge-weight optimization to maximize modularity using the Leiden community detection algorithm. We then use features of communities and nodes to classify both community and a node as fraud or not. We show that our methods achieve up to 9.92% lift in F1-score on internal data, which is significant at our scale, and up to 4.81% F1-score lift on two open datasets (Amazon, Yelp) vs. an XGBoost based baseline.

Keywords: Fraud detection · Graph learning · Community detection · Fraud rings identification · Graph neural network

1 Introduction

Similar to any e-commerce marketplace, every day, new loopholes are being exploited by fraudsters in hyper-local, online food delivery platforms. Fraudsters exploit business policies and services around food-issue claims, payments, coupons, etc., which have a material impact on the platforms' bottom line. Most often, these fraudsters operate in collusive groups. In many cases, when looked

© IFIP International Federation for Information Processing 2022
Published by Springer Nature Switzerland AG 2022
A. Holzinger et al. (Eds.): CD-MAKE 2022, LNCS 13480, pp. 150–167, 2022.
https://doi.org/10.1007/978-3-031-14463-9_10

at in isolation, it is easy to mistake the behavior of these fraudsters as not that dissimilar to normal, non-fraudulent patterns; it is only when looked at as a group that fraudulent patterns become apparent. Such fraud groups can cause more damage than individual fraudsters due to the collective knowledge they employ and can typically cover more surface area than an individual actor. As a result, identifying fraud rings is a crucial piece in a platform's overall risk mitigation plan.

For identifying frauds, graph-based approaches have been widely studied. Entities (like customers, sellers) in a marketplace and connections (like payment instruments, reviews) between them can be variously represented by homogeneous-, heterogeneous-, relational-, multi-, weighted graphs, and more. Nilforoshan [23] proposes a multi-view graph representation for mining fraud where multiple edges exist between nodes. Belle [2] proposes variants of graph representation learning such as traditional, inductive, and transductive for fraud detection in credit card transactions. While weighted and multi-view graphs solve for multiple edge types and weights, to the best of our knowledge, providing domain information to identify optimal weights has not been explored. Domain information is usually available in the form of, say, payment-instrument linkages being 'stronger' indicators of connectedness vs. linkages based on shared wi-fi addresses. Such fraud groups can cause more damage than individual fraudsters due to the collective knowledge they employ and can typically cover more surface area than an individual actor. As a result, identifying fraud rings is a crucial piece in a platform's overall risk mitigation plan.

GNNs play an important role in building graph-based machine learning (ML) approaches because of their ability to learn from graph structure. Graph Convolution Networks (GCN) have been applied in domains like opinion fraud [9,17] and insurance fraud [16]. For applications where graphs scale to millions of nodes with frequent updates and additions, scalability and run-time have been bottlenecks for GNN based approaches. On the other hand, community detection methods [26,36], which attempt to capture group structure by partitioning the graphs into communities, are typically more scalable. The authors of [4,13,30] have extensively studied GNN-based community formation and proposed methods to combine both approaches. In such methods, the number of communities to be formed has to be provided beforehand as a hyper-parameter, and this limits the attainability of optimal separation of communities.

Most fraud detection literature typically targets a single fraud detection task, like fake reviews or spam detection. However, in most real-world marketplaces, new fraud modus operandi (M.O) emerge all the time, and it is typically impractical, or even infeasible, to build M.O-specific detection models. It is our observation that a large swath of fraud is perpetrated by rings operating in collusion, constantly cooking up new M.Os. There is not much literature on methods or frameworks that can be extended across multiple M.Os being committed by similar sets of fraudsters (i.e., rings).

In this paper, we propose a novel generalizable framework to detect fraud rings using community detection on a graph whose edge weights are learned in a domain-aware manner. Our contributions are:

- We explicitly provide domain knowledge to community detection where optimal weights are algorithmically determined using weight bounds and edge priorities for fraud identification. This explicit feeding of information cannot be done in GNNs, as we expect them to learn this automatically, adding uncertainty.
- Our framework is modular – graph construction, community detection, downstream discriminator. We first convert a multi-edge graph into a weighted homogeneous graph which is then used by Leiden-based community detection. Community information is then used by downstream tasks to perform fraud ring detection. The node-to-community mapping can be used to develop rules and models for multiple M.Os using simple community feature aggregations.
- We perform extensive experiments on two public benchmark datasets (Yelp and Amazon Reviews) and an internal dataset demonstrating the effectiveness of the proposed framework, both in terms of run-time and F1-score performance. Our framework shows a 1.7% relative improvement in F1-score on Amazon, 4.8% on Yelp, and 9.92% on internal datasets.

The remainder of the paper is organized as follows. Section 2 outlines the existing literature in related works. Section 3 introduces the problem statement of fraud rings and challenges involved. Section 4 details the proposed framework. Section 5 demonstrates the experimentations conducted and the ablation study. We end in Sect. 6 by concluding the paper.

2   Related Work

2.1   Community Detection

While recent research has been backed by GNN based approaches, there has been significant research on non-GNN based approaches for community detection [31]. While GNN approaches focus on representation learning, traditional methods use various graph theory strategies. Traditional methods distribute graphs into communities and do not require the number of communities to be supplied beforehand. This is a major advantage as estimating the number of communities upfront is typically impossible. Modularity and Constant Potts Model (CPM) are well-known metrics to evaluate the quality of generated communities. Ghosh et al. [10] proposed a strategy to run the Louvain algorithm in a distributed computing environment. Even though Louvain, by nature, is not scalable, using these types of distributed implementations can process larger graphs, albeit at the cost of extra resources. You et al. [37] proposed a three-stage algorithm that includes central-node identification, label propagation, and community merging. Community detection can also be performed using optimization algorithms like Particle Swarm Optimization (PSO) [26] or semi-supervised approaches like label propagation [36]. The Leiden algorithm [33] introduced in 2019 by Traag et al. can run faster and find better partitions compared to other methods.

A majority of the recent literature on GNNs is based on the assumption of incorporating ring information within the embeddings. Luo et al. [20] attempted to identify communities in a heterogeneous graph by applying their Context Path based GNN model. While most community detection approaches focus on segregating graphs into non-overlapping communities, Shchur et al. [30] focus on generating overlapping communities using a GNN-based Neural Overlapping Community Detection (NOCD) model. Moreover, the NOCD model is also compared with non-deep models such as multi-layer perceptrons and free-variable models, studying the effectiveness of using deep network models. Bandyopadhyay et al. [1] attempted to solve community detection using GNN in an unsupervised approach. The authors integrated a self-expressive layer in GNN and designed a loss function to classify nodes into communities directly. Wang et al. [35] leverage the bipartite representation to perform community detection in heterogeneous graphs using GNNs. The attention mechanism is adopted for increased focus on prime nodes while segregating the graph. Jia et al. [14] presented a generative adversarial network based community detection framework solving overlapping community detection and graph representation learning.

Identifying edge priority has been an active area of research but focuses on learning weights rather than explicit inputs. Shang et al. [29] propose SACN (Structure-Aware Convolution Network) that uses an encoder-decoder architecture. The encoder consists of a weighted graph convolution network, which learns the network weights by utilizing the graph node structure, node attributes, and edge relation types. This generates accurate graph node embeddings capturing most of the important information from the graph. The decoder is a convolution network used for link prediction. In [12], the authors designed a fast algorithm known as MGFS (Multi-label Graph-based Feature Selection algorithm) that utilizes the Page Rank algorithm to estimate feature importance based on edge weights.

2.2   Graph-based Fraud Detection

Graph-based approaches for identifying frauds have been extensively studied in financial fraud [16,19,34] and opinion-fraud detection [9,15,24]. Wang et al. [34] proposed a SemiGNN variant solving the label uncertainty problem in fraud detection where only a small percentage is tagged as fraud with certainty. Liu et al. [19] present a GEM model utilizing graph embeddings and attention mechanisms to identify irregular malicious accounts on a financial platform. Dou et al. [9] proposed a CARE-GNN (CAmouflage-REsistant GNN) architecture emphasizing the identification of camouflaged fraudsters who had built ways to hide themselves among normal users. Liu et al. [18] present a DGFraud (Deep Graph Fraud) model that identifies frauds in social network graphs despite inconsistencies in context, features, or relations among the users. A few researchers [25, 28, 32] have also attempted to identify fraud community groups in medical, banking, and telecom sectors using traditional community detection such as label propagation and group mining. Sun et al. [32] proposed a person's similarity adjacency graph and Maximal Clique Enumeration-based approach to identify fraud in medical insurance.

Although using deeper networks or complex strategies might outperform during experimentations, these are often constrained by high inference time, complex retraining processes, and non-trivial deployment challenges, when applied in real-world applications. To the best of our knowledge, ours is the first novel and scalable end-to-end framework using traditional graph theory-based community detection, which outperforms GNNs in terms of both run-time and F1-score performance.

3   Fraud Detection Problem

3.1   Identifying Fraud Rings

A graphical representation of the user base of an e-commerce platform can have customers sharing stationary and non-stationary attributes. These customers can be grouped into communities, combined behavior of which can be used to identify fraud rings. Figure 1 illustrates how an entity's fraud status changes when looked at from the lens of its connections. Using only an entity's attributes might classify the entity as not-fraud. However, if the entity is connected to a number of high-risk entities, then the whole community and the entity at hand could be classified as fraudulent. It should be noted that it is not possible to constrain rings identified to be composed solely of fraudulent actors. It is typically a business decision to either add some post-processing to reduce such false positives or choose to live with it.

3.2   Incorporating Domain Knowledge

In e-commerce marketplaces, it is pretty likely that the customers can be connected to other customers. One can create connections based on identifiers like payments, wi-fi, etc., to behavioral aspects like similar buyers, areas, category preferences, etc. The importance of such connections varies based on the problem one is trying to solve. For example, behavioral relations can boost the quality of recommendations, and identifiers can help identify swindling tendencies. Further, in fraud rings identification, not all rings are equal. For example, for a hyper-local marketplace, a ring of entities connected by a common device(s) is a 'stronger' signal compared to rings of wi-fi addresses or broad geo-locations. It could be the opposite for a social network. It is crucial to encode this additional domain-specific knowledge.

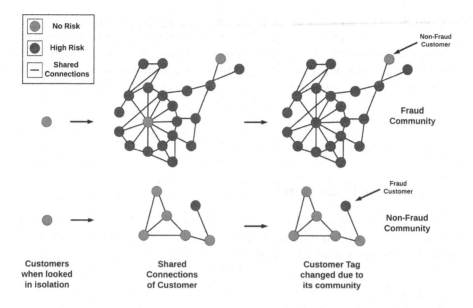

Fig. 1. An illustration of fraud rings identification problem.

Adding weights and domain knowledge to the graph helps enforce graph sub-structure by focusing on relationships that identify frauds with higher confidence and distilling the accidental or less relevant connections. The basic assumption is that "Not all the available information is useful or is of equal importance". Without weights, all the edges will have equal importance, which might increase false positives. Weights benefit the graph algorithms by providing domain knowledge information required to learn better representation from the neighborhood. These weights are learned automatically in GCNs but are taken as inputs in community detection. Explicitly setting weights to the edge relations provides us control over the amount of information to be used from the neighborhood.

3.3 Other Challenges in Fraud Detection

- **Concept Drift:** Fraudsters are constantly inventing new M.Os leading to the breaking of models trained on older data distributions. This phenomenon is known as concept drift. We tackle concept drift in two ways. Firstly, several M.O-agnostic features are employed which do not change with time, i.e., stationary attributes. Secondly, by abstracting community detection from fraud identification discriminator.
- **Scalability:** With millions of customers, it is challenging to represent them in a single graph or as multiple subgraphs. Further, it is computationally costly and time-consuming to train graphical ML models on large graphs. We rely on Leiden methods' fast local move approach to detect quality communities in much less time.

- **Cold Start:** A majority of the fraud detectors depend on entities' history or interactions on a platform to predict fraud. However, fraudsters continually create new accounts to commit frauds for which there will be no historical data. This is similar to cold-start problems in recommendation systems. To combat this, we match these new accounts with previously identified stationary attributes, linking them to a community. Since our framework tags an entire community as fraud or not, the same label is inherited by new accounts 'matched' to that community. We were thus able to uncover 20% more frauds on a daily basis.

4   Proposed Framework

In this section, we present our community-based fraud detection framework. Firstly, we provide an overview of the framework and its workflow. Then, we detail the different modules involved.

4.1   Overview

The proposed framework has three modules: Graph Construction (GC), Community Detection (CD), and Downstream Task (DT). The GC module molds the raw data into a graph representation. We then propose a way to convert a multigraph into a homogeneous one. The CD module processes this graph and segregates the nodes into possible rings based on their connectivity. This module also takes in the domain knowledge in the form of edge-weights optimized over the modularity metric. Finally, the DT module predicts communities as fraud rings and nodes as fraudulent customers. An illustration of the proposed framework is presented in Fig. 2.

4.2   Graph Construction

Graph Definition. We construct a graph where customers are the nodes, and stationary attributes are the edges between them. All edge types in our graph are undirected. Each edge type has a weight that signifies the importance of that attribute in identifying fraud. Since we intend to use off-the-shelf community detection algorithms which require input graphs to be homogeneous, we make a simplifying assumption of treating our edge types as homogeneous (our nodes are already homogeneous).

Graph Representation. A multigraph is defined as $G(V, E(e, t), W(t, w))$,
    where
        V is a set of vertices representing customers with $\|V\| = n$,
        E is the set of edges with $\|E\| = m$, each edge e having a relation type attribute $t \in T' = \{1, 2, \ldots, T\}$ where T' is the set of possible edge relation types, and

W = $(w_1, w_2 ..., w_T)$ is a weight vector that maps each edge relation type t to a weight w_t. By default, all edge relation types have unit weight.

The multigraph (G) is converted into a homogeneous graph (G'), by merging multiple edges between the same two nodes as one edge and summing the edge weights.

G' is defined as $G'(V, E'(e, m), W')$,

where

V is the same set of vertices representing users with $\|V\| = n$,

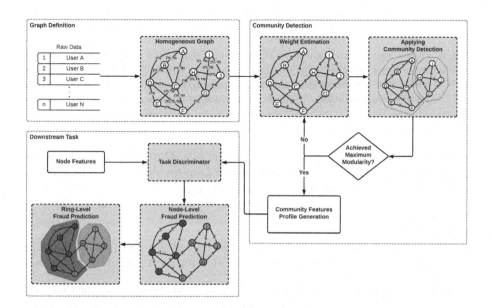

Fig. 2. Proposed community-based fraud detection framework.

E' is the set of edges with $\|E'\| = m'$, each edge e having a merge set type attribute u ∈ P(T'), power set of T'

W' = $(w'_1, ..., w'_{2n})$ is a new weight vector, where w'_u is the sum of weights of all edge types in the merge set u.

We experimented with summation, averaging, and multiplication as weight aggregation techniques, out of which, summation worked best (as indicated by goodness-of-fit in downstream tasks) in our experiments. Figure 3 illustrates the conversion from a multigraph to a homogeneous graph.

4.3   Community Detection

Weighted Community Detection. In a graph, a community is defined as a set of nodes that can be grouped together such that each set of nodes is densely connected internally, and loosely connected with the rest of the nodes. Several graph algorithms exist for community detection, which evolved over time from

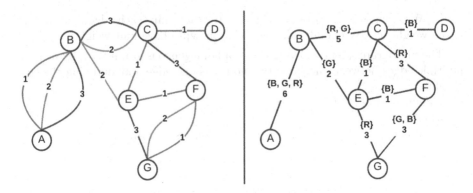

Fig. 3. Left: A sample multigraph representation. Edge colors and numbers represent different edge types and their edge weights respectively. Right: The homogeneous graph derived where edge attributes store the merged information such as edge types involved and their weights aggregation. A simple summation of edge weights for aggregation is illustrated here.

Newman [22], Louvain [3] to Leiden algorithm [33]. Our framework adopts the Leiden algorithm (LDN), which builds on the Louvain algorithm. LDN employs a three-step process for segregating communities:

1. local moving of nodes,
2. refinement of the partition, and
3. aggregation of the network based on the refined partition, using the non-refined partition to create an initial partition for the aggregate network.

LDN supports two objective functions known as Modularity and the Constant Potts Model (CPM). Modularity is a measure of how well a graph is partitioned into communities. It tries to maximize the difference between the actual number of edges in a community and the expected number of such edges and is defined as follows:

$$H = \frac{1}{2m} \sum_c \left(e_c - \gamma \frac{K_c^2}{2m} \right)$$

Here, e_c denotes the actual number of edges in community c. $\frac{K_c^2}{2m}$ denotes the expected number of edges, where K_c is the sum of the degrees of the nodes in community c and m is the total number of edges in the network. γ is the resolution parameter that ranges in $[0, 1]$. Higher resolution leads to more communities, while lower resolution leads to fewer communities.

Our framework's novelty is the inclusion of domain knowledge in community detection. The various edge types between the customers can differ in the information they convey with respect to a task. For example, to suggest friends in a social graph, subscribing to a common page is likely a more important edge than clicking on the same ad. A similar analogy applies to fraud detection in e-commerce graphs where the edge weight of stationary attributes could be different. Prior domain knowledge is required to define these weights. Weighted community detection can then exploit this domain knowledge to better segregate a graph, compared to an unweighted or domain-agnostic method. The brute-force way would be to take the best guesstimate and pick weights as constant values such that order is maintained between the edges type. The problem with this approach is that weights might not be optimal. The proposed way picks inspiration from constrained optimization. We first define upper and lower bounds on weights for each edge type. Then, we use a relative priority of edge types as constraints over these weights. The bounds can overlap among edge types, but estimated weight combinations should follow priority constraints. For example, consider two edge types with bounds as $[0,10]$ and $[5,10]$ respectively and edge_type_1 with a lower priority than edge_type_2. The set of weight combinations considered during optimization have edge_type_1_weight $<$ edge_type_2_weight with $[5{<}6,4{<}8,1{<}10]$ as valid and $[8{<}5, 9{<}5]$ as invalid example combinations.

Community Profile. Each node is represented by a set of features (F) derived from the node's domain behavior. Key group indicators (KGI) are a subset of F that can be directly influenced by a group context. Community profile (CP) is defined by the combined representation of the member node. A CP vector is an aggregate of KGI feature values corresponding to each node in the community. Figure 4 explains CP in a fraud reviews detection problem.

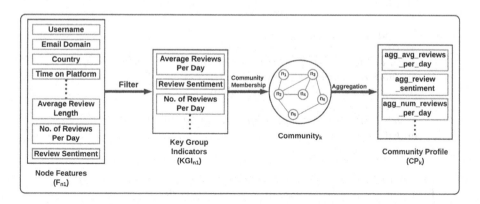

Fig. 4. Community profile

Proposed Algorithm. The CD module comprises of the below step by step process:

1. Define weight bounds and edge type priorities for the input homogeneous graph
2. Iteratively select combinations of weights using optimization methods like grid search.
3. Apply weighted community detection with weights from step 2 and modularity as the optimization metric

4. Output the community separation with maximum modularity as the best separation
5. For each community in the best separation, aggregate node features to create a community profile

The community detection process is presented as Algorithm 1.

Algorithm 1. Domain Aware Community Detection

Input: G' = (V, E'(e, m), W): A homogeneous graph with equal uni-weighted edge types

Output: A mapping of each node in V to a community and a community features vector representation for each community formed

   procedure COMMUNITYDETECTION(Graph G')

       $max\_modularity \leftarrow 0$

       $best\_communities\_seperation \leftarrow None$

       $ub = (u_1, u_2, ..., u_t)$

       $lb = (l_1, l_2, ..., l_t)$

       $constraint = \{n\}$ #example $\rightarrow n_r > n_g$ & $n_g \geq n_r$ & $n_b = n_r$

       $weight\_combinations \leftarrow weight\_estimation(ub, lb, constraint)$

       for each w in $weight\_combinations$ **do**

          for each e in E' **do**

             $E" + = e' = sum(m, w)$

          end for

          $G" \leftarrow graph(E")$

          $partitions \leftarrow leiden\_community\_detection(G")$

          $communities \leftarrow group\_partitions\_by\_community\_id(partitions)$

          $modularity \leftarrow evaluate(communities)$

          if $modularity > max\_modularity$ **then**

             $max\_modularity \leftarrow modularity$

             $best\_communities\_separation \leftarrow communities$

          end if

       end for

       $community\_features \leftarrow dict\{community\_id : aggregated\_features\}()$

       for community in $best\_communities\_seperation$ **do**

          $aggregated\_features \leftarrow vector()$

          for each node in community **do**

             Aggregate individual node features within a community as required and add the to the aggregated_features vector

             $aggregated\_features \leftarrow addup\_node\_features()$

             $community\_features[community\_id] \leftarrow aggregated\_features$

          end for

       end for

   return $partitions, community\_features$

   end procedure

4.4   Downstream Task

The DT module consists of a discriminator which takes node features and community profiles as inputs. In GCN-based methods, the task information is provided during the optimization process, and hence, the learned node-embeddings are tuned to the respective M.Os/tasks. In the proposed framework, the communities identified in the previous step(s) can be used across M.Os. Community 'profile' consists of aggregated features across all possible identified frauds. For example, for an M.O involving order cancellations, the incidence of cancellation in the community can be used. For M.Os around fraudulent claims, the incidence of claims at a community level can be used. Since the communities' profile vector already encapsulates the incidence of various fraudulent actions, it can help in creating a variety of lightweight discriminators with little to no future feature engineering. For example, a simple rule like flagging all communities with incidences of M.Os above a certain threshold can be employed to identify rings. This is especially useful for emerging M.Os or when it is challenging to get labeled data. For M.Os where sufficient labeled data is available, supervised methods with node and community features as inputs can be used.

5   Experiments

In this section, we demonstrate the effectiveness of our framework on real-world fraud detection problems, namely, opinion fraud and financial fraud. We also compare and contrast the framework in unweighted and weighted variants, using grid-search and PSO (particle swarm optimization) methods for weights estimation, in addition to GCN-based and Leiden community detection. We use the end-to-end run-time of the framework and the F1-score of the downstream task as the comparison metrics.

Table 1. Graph statistics of reviews datasets for opinion fraud detection

Dataset	#Nodes	#Edges	Degree	Fraud percentage
Amazon	11,944	4,398,392	739.71	14.5%
Yelp	45,954	3,846,979	167.47	9.5%
Our graph	~3.9M	~4.9M	2.48	< 5%*

* to preserve confidentiality, we do not reveal the exact number

5.1   Experimental Setup

Datasets. For opinion fraud detection, two open datasets are considered: restaurant-review spam data from Yelp [27] and product-review fraud data from Amazon [38]. Both have labels for each review/user being either fraud (spam) or benign (genuine). Both of these datasets can be represented as multi graphs, with one node type and multiple edge types.

Yelp dataset has reviews as nodes and 32 handcrafted node features with the following three relations:

1. R-U-R: links different reviews posted by the same user
2. R-S-R: links different reviews under the same product with the same star rating
3. R-T-R: links different reviews under the same product posted in the same month

Amazon dataset has users as nodes and 25 handcrafted node features with the following three relations:

1. U-P-U: links different users reviewing at least one same product
2. U-S-U: links different users having at least one same star rating within one week
3. U-V-U: links different users with top 5% mutual review text TF-IDF similarities

For benchmarking on our internal dataset, we tackle a M.O related to cash transactions where the graph is built on two months' worth of cash-transacting customers. Our graph has customers as nodes and 60 node features with our relations (for confidentiality reasons, we cannot reveal the exact relationship types).

As mentioned in Sect. 4.2, all datasets are converted to homogeneous graphs. Table 1 shows the statistics.

Methods Compared. We compare our proposed methods (domain-aware, weighted community-detection based) against GCN methods as shown in Table 2. The baseline is an XGBoost classifier which also serves as the discriminator for all non-baseline methods.

Table 2. Methods used for ablation

Name	Description	Community detection method	Input to discriminator
Baseline	XGBoost	-	Node features
GCN	Graph Convolutional Network	-	Node features & Graph embeddings
DMoN	Deep Modularity Network [17]	Modularity-based GCN	Node features & Graph embeddings
UWL	Unweighted community detection	Leiden	Node features & Graph embeddings
OWL*	Domain-aware weighted community detection using PSO optimization	Leiden	Node features & Graph embeddings
WL*	Domain-aware weighted community detection using grid search	Leiden	Node features & Graph embeddings

* our proposals

5.2   Experimental Settings

For GCN variants, we implemented Deep Graph Infomax [18] to learn node representations. A GCN with 2-layer and 128-dimension embeddings was used for comparison methods across all three datasets. For GCN-based community detection using DMoN, a 2-layer network each of 32 dimensions was trained for all datasets. Both GCN & DMoN were trained for 100 epochs with Adam optimizer with a learning rate of 0.001 and a dropout of 0.5. The choice of architectures for GCN and DMoN was primarily driven by the computing resources required. In the case of the XGBoost discriminator, hyper-parameter settings were kept the same for all the comparison methods but were specific to the dataset. To achieve a fair comparison, the classification threshold of the XGBoost discriminator was adjusted so that the fraud coverage (percentage of samples tagged as fraud by a model) was the same for all variants and equal to the dataset's fraud percentage. This is called the threshold-moving strategy in the literature [6].

5.3   Implementation

Graphs were constructed and maintained using networkx [11] and igraph [7] libraries. For PSO, the global optimization variant was adopted from pyswarms [21]. We used the Leiden implementation from igraph with default parameters. For GCN and XGBoost implementations, Stellargraph [8] and XGBoost [5] packages were used respectively. All models were trained on an AWS m4.4xlarge CPU instance.

Table 3. Performance comparison datasets

Comparison method	Amazon			Yelp			Our graph	
	F1 score			F1 score				
		Relative improvement	Time taken		Relative improvement	Time taken	Relative improvement in F1-score*	Time taken
Baseline	0.8383	-	2 s	0.7751		2 s	0%	2 s
GCN	0.8423	0.48%	3.5 min	0.7958	2.67%	2.65 min	3.58%	35 min
DMoN	0.832	−0.75%	10.2 min	0.7928	2.28%	8 min	−0.64%	48 min
UWL	0.8406	0.27%	8.6 s	0.8279	6.81%	2 s	2.47%	44.5 s
OWL**	0.839	0.08%	9.5 s	0.792	2.18%	2 s	2.94%	44.5 s
WL**	0.8532	1.78%	9.5 s	0.8124	4.81%	2 s	9.92%	44.5 s

* to preserve confidentiality, we only report relative improvement numbers for our data

** our proposals

5.4   Experimental Results

Table 3 shows the F1 scores and time taken on the CPU of each dataset. As previously mentioned, the baseline method uses only node features, while the other methods use both node and neighborhood information. On the Amazon dataset, only our WL method shows a pragmatically meaningful improvement (1.78%). On the Yelp dataset, all methods handily beat the baseline, demonstrating the usefulness of neighborhood information. Our WL method trails UWL on the Yelp dataset. We hypothesize that this is primarily due to our limited knowledge of Yelp's domain which has, in turn, had a direct bearing on and affects weight optimization and the quality of communities formed. On our dataset, DMoN under-performs the baseline signifying that, in larger graphs, limiting the number of communities can negatively affect downstream performance. Both OWL (+2.94%) and WL (+9.92%) outperform UWL (+2.47%), indicating the importance of domain-aware weights. Between OWL and WL, we hypothesize that PSO was not able to optimize better by using only modularity as the objective function and hence could not outperform the grid-search-based WL.

On the computation time front, while for smaller graphs like Amazon and Yelp, the difference is in seconds, for larger graphs like ours with 3 M nodes, LDN concludes in less than a minute, while GCN methods take at least half an hour. We only used two months of data for these experiments; expanding this horizon will have a direct bearing on graph size. Hence, we hypothesize that for even larger graphs, training GCN will take much longer.

We also investigated the temporal stability of the communities formed. We constructed graphs over a moving time window of two months at a weekly level. We then compared the movements of customers' assignments from one community to another in every iteration and found a 3–5% movement which is an acceptable threshold for us. As previously claimed, we were also able to create a rule-based classifier for an unseen-before M.O using the results from the community module within a few days (as opposed to weeks/months if we had to build models from scratch for this M.O). We were also able to change the threshold of these rules based on changing fraud behavior with no change in the underlying graph or community modules. This framework is currently deployed in production, inferencing millions of transactions per day, with the graph and community modules being updated weekly.

6   Conclusion and Future Work

In this paper, we proposed a novel end-to-end fraud detection framework to identify fraud rings. To the best of our knowledge, this is the first attempt at a scalable graph-based system utilizing domain knowledge as weighted edge priorities in Leiden community detection. Experiments were conducted on large-scale open and internal fraud datasets demonstrating the effectiveness of the proposed framework using F1 score and CPU run-times.

As an extension, we plan to experiment with different objective functions and strategies that can potentially outperform grid searching. On the community detection front, we want to explore how we can extend this work to handle overlapping communities. Our current implementation uses stationary attributes for edges. Given the dynamic nature of fraud, it is important to explore ways to incorporate non-stationary attributes, which can potentially help make detections resistant to changing fraud patterns. In this work, we also made the simplifying assumption of converting heterogeneous graphs into homogeneous ones. We would like to explore if there are additional benefits to be derived by researching ways to directly use heterogeneous graphs instead.

References

1. Bandyopadhyay, S., Peter, V.: Unsupervised constrained community detection via self-expressive graph neural network. In: Uncertainty in Artificial Intelligence, pp. 1078–1088. PMLR (2021)
2. Van Belle, R., Mitrović, S., De Weerdt, J.: Representation learning in graphs for credit card fraud detection. In: Bitetta, V., Bordino, I., Ferretti, A., Gullo, F., Pascolutti, S., Ponti, G. (eds.) MIDAS 2019. LNCS (LNAI), vol. 11985, pp. 32–46. Springer, Cham (2020). https://doi.org/10.1007/978-3-030-37720-5_3
3. Blondel, V.D., Guillaume, J.L., Lambiotte, R., Lefebvre, E.: Fast unfolding of communities in large networks. J. Statist. Mech. Theory Exp. **2008**(10), P10008 (2008)
4. Cao, C., Li, S., Yu, S., Chen, Z.: Fake reviewer group detection in online review systems. In: 2021 International Conference on Data Mining Workshops (ICDMW), pp. 935–942. IEEE (2021)
5. Chen, T., Guestrin, C.: Xgboost: a scalable tree boosting system. In: Proceedings of the 22nd ACM SIGKDD International Conference on Knowledge Discovery and Data Mining, pp. 785–794 (2016)
6. Collell, G., Prelec, D., Patil, K.R.: A simple plug-in bagging ensemble based on threshold-moving for classifying binary and multiclass imbalanced data. Neurocomputing **275**, 330–340 (2018)
7. Csardi, G., Nepusz, T., et al.: The igraph software package for complex network research. Int. J. Complex Syst. **1695**(5), 1–9 (2006)
8. Data61, C.: Stellargraph machine learning library (2018). https://github.com/stellargraph/stellargraph
9. Dou, Y., Liu, Z., Sun, L., Deng, Y., Peng, H., Yu, P.S.: Enhancing graph neural network-based fraud detectors against camouflaged fraudsters. In: Proceedings of the 29th ACM International Conference on Information & Knowledge Management, pp. 315–324 (2020)
10. Ghosh, S., et al.: Distributed Louvain algorithm for graph community detection. In: 2018 IEEE International Parallel and Distributed Processing Symposium (IPDPS), pp. 885–895. IEEE (2018)
11. Hagberg, A., Swart, P., S Chult, D.: Exploring network structure, dynamics, and function using network. Technical report, Los Alamos National Lab. (LANL), Los Alamos, NM (United States) (2008)
12. Hashemi, A., Dowlatshahi, M.B., Nezamabadi-Pour, H.: MGFS: a multi-label graph-based feature selection algorithm via PageRank centrality. Expert Syst. App. **142**, 113024 (2020)
13. He, D., Song, Y., Jin, D., Feng, Z., Zhang, B., Yu, Z., Zhang, W.: Community-centric graph convolutional network for unsupervised community detection. In: Proceedings of the Twenty-Ninth International Conference on International Joint Conferences on Artificial Intelligence, pp. 3515–3521 (2021)

14. Jia, Y., Zhang, Q., Zhang, W., Wang, X.: Communitygan: community detection with generative adversarial nets. In: The World Wide Web Conference, pp. 784–794 (2019)
15. Li, A., Qin, Z., Liu, R., Yang, Y., Li, D.: Spam review detection with graph convolutional networks. In: Proceedings of the 28th ACM International Conference on Information and Knowledge Management, pp. 2703–2711 (2019)
16. Liang, C., Liu, Z., Liu, B., Zhou, J., Li, X., Yang, S., Qi, Y.: Uncovering insurance fraud conspiracy with network learning. In: Proceedings of the 42nd International ACM SIGIR Conference on Research and Development in Information Retrieval, pp. 1181–1184 (2019)
17. Liu, Y., et al.: Pick and choose: a GNN-based imbalanced learning approach for fraud detection. In: Proceedings of the Web Conference 2021, pp. 3168–3177 (2021)
18. Liu, Z., Dou, Y., Yu, P.S., Deng, Y., Peng, H.: Alleviating the inconsistency problem of applying graph neural network to fraud detection. In: Proceedings of the 43rd International ACM SIGIR Conference on Research and Development in Information Retrieval, pp. 1569–1572 (2020)
19. Liu, Z., Chen, C., Yang, X., Zhou, J., Li, X., Song, L.: Heterogeneous graph neural networks for malicious account detection. In: Proceedings of the 27th ACM International Conference on Information and Knowledge Management, pp. 2077–2085 (2018)
20. Luo, L., Fang, Y., Cao, X., Zhang, X., Zhang, W.: Detecting communities from heterogeneous graphs: A context path-based graph neural network model. In: Proceedings of the 30th ACM International Conference on Information & Knowledge Management, pp. 1170–1180 (2021)
21. Miranda, L.J.: PySwarms: a research toolkit for particle swarm optimization in python. J. Open Source Softw. **3**(21), 433 (2018)
22. Newman, M.E.: Fast algorithm for detecting community structure in networks. Phys. Rev. E **69**(6), 066133 (2004)
23. Nilforoshan, H., Shah, N.: Slicendice: mining suspicious multi-attribute entity groups with multi-view graphs. In: 2019 IEEE International Conference on Data Science and Advanced Analytics (DSAA), pp. 351–363. IEEE (2019)
24. Noekhah, S., binti Salim, N., Zakaria, N.H.: Opinion spam detection: using multi-iterative graph-based model. Inf. Process. Manage. **57**(1), 102140 (2020)
25. Peng, L., Lin, R.: Fraud phone calls analysis based on label propagation community detection algorithm. In: 2018 IEEE World Congress on Services (SERVICES), pp. 23–24. IEEE (2018)
26. Rahimi, S., Abdollahpouri, A., Moradi, P.: A multi-objective particle swarm optimization algorithm for community detection in complex networks. Swarm Evol. Comput. **39**, 297–309 (2018)
27. Rayana, S., Akoglu, L.: Collective opinion spam detection: bridging review networks and metadata. In: Proceedings of the 21th ACM SIGKDD International Conference on Knowledge Discovery and Data Mining, pp. 985–994 (2015)
28. Sarma, D., Alam, W., Saha, I., Alam, M.N., Alam, M.J., Hossain, S.: Bank fraud detection using community detection algorithm. In: 2020 Second International Conference on Inventive Research in Computing Applications (ICIRCA), pp. 642–646. IEEE (2020)
29. Shang, C., Tang, Y., Huang, J., Bi, J., He, X., Zhou, B.: End-to-end structure-aware convolutional networks for knowledge base completion. In: Proceedings of the AAAI Conference on Artificial Intelligence, vol. 33, pp. 3060–3067 (2019)
30. Shchur, O., Günnemann, S.: Overlapping community detection with graph neural networks. arXiv preprint arXiv:1909.12201 (2019)

31. Souravlas, S., Anastasiadou, S., Katsavounis, S.: A survey on the recent advances of deep community detection. Appl. Sci. **11**(16), 7179 (2021)
32. Sun, C., Yan, Z., Li, Q., Zheng, Y., Lu, X., Cui, L.: Abnormal group-based joint medical fraud detection. IEEE Access **7**, 13589–13596 (2018)
33. Traag, V.A., Waltman, L., Van Eck, N.J.: From Louvain to Leiden: guaranteeing well-connected communities. Sci. Rep. **9**(1), 1–12 (2019)
34. Wang, D., et al.: A semi-supervised graph attentive network for financial fraud detection. In: 2019 IEEE International Conference on Data Mining (ICDM), pp. 598–607. IEEE (2019)
35. Wang, L., Li, P., Xiong, K., Zhao, J., Lin, R.: Modeling heterogeneous graph network on fraud detection: a community-based framework with attention mechanism. In: Proceedings of the 30th ACM International Conference on Information & Knowledge Management, pp. 1959–1968 (2021)
36. Yang, G., Zheng, W., Che, C., Wang, W.: Graph-based label propagation algorithm for community detection. Int. J. Mach. Learn. Cybern. **11**(6), 1319–1329 (2019). https://doi.org/10.1007/s13042-019-01042-0
37. You, X., Ma, Y., Liu, Z.: A three-stage algorithm on community detection in social networks. Knowl. Based Syst. **187**, 104822 (2020)
38. Zhang, S., Yin, H., Chen, T., Hung, Q.V.N., Huang, Z., Cui, L.: GCN-based user representation learning for unifying robust recommendation and fraudster detection. In: Proceedings of the 43rd international ACM SIGIR Conference on Research and Development in Information Retrieval, pp. 689–698 (2020)

Capabilities, Limitations and Challenges of Style Transfer with CycleGANs: A Study on Automatic Ring Design Generation

Tomas Cabezon Pedroso[1], Javier Del Ser[2,3], and Natalia Díaz-Rodríguez[4(✉)]

[1] Carnegie Mellon University, 5000 Forbes Avenue, Pittsburgh, PA 15213, USA
tcabezon@andrew.cmu.edu
[2] TECNALIA, Basque Research and Technology Alliance (BRTA),
48160 Derio, Spain
javier.delser@tecnalia.com
[3] University of the Basque Country (UPV/EHU), 48013 Bilbao, Spain
[4] Department of Computer Sciences and Artificial Intelligence,
Andalusian Research Institute in Data Science and Computational Intelligence
(DaSCI), CITIC, University of Granada, Granada, Spain
nataliadiaz@ugr.es

Abstract. Rendering programs have changed the design process completely as they permit to see how the products will look before they are fabricated. However, the rendering process is complicated and takes a significant amount of time, not only in the rendering itself but in the setting of the scene as well. Materials, lights and cameras need to be set in order to get the best quality results. Nevertheless, the optimal output may not be obtained in the first render. This all makes the rendering process a tedious process. Since Goodfellow et al. introduced Generative Adversarial Networks (GANs) in 2014 [1], they have been used to generate computer-assigned synthetic data, from non-existing human faces to medical data analysis or image style transfer. GANs have been used to transfer image textures from one domain to another. However, paired data from both domains was needed. When Zhu et al. introduced the CycleGAN model, the elimination of this expensive constraint permitted transforming one image from one domain into another, without the need for paired data. This work validates the applicability of CycleGANs on style transfer from an initial sketch to a final render in 2D that represents a 3D design, a step that is paramount in every product design process. We inquiry the possibilities of including CycleGANs as part of the design pipeline, more precisely, applied to the rendering of ring designs. Our contribution entails a crucial part of the process as it allows the customer to see the final product before buying. This work sets a basis for future research, showing the possibilities of GANs in design and establishing a starting point for novel applications to approach crafts design.

© IFIP International Federation for Information Processing 2022
Published by Springer Nature Switzerland AG 2022
A. Holzinger et al. (Eds.): CD-MAKE 2022, LNCS 13480, pp. 168–187, 2022.
https://doi.org/10.1007/978-3-031-14463-9_11

Keywords: Deep learning · Generative adversarial networks · Automatic design · Image-to-Image translation · Jewelry design · CycleGAN

1   Introduction

With the advances on artificial intelligence and deep learning (DL), the capabilities of computation in the field of design and computational creativity have spun. Machines have gone beyond doing what they are programmed to do, and debates spur questioning the creativity of models that, in any case, will not replace, but can definitely save time and assist designers do what they do best, more efficiently, and focusing on what really requires their expertise and skillful effort. Works in the intersection of design and computation are growing and are more relevant than ever. In the last years, engineers, researchers or artists have begun to explore the possibilities of artificial intelligence for creative tasks that can vary from the AI generated music of Arca that sounds in the MOMA's lobby[1] to the drawings by AARON computer program that can be visited at TATE Museum[2].

This work rises in this same intersection of design and technology, design and engineering, design and computation. The aim of this paper is to explore new areas and applications in which computers will change the way we consider design and the role that computers have on it. When this statement is made, often the fear of computers stealing people's jobs arises, nevertheless, this is not how we conceive this intersection of computers and design. While algorithms will spend time doing repetitive work, designers will be able to focus on what really matters: the users, the emotional feeling needs, innovations, or other needs... The objective is thus to make the most out of it for all agents taking part in the design process, from computer programs to designers. We believe the tools at our disposal cannot determine what we are capable of creating. Instead, AI, as another tool more, should serve to develop new ideas and not to limit the ones that the designer already has.

This work is organized in two parts, a theoretical one and a practical one, both complementary. The first one is motivated by the recent arrival of Generative Adversarial Networks (GAN) that since they were first introduced few years ago, in 2014 [5], have experimented and exponential growth and development. This research will be focus on the study and comprehension of the theory that supports GANs and their components. In the second part of the work, taking into account all of the above, a new tool of image generation is applied to an actual design problem, in this case, the rendering of an example of the XYU ring (finger, jewelry area). This tool will consist on a CycleGAN that taking as an input the sketch of the shape of the ring will generate a 3D object representation or a rendered image of it.

Although the steps in the design process can differ among authors, the whole process consists of going from a virtual concept or idea to the materialization in a concrete product[3]. This process starts with an initial brainstorming and,

---

[1] *Arca will use AI to soundtrack NYC's Museum of Modern Art*, https://www.engadget.com/2019-10-17-arca-ai-soundtrack-for-nyc-moma.html.

[2] *Untitled Computer Drawing*, by Harold Cohen, 1982, Tate. (n.d.), https://www.tate.org.uk/art/artworks/cohen-untitled-computer-drawing-t04167.

[3] *Design Thinking*, https://hbr.org/2008/06/design-thinking.

later, some of the concepts are developed, prototyped and, only after evaluation, the final product is selected. Computers have become fundamental in these last steps allowing designers not only to materialize their ideas with 3D objects and renders but also to show the clients how the final products look like. Actually, the famous furniture seller Ikea reaches their clients with the yearly catalogs, full of not real images but renderings of it[4]. Our ML model aims to input new tools for this last part of the design process.

While the proposed GANs have been used in a broad set of applications, we limit the scope of this paper to assess a potential technology impact that these models could have in automatic design. Although realistic images could be generated by other means, such as rendering, mocking up or photographing, in this work, the objective is not getting the output images themselves, but rather visually assessing the possibilities and limitations of paired GANs. More particularly, we will use GANs to assess the generative and creative capabilities to construct realistic and physically plausible designs. The concrete example we take is the rendering of a sketch of the XYU ring example. To approach this issue, we need to understand how GANs create images.

The rest of this paper is organized as follows. Section 2 briefly describes the state of the art on design tools and generative models for computational creativity with respect to design. Section 3 presents the proposed design model pipeline, Sect. 4 shows and examines critically the obtained results. Finally, Sect. 5 concludes with insights for further research and development.

2   Related Work

A large body of works has emerged in the literature exploiting the highly performing abilities of generative models such as GANs. We briefly discuss those more closely related to automatic computational design generation.

With respect to neural rendering models, some approaches produce photorealistic renderings given noisy or incomplete 3D or 2D observations. In Thies et al. [2], incomplete 3D inputs are processed to yield rich scene representations using neural textures, which regularize noisy measurements. Similar to our work, Sitzmann et al. [3] aggregate and encode geometry and appearance into a latent vector that is decoded using a differentiable ray marching algorithm. In contrast with our work, these methods either require 3D information during training, complicated rendering priors or expensive inference schemes. In [4] they present a way to learn neural scene representations directly from images, without 3D supervision, which permits to infer and render scenes in real time, while achieving comparable results to models requiring minutes for inference.

Since the introduction of generative adversarial networks (GANs) [1] and its spread use to generate data –from images to sound, music or even text–, a *zoo* of GANs has emerged. In the plethora of existing models we focus on style transfer models that generally consist of translating images from one domain to a different

---

[4] *Why IKEA Uses 3D Renders vs. Photography for Their Furniture Catalog*, https://www.cadcrowd.com/blog/why-ikea-uses-3d-renders-vs-photography-for-their-furniture-catalog.

one, where some dimension or data generating factor should be preserved. Style transfer was proposed by [5] as a neural algorithm able to disentangle content from style from an artistic image, and recombine these elements being taken from arbitrary images.

Among the most popular models for style transfer there are models that use paired datasets to perform image-to-image translation [6]. Image-to-image translation models can be used to generate street imaging from semantic segmentation masks (DCGAN [7], Pix2pixHD [8], DRPAN [9], SPADE [10], or OASIS [11]); however the need for paired data makes data collection tedious and costly.

When high-resolution photorealism is a priority despite the computational cost, models such as Pix2pixHD [8] have demonstrated to generate accurate images that are both physically-consistent and photorealistic (e.g., to visualize the impact of floods or ice melt [12,13]).

Alternative to GANs also exist to learn the distribution of possible image mappings more accurately [14], e.g., normalizing flows [15,16] or variational autoencoders [17], although they have shown this happens in detriment of the result realistic effect [18,19].

3   Proposed Model for Automatic Ring Design Generation

In this section we present the motivating design problem and the model proposed to achieve this objective in an automatic manner.

3.1   Practical Use Case: Designing XYU Rings: Traditional Ring Design Pipeline

XYU is not only a ring but an algorithm to create rings[5] [20]. This algorithm uses splines to generate infinite ring possibilities. The starting point is set by the user, who specifies the number of splines and the length and thickness of the ring. The control points of the splines are randomly selected and adapted to make them continuous on the ring. If the user does not like the result, the algorithm can be run again until an aesthetic shape is achieved (Fig. 1).

The XYU ring algorithm allows the users to design their own 3D ring example based on their preferences, which are used by the algorithm to generate random rings. Each time the XYU ring code is run, a different and unique ring is produced. This procedure permits to personalize each of the XYU ring examples. Once the user finds a ring he/she likes, this is automatically modelled on *Maya* computer graphics application and the 3D object is sent to the jeweler. Each ring is 3D printed and cast, so each piece is unique.

The description of the actually used programs and different steps followed to go from the initial starting data to the final design of the ring are shown in Fig. 2.

---

[5] The XYU ring project is key to understanding this work. More information in https://tomascabezon.com/. XYU is the name of this project, it is not an acronym, but the name of this ring composed of 3 randomly chosen letters.

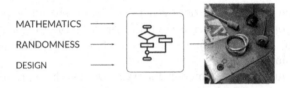

Fig. 1. Original conceptualisation schema for traditional ring design processes.

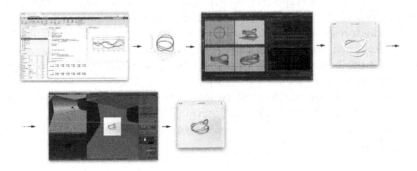

Fig. 2. Traditional pipeline and steps followed. On top Matlab program screenshot where the algorithm is run. In the middle, the Maya application with the 3D object of the ring. On the bottom, Blender software and the rendered image.

The algorithm is run on Matlab, the ring is later automatically translated into Maya Embedded Language (MEL) language where the ring is 3D modelled and an .obj/.stl file is created. Finally, the realistic images of the ring are rendered using the Blender software (Fig. 3).

To generate the 3D object and send the .obj file to the jeweler who will 3D print and cast it, *Maya 3D* application is used. To do so, the information of the ring is passed to Maya using the Maya MEL coding language. This is, the output of the Matlab algorithm is a .txt file with the instructions of the curves that generate the different brands of the ring, the splines, as well as the circumferences that will be extruded along the splines to form the 3D object. Therefore, the information to generate the ring in 3D is passed as coded instructions to Maya.

The last stage of the process corresponds to the rendering of the final product. In this last step, Blender rendering program is used to generate realistic images on the ring and to show the final product to the customer (Fig. 4).

Rendering is the process of turning a 3D scene into a 2D image. A 3D scene is composed of various elements apart from the object we want to render, such as the background, the camera, the materials and the light. This step is the most tedious part of the XYU ring generation: the rendering of images not only takes a long time to be calculated, but scenes need to be arranged and the images do not always render as expected the first time. As a matter of fact, companies that create whole animation films by rendering each of the video frames of their films, such as Pixar, have rendering directors to optimize this process.

Fig. 3. Schema of how each of the brands of the ring are 3D modelled on Maya 3D modelling software. The points that compose the curve as well as the circle that will be extruded along this curve are passed to the Maya program using Maya MEL coding language.

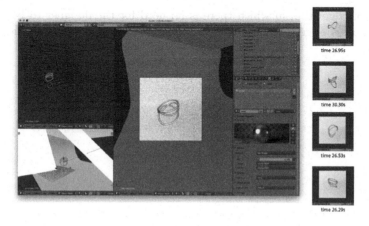

Fig. 4. Rendering settings and image rendering duration times for a set of 1000×1000 pixel size sample images.

To calculate how long it takes to render an image, on top of the time needed to compute the color of each pixel, which is not the longest one of the process, the scene setting time, the lighting configuration and the material generation and selection times should be added. A properly rendered image of an XYU ring would take, in total, around an hour in the making (Fig. 5).

As it was seen, the original idea of completely automatizing the whole process of the XYU ring generation by the user was not achieved, as intermediate external programs need to be used in the process by the designer. Figure 6 shows the actual process intermediate steps needed for the 3D object generation and rendering.

3.2 Proposed Pipeline: Automatic Design Rendering Through Generative Models for Image-to-Image Translation

The initial process in which the generation of rings has been completely automatized has not been achieved yet. This limitation was the starting point that motivates this work. Therefore, we propose a new approach for this ring design

Fig. 5. Current traditional design pipeline schema from left to right: the algorithm run in Matlab, the 3D object modelled in Maya, rendered image in Blender and final product after being made by the jeweler. This process is slow and requires human intervention to render the 3D model.

generation algorithm in which there is no need for the designer to be involved in the process of generating rendered images of the ring to show to the user.

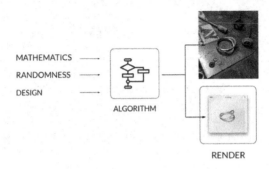

Fig. 6. Proposed model design pipeline. The algorithm generates not only the ring 3D object, but also its rendered images. The traditional rendering process is skipped, because it is automatically done by the CycleGAN within the full algorithm. The interesting aspect is that this allows: 1) the end user to instantaneously visualize the designed sample, which would not be possible without intervention from the designer to render it. 2) automating the full process.

It is important to note that the process for the jeweler does not change, as the 3D model file (e.g. .obj, .stl) is in both cases created for it to be 3D printed and later cast by the jeweler. This means the ring sketch is generated by the Matlab algorithm, which generates the different spline curves that compose the ring. These 3D spline curves are plotted in 2D and this image is used as input sketch for the CycleGAN. Therefore, since the algorithm also produces the rendering of the 3D models, the end user benefits from the CycleGAN by running it, choosing the preferred 3D model, and printing it in 3D, which means the full process is amenable to be automated.

3.3 CycleGAN as a Generative Model Trained on Unpaired Images

CycleGANs [21] are generative models that are trained on unpaired sets of images in tuple format. They are used to translate between image *styles* or domains.

Some examples can be seen in computer vision applications *translating* or transforming images from one domain into another (e.g. *horse2zebra*, *apple2orange*, *photo2Cezanne*, *winter2summer*... and vice versa).

In this work, we propose an innovative use of CycleGANs that consists of speeding up the last parts of the process of the design of XYU rings, i.e., their presentation. To achieve this, a *Sketch2Rendering* CycleGAN is trained. It will get a sketch image as an input and will apply a style transfer to generate a rendered image of the ring. In the following image this CycleGAN can be seen. Our hypothesis is that training a CycleGAN that would render the generated rings would not only ease this process and reduce its time but also automatize and speed up the process (Fig. 7).

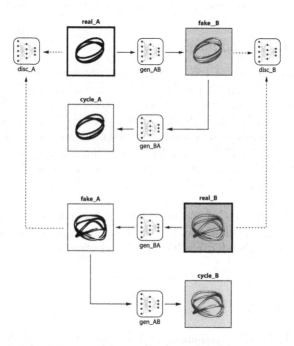

Fig. 7. Proposed CycleGAN where *gen* and *disc* prefixes stand for generator and discriminator, respectively. Input training images are the ones with a thicker frame.

One reason to choose CycleGANs against other *image2image* translation models is their ability to create infinite samples for a given input. The variational autoencoder (VAE) module that is part of the CycleGAN is responsible for this feature, and one of the reasons behind our choice of this architecture. Against other models that are deterministic and thus, limited to produce a single unique output for a given input, a CycleGAN allows to produce a one-to-many output pairings for a given input image.

We propose a CycleGAN based on the one proposed by Zhu et al. as an unpaired Image-to-Image Translation using Cycle-Consistent Adversarial Networks [21]. The CycleGAN involves the automatic training of image-to-image translation models without paired examples. This capability is very suitable for this application, as there is no need for paired image datasets, since this is usually challenging and time consuming to obtain.

4   Results and Analysis

In this section we present the produced designs by the CycleGAN architecture and provide some visual galleries to analyze them. Models, scripts and notebooks used to produce these results have been made publicly available[6].

4.1   *Sketch2Rendering* image results

In Fig. 8 some examples of the style transfer performed by the once trained *Sketch2Rendering* CycleGAN are shown. For comparison purposes, some of the ring sketches used to train the data have been modelled and rendered using the traditional procedure. In the following image, the outputs of the CycleGAN are show next to what they could be some expected rendered images, using the *Maya* modelling program and *Blender* rendering program. Although the data consists of unpaired images, in Figure 8, images are show as pairs of the input sketch and the generated image by the Sketch2Rendering CycleGAN and the rendered image using Blender.

In Figs. 9 and 10, 360° of the same XYU ring can be seen. On the left, the input (sketch) image is shown and, next to it, an expected rendered image of the 3D object using Blender. On its right, the output of the *Sketch2Rendering* CycleGAN.

5   Discussion

After having shown the possibilities of the applications of style transfer with CycleGANs for rendering purposes, in this section, the artefacts found during the training and testing, as well as some of the limitations found on the model will be considered. Although the Sketch2Rendering model can achieve reasonable results in some cases, there are areas for improvement in future works. As it can be seen in the following lines, the results are far from uniformly positive and there are still some challenges and improvements to be done before good quality realistic images of rings are generated by the CycleGAN. The following describes some artefacts found both while training and testing the model.

---

[6] https://github.com/tcabezon/automatic-ring-design-generation-cycleGAN,
https://tcabezon.github.io/automatic-ring-design-generation-cycleGAN/.

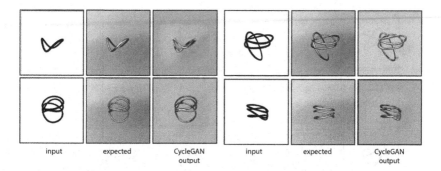

Fig. 8. Different rings in the sketch domain and the rendered domain. The expected rendering were generated with Blender, while the other using the Sketch2Rendering CycleGAN.

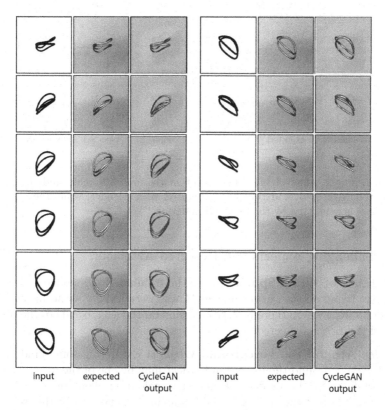

Fig. 9. Different views of the same object in the sketch domain and the rendered domain. The expected rendering was generated with Blender, while the other was generated by the Sketch2Rendering CycleGAN.

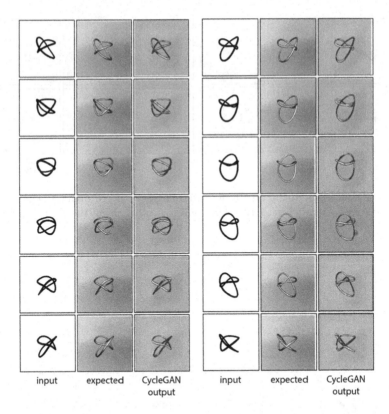

input   expected   CycleGAN   input   expected   CycleGAN
                       output   output

Fig. 10. Different views of the same object in the sketch domain and the rendered domain. The expected rendering was generated with Blender, while the one on the right of it was generated by the Sketch2Rendering CycleGAN.

5.1 Challenging Aspects and Detected Artefacts

Appearance of White Spots: During the training of the model, white blurry spots were found on the output images. These were usually found in the edge of the ring in areas where there is a strong shine on the ring, or where the different bands of the ring intersected.

***Aureoles* Around the Foreground Object:** Exhibiting some kind of background *aureole* around the foreground object may be one of the most commonly found artefacts in the model outputs. The colour gradient or aureole shows around the edges of the ring. We could explain this effect to be due to the different lighting settings, since there is non uniform background color in the full training dataset. Actually, the rendered images created using the Blender program show noise in the background, as if a *Photoshop Film Grain* filter would have been applied to the background. This is due to the renderization parameters on Blender. In order to accelerate the renderization process, the number of calculation steps for the color of each pixel was reduced when the dataset was created. In order to verify whether this is the actual cause of this artefact, in future works, better-quality datasets should be created, not only for the rings themselves but also for the backgrounds.

Checkerboard Patterns: Checkerboard effect is a common and one of the most typical artefacts in GANs. The reason for this checkerboard-like pattern in images is due to the upsampling process of the images from the latent space, which becomes visible in images with strong colours. This artefact appears as a consequence of the ability of deconvolutions (i.e., transposed convolutions) to easily show an uneven overlap that adds *more of the metaphorical paint in some places than others* [22]. Since there are some solutions proposed for solving such artefact [22], these may be some of the first strategies to be applied in future works to improve the *pixelated*-looking effect on generated images (Fig. 11).

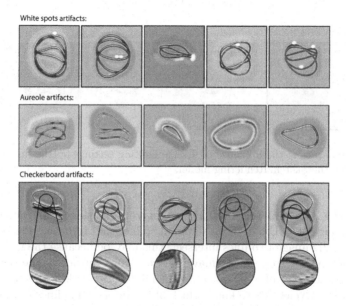

Fig. 11. Examples of generated samples by the Sketch2Rendering model exhibiting the different artefacts. Examples of the white spots artefact on the top, the aureole artefact in the middle, and the checkerboard pattern on the bottom.

5.2 Model Limitations

Apart from the artefacts described in the previous section, some limitations of the actual model have been found. Solving these would require further development of the model itself, for example, using paired data for the lighting setting to help guide the model to infer where the light comes from, or change how lines in the sketch intersect, for the model to better disentangle which one is on top of the other.

Coping with Arbitrary Lighting Settings: In the following image, different lighting settings were used in the rendering of the same object with the same materials. Therefore, different images were created. Even if it may not be as realistic, in order to improve the training of the CycleGAN it could be beneficial for the model to be trained using always the same rendering settings, so that the CycleGAN is able to more sharply learn the rendering style.

Another solution, as previously mentioned, could be using other prior information or labeled data, e.g., adding information about the light position and direction, so the network can learn the differences. However, this would complicate the construction of datasets, since really precise information would be required to make sure that all the information about the lighting settings is included in the labels, which is something to be avoided, due to an increased cost in time and effort of the annotation process. Furthermore, this additional input would require some effective information fusion strategy for the model to leverage this information adequately (Fig. 12).

Fig. 12. Examples of the influence of diverse lighting settings on the final renderings generated by the Sketch2Rendering model.

Learning to Account for a 3D Perspective: The sketch input image of the ring is an image of a 3D plot of the different splines that form the ring. Therefore, when in the 2D image two lines intersect, this may be because the lines actually intersect in the 3D space, or it may just be a consequence of the perspective. When creating a plot of a 3D object, some information is lost and thus, there is no way for the CycleGAN to know which of the intersecting lines in the image is on top of which or whether they are actually intersecting. A good way to solve this could be to use a different representation for intersecting lines and those that are not, for example, the diagrams used for knot representation in the study of mathematical knots [23] could be used. That way the model will learn when to render both of the splines together, or when they are not intersecting. In Fig. 13 an example of this can be seen, on the top, the splines actually intersect; while in the bottom, both of the splines are separated.

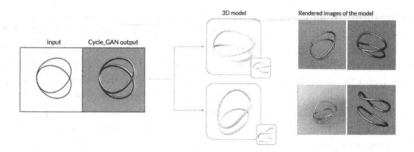

Fig. 13. Example of two lines intersecting in the 2D image, and two different 3D examples of the actually intersecting and non intersecting cases.

6   Conclusions and Future Work

After presenting the obtained results through the proposed and trained Cycle-GAN architecture and having discussed the problems encountered, some conclusions around the initially set objectives can be drawn.

First of all, we can assert that the Sketch2Rendering model can achieve compelling results for the rendering of the design through style transfer. That was the initial objective; nevertheless, as it was seen, there are areas for improvement in future works before high quality realistic images of the ring examples are generated by the CycleGAN. However, the results obtained exceed the initial expectations for this work. Indeed this new model supposes a new approach for the XYU ring algorithm and even though perfect results were not obtained, the reduction of the time and the allowance of a complete automation of the different ring design generation, makes this work an exemplary starting point for future research and improvements.

Secondly, this work shows the possibilities lying in the intersection of computation and design, which allows designers to focus on what really matters, while the algorithms do the repetitive non creative work. The rendering style transfer supposes going from the rendering of images that could take up to one hour on the making, to renders generated by the CycleGAN in seconds.

Therefore, it can be concluded that having developed a software that is capable of transferring the rendering style to the initial sketches of the ring, the research objective was achieved. The contribution of this work to the XYU ring design generation algorithm supposes an inflexion point for the way rings are shown to the end user, who now would be able to see real time rendered images of the ring that is being generated while interacting with the algorithm. Although we succeeded at the objective of validating image-to-image translation frameworks for automatic design rendering, some problems were encountered during the development of this work and discussed. We hope the research community finds the potential avenue of future works motivating for the exciting field of computational creativity and AI-assisted design to thrive.

In the future other types of GANs could be trained, e.g., models for higher resolution such as Pix2Pix, BiGAN or StyleGAN, and train them with paired data when available, to compare the gain in quality with this model's results. This quantitative comparison with other methods could help decide whether more dataset agnostic models that do not require paired data (such as the Cycle-GANs used in this work), are the best approach for this problem, or instead preparing paired data to train an image-to-image translation GAN is worth the time and effort. Future works should also assess some mathematical notions of correctness, and practical implications they could have for jewelry design.

Acknowledgments. Díaz-Rodríguez is supported by IJC2019-039152-I funded by MCIN/AEI/10.13039 /501100011033 by "ESF Investing in your future" and Google Research Scholar Program. Del Ser is funded by the Basque Government ELKARTEK program (3KIA project, KK-2020/00049) and research group MATHMODE (T1294-19).

A   Appendix: Supplementary Materials

A.1   Datasets

Some randomly selected .jpg images from the different datasets generated for this work are shown in this section. The aim is to show the diversity of images that have been used for the purpose of training the CycleGAN.

Sketch2Rendering: 179 sketch images and 176 rendered images of the training dataset were used for the training. The images were scaled to 400×400 pixels when loaded. The Sketch dataset is composed of .jpg images. These, in Fig. 14, have been generated using Matlab and the XYU ring algorithm. These images are a 3D plot of the splines that compose each of the rings, all with the same line thickness. The thickness was varied to show different ring thicknesses.

Rendered Dataset: The images in Fig. 14 have been created using the Blender rendering software. As it can be seen, although the background color has always been the same blue (#B9E2EA), the lighting setting has changed, as well as the camera position and orientation, and thus, different shadows and lights can be appreciated across the dataset.

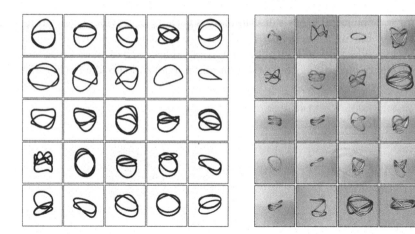

Fig. 14. On the left, random images of wire sketches created in Matlab language and used to train the CycleGAN model (domain A). On the right, random images of the rendered set of images, created using Blender rendering software, and used to train the CycleGAN model (domain B).

A.2   CycleGAN Model

To achieve a cycle consistency among two domains, a CycleGAN requires two generators: the fist generator (G_{AB}) will translate from domain A to B, and the second generator (G_{BA}) will translate from domain B back to A. Therefore, there will be two losses, one forward cycle consistency loss and other backward cycle consistency loss. These mean that $x* = G_{AB}(G_{BA}(x))$ and $y* = G_{BA}(G_{AB}(y))$.

CycleGAN Generator Architecture: The generator in the CycleGAN has layers that implement three stages of computation:

1. The first stage encodes the input via a series of convolutional layers that extract image features.
2. the second stage then transforms the features by passing them through one or more residual blocks.
3. The third stage decodes the transformed features using a series of transposed convolutional layers, to build an output image of the same size as the input.

The residual block used in transformation stage 2 consists of a convolutional layer, where the input is added to the output of the convolution. This is done so that the characteristics of the output image (e.g., the shapes of objects) do not differ too much from the input. Figure 16 shows the proposed architecture with example paired images as input.

CycleGAN Discriminator Architecture: The discriminator of the Cycle-GANs is based in the PatchGAN architecture [6]. The difference between this architecture and the usual GAN's discriminators is that the CycleGAN discriminator, instead of having a single float as an output, it outputs a matrix of values. A PatchGAN architecture will output a matrix of values, each of them between 0 (fake) and 1 (real), classifying the corresponding portions of the image (Fig. 15).

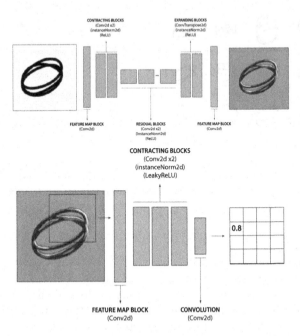

Fig. 15. Generator (upper) and Discriminator (below) architectures. Example classifying a portion of the image in the PatchGAN architecture, part of the CycleGAN discriminator. In this example 0.8 is the score the discriminator gave to that patch of the image (i.e., this patch looks closer to a real image (1)).

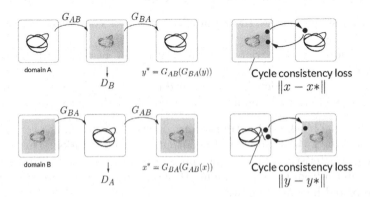

Fig. 16. Proposed CycleGAN model to learn an unsupervised Sketch2Rendering mapping.

Losses: The objective of CycleGANs is to learn the mapping between domains X and Y given training examples $x_i \in X$ and $y_i \in Y$. The data distributions are $x \sim p_{data}(x)$ and $y \sim p_{data}(y)$. As shown in Fig. 16, the model includes two mappings, one learned by each generator, $G_{AB} : X \rightarrow Y$ and $G_{BA} : Y \rightarrow X$.

Apart from these generators, the model has two discriminators, one for each domain. D_X will learn to distinguish between real images x and fake images $x* = G_{BA}(y)$, while discriminator D_B will learn to distinguish between real images y and fake images $y* = G_{AB}(x)$. The objective functions will therefore contain two different losses, the adversarial losses [1] that will measure whether the distribution of the generated images match the data distribution in the target domain, and the cycle consistency losses [21], that will make sure that G_{AB} and G_{BA} do not contradict each other.

Cycle Consistency Loss: It can be expressed as $||x - x^*||$ or $||y - y^*||$, depending on which of the styles we consider as the starting point, where x^* and y^* represent the fake images generated by the generators. These equations ensure that the original image and the output image, after completing the cycle, i.e., the twice-translated image, are the same. This loss function is expressed as:

$$\mathcal{L}_{cyc}(G_{AB}, G_{BA}) = \mathbb{E}_{x \sim p_{data}(x)}[|||G_{BA}(G_{AB}(x)) - x||_1]$$
$$+ \mathbb{E}_{y \sim p_{data}(y)}[|||G_{AB}(G_{BA}(y)) - y||_1] \quad (1)$$

Adversarial Loss: Apart from the cycle consistency loss, CycleGANs also use adversarial loss to train. As in traditional GAN models, the adversarial loss measures whether the generated images look real, i.e., whether they are indistinguishable from the ones coming from the same probability distribution learned from the training set [1]. For the mapping $G_{AB} : X \rightarrow Y$ and the corresponding discriminator, we express the objective as:

$$\mathcal{L}_{GAN}(G_{AB}, D_B, X, Y) = \mathbb{E}_{x \sim p_{data}(x)}[log D_B(y)]$$
$$+ \mathbb{E}_{x \sim p_{data}(x)}[log(1 - D_B(G_{AB}(x)))] \quad (2)$$

Every translation by the G_{AB} generator will be checked by the D_B discriminator, and the output of generator G_{BA} will be assessed and controlled by the D_A discriminator. Every time we translate from one domain to another, the discriminator will test if the output of the generator looks real or fake. Each generator will try to *fool* its adversary, the discriminator. While each generator tries to minimize the objective function, the corresponding discriminator tries to maximize it. The training objectives of this loss are $\min_{G_{AB}} \max_{D_B} \mathcal{L}_{GAN}(G_{AB}, D_B, X, Y)$ and $\min_{G_{BA}} \max_{D_A} \mathcal{L}_{GAN}(G_{BA}, D_A, X, Y)$.

Identity Loss: The identity loss measures if the output of the CycleGAN preserves the overall color temperature or structure of the picture. Pixel distance is used to ensure that ideally there is no difference between the output and the input, this ensures that the CycleGAN only changes the parts of the image when it needs to.

Model Training: The full objective of the CycleGAN is reducing these three loss functions. Actually, Zhu et al. show that training the networks with only one of the functions doesn't arrive to high-quality results. In the formula, we can see that both the identity loss and cycle consistency functions are weighted by λ_{ident} and λ_{cyc}, respectively. These scalars control the importance of each of the losses in the training. In our case, following the values for these parameters proposed in the original paper [21], λ_{cyc} will be 10, and λ_{ident} will be 0.1, as this last function only controls the tint of the background of the input and output images; and as our dataset is composed of the same colors, it does not suppose a large influence.

$$\mathcal{L}(G_{AB}, G_{BA}, D_A, D_B) = \mathcal{L}_{GAN}(G_{AB}, D_B, X, Y) + \mathcal{L}_{GAN}(G_{BA}, D_A, X, Y)$$
$$+ \lambda_{cyc}\mathcal{L}_{cyc}(G_{AB}, G_{BA}) + \lambda_{ident}\mathcal{L}_{ident}(G_{AB}, G_{BA}) \quad (3)$$

A.3   CycleGAN Training details

The networks were trained from scratch with a starting learning rate of 0.0002 for 100 epochs, after this, it was trained for 100 epochs more with a learning rate of 0.00002, as suggested by Zhu et al. in [21]. Following this procedure, the objective loss function of the discriminator D was divided by 2, which slows down the rate at which D learns compared with the generator G.

For the generator and discriminator we adopt the same architectures as the ones proposed by Zhu et al. [21], with the difference that for the first and last layers in the generator, we used a padding of 3 due to the input image size of our dataset.

References

1. Goodfellow, I., et al.: Generative adversarial nets. Adv. Neural Inf. Process. Syst. **27**, 1–9 (2014)
2. Thies, J., Zollhöfer, M., Theobalt, C., Stamminger, M., Nießner, M.: Ignor: Image-guided neural object rendering. arXiv preprint arXiv:1811.10720 (2018)
3. Sitzmann, V., Thies, J., Heide, F., Nießner, M., Wetzstein, G., Zollhofer, M.: Deepvoxels: learning persistent 3d feature embeddings. In: Proceedings of the IEEE/CVF Conference on Computer Vision and Pattern Recognition, pp. 2437–2446 (2019)
4. Dupont, E., et al.: Equivariant neural rendering. arXiv preprint arXiv:2006.07630 (2020)
5. Gatys, L.A., Ecker, A.S., Bethge, M.: A neural algorithm of artistic style. arXiv preprint arXiv:1508.06576 (2015)
6. Isola, P., Zhu, J.-Y., Zhou, T., Efros, A.A.: Image-to-image translation with conditional adversarial networks. In: 2017 IEEE Conference on Computer Vision and Pattern Recognition (CVPR) (2017)
7. Radford, A., Metz, L., Chintala, S.: Unsupervised representation learning with deep convolutional generative adversarial networks (2016)

8. Wang, T.-C., Liu, M.-Y., Zhu, J.-Y., Tao, A., Kautz, J., Catanzaro, B.: High-resolution image synthesis and semantic manipulation with conditional GANs. In: Proceedings of the IEEE Conference on Computer Vision and Pattern Recognition (CVPR), pp. 8798–8807 (2018)
9. Wang, C., Zheng, H., Yu, Z., Zheng, Z., Gu, Z., Zheng, B.: Discriminative region proposal adversarial networks for high-quality image-to-image translation. In: Proceedings of the European Conference on Computer Vision (ECCV), September 2018
10. Park, T., Liu, M.-Y., Wang, T.-C., Zhu, J.-Y.: Semantic image synthesis with spatially-adaptive normalization. In: Proceedings of the IEEE Conference on Computer Vision and Pattern Recognition (2019)
11. Schönfeld, E., Sushko, V., Zhang, D., Gall, J., Schiele, B., Khoreva, A.: You only need adversarial supervision for semantic image synthesis. In: International Conference on Learning Representations (2021)
12. Lütjens, B., et al. Physically-consistent generative adversarial networks for coastal flood visualization. arXiv preprint arXiv:2104.04785 (2021)
13. Lütjens, B., et al.: Physics-informed GANs for coastal flood visualization. arXiv preprint arXiv:2010.08103 (2020)
14. Casale, F.P., Dalca, A., Saglietti, L., Listgarten, J., Fusi, N.: Gaussian process prior variational autoencoders. In: Advances in Neural Information Processing Systems, pp. 10369–10380 (2018)
15. Rezende, D., Mohamed, S.: Variational inference with normalizing flows. In: Bach, F., Blei, D., (eds.) Proceedings of the 32nd International Conference on Machine Learning (ICML), vol. 37, pp. 1530–1538 (2015)
16. Lugmayr, A., Danelljan, M., Van Gool, L., Timofte, R.: Learning the super-resolution space with normalizing flow. In: ECCV, Srflow (2020)
17. Kingma, D.P., Welling, M.: Auto-encoding variational Bayes. In: Proceedings of the 2nd International Conference on Learning Representations (ICLR) (2014)
18. Dosovitskiy, A., Brox, T.: Generating images with perceptual similarity metrics based on deep networks. In: Advances in Neural Information Processing Systems, vol. 29, pp. 658–666. Curran Associates Inc (2016)
19. Zhu, J.-Y., Zhang, R., Pathak, D., Darrell, T., Efros, A.A., Wang, O., Shechtman, E.: Toward multimodal image-to-image translation. In: Advances in Neural Information Processing Systems (NeurIPS), vol. 30, pp. 465–476 (2017)
20. Pedroso, T.C.: Utilización de métodos aleatorios en la generación de formas geométricas. Master's thesis, Universidad Politécnica de Madrid, Spain (2003). https://oa.upm.es/69208/
21. Zhu, J.-Y., Park, T., Isola, P., Efros, A.A.: Unpaired image-to-image translation using cycle-consistent adversarial networks. In: 2017 IEEE International Conference on Computer Vision (ICCV) (2017)
22. Odena, A., Dumoulin, V., Olah, C.: Deconvolution and checkerboard artifacts. Distill 1(10), e3 (2016)
23. Murasugi, K., Kurpita, B.: Knot Theory and Its Applications. Springer, Boston (1996). https://doi.org/10.1007/978-0-8176-4719-3

Semantic Causal Abstraction for Event Prediction

Sasha Strelnikoff[(✉)], Aruna Jammalamadaka, and Tsai-Ching Lu

Information Systems and Sciences Laboratory, HRL Laboratories, LLC,
Malibu, CA 90265, USA
{sstrelnikoff,ajammalamadaka,tlu}@hrl.com
https://www.hrl.com/laboratories

Abstract. Causal models promise many benefits if applied correctly to machine learning tasks. However, in order to leverage fine grained causal information, it is often useful to reduce the complexity of the causal connections by producing an abstracted version of the graph. In this paper, we introduce semantic causal abstractions, a scheme for constructing abstracted causal graphs in order to provide a domain-independent approximation to formal causal abstraction for unstructured textual data. We then analyze the effects that this type of abstraction has on the performance of a causal graph-based prediction model under multiple semantic representations. Our experiments on two stock prediction tasks provide evidence for the efficacy of semantic causal abstraction to improve prediction performance and give insight into the consistency of the optimal semantic causal abstraction levels across tasks.

Keywords: Causal abstraction · Semantic representations · Stock trend prediction · Stock volatility prediction

1 Introduction

Prediction under any sufficiently complex causal system is immensely difficult. For example, the weather could in principle be analyzed by understanding the causal interactions between each of the individual particles in the atmosphere, however, in practice, the size of the causal graph renders it infeasible for humans to reason about or simulate directly. As a result, it becomes necessary to model a macroscopic representation of the causal graph by analyzing the interactions between large-scale pressure systems. This example provides an illustration of causal abstraction, which formally defines a process of constructing a "macro" causal model which is a distilled form of a corresponding "micro" causal model. While there are several theoretical works defining causal abstraction, few of these focus on the interaction between causal abstraction and prediction performance, despite evidence that appropriate causal model abstractions improve accuracy of predictions made from these models [9]. Moreover, to the knowledge of the

© IFIP International Federation for Information Processing 2022
Published by Springer Nature Switzerland AG 2022
A. Holzinger et al. (Eds.): CD-MAKE 2022, LNCS 13480, pp. 188–200, 2022.
https://doi.org/10.1007/978-3-031-14463-9_12

authors', there are no previous works addressing the problem of causal *language* abstraction. In this study, we address this gap by jointly analyzing the effects of semantic feature representations and semantically driven causal abstraction on prediction performance.

The application of causal systems to machine learning models has gained much interest recently due to a number of advantages they may provide. For example, since causality provides the direct mechanisms by which events occur, it is hypothesized that causality-based systems may be less prone to overfitting [20]. Indeed, causal relations have been successfully utilized in multiple domains to improve model robustness to rare, adversarial, and out-of-distribution examples [12, 24, 25], and causal structures have also proved useful for domain adaptation [14, 26]. Causally derived structures are also desirable for humans to reason about and understand systems. For example, [23] outlined a method for automatically producing human-interpretable causal graphs for summarizing information contained in large corpora. Recently, causality-based methods were used to produce explainable reinforcement learning agents [13] and image classification models [7]. As is evidenced by these studies, it is clear that there is a potential for numerous benefits to be had by developing causality-based models for machine learning tasks. In this work, we are particularly interested in abstracting *textually derived* causal models for classification. In order to make use of the abstracted phrasal nodes in the causal graph, we adopt semantic feature representations and observe the effect of the choice of representation on the prediction performance as a function of semantic causal abstraction (referred to as abstraction throughout).

After outlining related works in Sect. 2, we propose a pipeline for producing semantically abstracted causal graphs from textual data and introduce a neural-based classification model for utilizing the abstracted causal graph for downstream tasks in Sect. 3. In Sect. 4, by considering both stock price trend and volatility prediction tasks as a case study, we jointly analyze the impact of both the semantic representation and semantic causal abstraction of relevant textual information on prediction performance, which produces evidence in support of the following claims:

- Semantic causal abstraction can improve model prediction performance.
- The optimal level of semantic causal abstraction depends on the semantic feature representation, but is moderately consistent across prediction tasks.
- Multiple locally optimal levels of semantic causal abstraction may exist for a particular task.

Finally, we conclude in Sect. 5 and outline possible future directions of inquiry in this domain.

2   Related Work

Stock Value Prediction From Text – Financial text has been used as a means of stock market prediction for some time. Textual sentiment is a commonly used feature for stock prediction [3, 17, 22]; however, distilling the rich semantic meaning of text to its sentiment is likely to remove other predictive features of the text (e.g. relational or event information). As a result, more sophisticated methods have been developed in order to make use of the additional information contained in the textual data. For example, an event-driven approach to stock prediction was used in [5], wherein structured events of the form (*Actor, Action, Object*) are extracted and generalized in order to create dense feature representations for a collection of documents. Another approach is presented in [27], which shows superior performance by extracting the causal phrases from text in order to generate novel embeddings designed to capture the corresponding causal relations. While these methods represent the causal relations at the embedding stage, they all aggregate the individual event representations in order to make predictions. Because of the reliance on a single vector representation of the textual information to make predictions, none of these methods make explicit use of the relational information present in the data. Here we overcome this limitation by implementing a graph neural network (GNN) architecture, which enables us to explicitly incorporate the causal graph into the prediction model.

Causal Abstraction – Broadly, causal abstraction provides a scheme for constructing a "macro" causal model that is derived from a corresponding "micro" causal model. Empirically, [8] shows that in certain systems there exists an optimal, non-trivial, level of causal abstraction, where "optimality" is formally defined to capture how effectively causes produce effects in the system, and how selectively causes can be identified from effects. Here, we take inspiration from [1] which introduces a form of causal abstraction known as constructive abstraction. A constructive abstraction of a fine-grained causal model is a coarse causal model in which the variables define a partition of the fine-grained causal model while preserving the original causal relationships. In order to adopt this to a domain-independent text-based setting, we propose a graph coarsening scheme based on the semantic similarity between causal phrases in the text, which enables us to approximate constructive abstraction in both a general and text-driven setting.

3   Methods

In this section, we introduce our financial news dataset, describe the our causal extraction methodology, semantic representation scheme, and finally outline our semantic causal abstraction pipeline and GNN model.

3.1   Dataset

The dataset used throughout this work is a publicly available Kaggle dataset[1] derived from financial news articles published by a variety of financial media companies. The dataset comprises of 21 publishers in total, with the following

---

[1] This dataset is publically available and can be accessed at https://www.kaggle.com/miguelaenlle/massive-stock-news-analysis-db-for-nlpbacktests.

top three most frequently represented publishers: Seeking Alpha (49%), Zacks (24%), and GuruFocus (12%). In this work, we focus only on the financial news headlines for articles published between 01/01/2017 and 06/11/2020 (the latest date in the dataset).

3.2 Causal Phrase Parsing

In order to extract cause-effect relations from text, we employ the pre-trained model from [10], which we refer to as *Kao2020*. This model consists of a Viterbi decoder built on top of a fine-tuned BERT classifier [4]. It ranked first in the cause-effect detection portion of the FinCausal-2020 task [15], wherein participants are tasked to design a model to identify causal language and extract the causal elements contained in financial news text[2]. As a means of assessing the efficacy of this method to identify causal text within our dataset, we randomly sample 100 news headlines and then assign ground-truth labels to each sample by manual inspection. After performing inference using *Kao2020* on these headlines, we obtain precision, recall, and F1 scores of 0.83, 0.48, and 0.61 respectively. The relatively high estimated precision score of this method indicates that we should expect to obtain high-quality causal pairs, however it should be noted that, due to the low estimated recall score, we may incur a loss of important causal information.

3.3 Semantic Representations

Throughout this work, we use the term *semantic representation* to describe the result of any method which aims to produce a quantitative vector representation of its input phrase with the aim of preserving some feature of its semantic meaning. Semantic representations provide us with the ability to quantify qualitative (textual) causal relationships, which is an essential step for causal abstraction. There are numerous semantic representation methodologies; however, they generally classify into semantic network and semantic space approaches. Semantic networks aim to represent the meaning of a lexical unit node in a graph whose edges encode semantic relations between the units [19]. For example, in this paper we consider WordNet [16]. This is a popular word-based semantic network scheme, which we use to map each noun in the cause and effect phrases to its WordNet hypernym. This construction has the benefit of producing a dense vector representation of each of the phrases, while preserving the high-level semantic content implied by the nouns. VerbNet is an analogously popular verb-based semantic network introduced in [21]. VerbNet is a lexicon of verbs and groups of verbs organized according to their shared syntactic linking behavior. Similarly to the application of WordNet, we produce dense vector representations of the verbs in each of the phrases by mapping each verb to its corresponding group in VerbNet. The Universal Sentence Encoder (USE) [2] provides an example of

---

[2] This dataset is restricted for use to the participating members of the competition only.

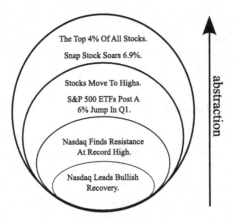

Fig. 1. Visual illustration of the result of macro-node abstraction. This is a selected example generated from our dataset using the DBSCAN-based causal abstraction scheme outlined in Sect. 3.4. Each ring includes all of the phrases in the smaller rings, making it a less specific (more abstracted) collection of phrases, corresponding to a higher value of ϵ.

a semantic space representation approach. USE is a transformer-based encoder capable of natively producing a dense representation of the semantic information contained in sentence length text, which makes it a suitable choice for our problem domain.

3.4   Semantic Causal Abstraction

The following section outlines our method for semantic causal abstraction. Conceptually, we aim to utilize the semantic similarity between cause and effect phrases within the original ("micro-node") causal graph in order to produce macro-nodes composed of semantically related phrases. We additionally use the edge information from the original causal graph in order to induce the edge relations in the macro-node graph. See Fig. 3 for an example of this coarsening on a small subset of our data.

Formally, if we let N denote the number of cause-effect phrase pairs extracted from our input documents through the methods outlined in Sect. 3.2, we have tuples of strings (C_i, E_i) containing the cause and effect phrases respectively, for $i = 1, ..., N$. We refer to this dataset as $\mathcal{D} = \{(C_i, E_i)\}_{i=1}^{N}$. This relational data defines our original causal graph $\mathcal{G} = (\mathcal{V}, \mathcal{E})$ wherein each of the $2N$ nodes of the graph corresponds to one of the cause or effect phrases in $\mathcal{D}$ with directed edges from each cause to its corresponding effect. Notationally, $\mathcal{V} = \{C_i\}_{i=1}^{N} \cup \{E_i\}_{i=1}^{N}$ and $\mathcal{E} = \mathcal{D}$.

At this stage we have a bipartite graph in which each extracted cause is connected only to it's corresponding effect. In order to cluster $\mathcal{G}$ and form an abstracted version of the graph, we first need to quantify the semantic similarity of cause/effect nodes in $\mathcal{G}$. We utilize the USE embedding, $Emb : \{\text{Strings}\} \rightarrow$

$\mathbb{R}^{512}$ as a means to capture the semantic similarity, sim, between phrases $p_1, p_2 \in \mathcal{D}$ as

$$sim(p_1, p_2) = Emb(p_1) \cdot Emb(p_2). \tag{1}$$

This choice of similarity metric is made because USE can natively encode sentence-level semantic meaning, with performance comparable with state-of-the-art methods [18]. This similarity function is then used to apply the DBSCAN algorithm [6] to construct a partition, $\mathcal{A}$, of semantically similar phrases in $\mathcal{D}$. DBSCAN is a popular density-based non-parametric clustering algorithm. The characteristics of the clustering it produces are primarily determined by two parameters, ϵ, controlling the density required for points to be within the same class, and N_{min}, specifying the number of dense points required to define a class. We choose DBSCAN clustering for two reasons:

1. Density-based clustering allows for an interpretable measure of phrase abstraction; larger values of ϵ correspond to larger classes of semantically related phrases and therefore a more abstract representation of the causal graph.
2. The non-parametric nature of DBSCAN is suitable for semantic clustering since we do not have any prior knowledge about the distribution of points within the USE-embedding space.

Figure 1 provides an illustration of the causal abstraction outlined above for a single node in the graph, wherein the size of the ring corresponds to ϵ, and therefore larger rings depict higher levels of abstraction. As the abstraction levels increase, we can see that the single phrase *"Nasdaq Leads Bullish Recovery"* broadens to include a re-phrasing with a similar meaning (*"Nasdaq Finds Resistance At Record High"*) in the second ring, again in the third ring to describe the concept of stocks performing well, and again in the final ring to include broad positive sentiment about stocks.

Given derived node partitioning, $\mathcal{A}$, we then define $C_i^{\mathcal{A}}$ and $E_i^{\mathcal{A}}$ to be the classes in $\mathcal{A}$ corresponding to C_i and E_i respectively. This in turn induces a coarsened version of $\mathcal{D}$ as $\mathcal{D}^{\mathcal{A}} = \{(C_i^{\mathcal{A}}, E_i^{\mathcal{A}})\}_{i=1}^{N}$ and a corresponding coarsened version of $\mathcal{G}$, which we denote by $\mathcal{G}^{\mathcal{A}} = (\mathcal{V}^{\mathcal{A}}, \mathcal{E}^{\mathcal{A}})$. Note, there is a directed edge between two macro-nodes in the abstracted graph $\mathcal{G}^{\mathcal{A}}$ exactly when there is a directed edge between any of the constituent micro-nodes for those macro-nodes.

3.5 Prediction Model

If our semantic causal abstraction method results in an appropriate causal abstraction, we expect an improvement in the prediction accuracy of a model which leverages these causal relationships [9]. The following is a description of that model, which incorporates the semantic representation information of the nodes in $\mathcal{G}$. Note, for ease of notation, for every cause C_i and effect E_i, we take each of the semantic representations discussed in Sect. 3.3 to be denoted by $\mathbf{c}_i$ and $\mathbf{e}_i$ respectively, where the particular semantic representation is assumed from the context. We define the node feature of each node C_i (or E_i) in $\mathcal{G}$ to be the corresponding semantic representation, c_i (or e_i). Then, for each node V^A of $\mathcal{G}^A$, we define its node feature, v^A, to be the mean of the features of the nodes contained in the class $V^A \subseteq V$.

We leverage the structure of $\mathcal{G}^A$ for classification by employing a graph convolutional network (GCN). In particular, we make use of the popular GCN architecture introduced in [11]. Explicitly, for a GCN with L layers, the hidden representation of node i at layer $l + 1 \leq L$ is given by

$$h_i^{l+1} = \sigma \left(b_i^l + \sum_{j \in \mathcal{N}(i)} \frac{W^l h_j^l}{\sqrt{|\mathcal{N}(i)||\mathcal{N}(j)|}} \right), \tag{2}$$

where b_i^l, W^l are the learned bias and weight terms, $\mathcal{N}(i)$ is the collection of neighbors of node i, and σ is an element-wise Rectified Linear Unit (ReLU) function. In order to produce a classification output, p, we average the features at the final layer and apply a softmax function,

$$p = \text{softmax} \left(\frac{1}{|\mathcal{V}^A|} \sum_{i=1}^{|\mathcal{V}^A|} h_i^L \right). \tag{3}$$

4   Experiments

In the following experiments, we aim to empirically analyze the effect that the semantic representation and semantic causal abstraction have on the predictive power of our model applied to financial tasks. In particular, we observe the Area Under the Receiver Operating Characteristic (AUROC) as a function of semantic representation and semantic causal abstraction.

4.1   Setup

For all of the following experiments, we fix $N_{min} = 3$. In order to test a variety of abstraction levels, we consider the set of distances Δ between each point in our data set and its N_{min}-neighbor. From this, we take ϵ to be equally spaced values between the 10th and 90th percentile of Δ. Including the case of not performing any causal abstraction, $\epsilon = 0$, this provides five values, $\epsilon = 0, 0.09, 0.17, 0.22, 0.27$. In order to compare the semantic representations with one another, it is desirable to standardize their vector representations. First, because each of the vector representations have different lengths (512, 2591, and 223, for the USE, WordNet, and VerbNet representations respectively) we project each onto the top k corresponding principal components and repeat each experiment for $k = 16, 32, 64, 128$. We additionally restrict the range of values for the representations by normalizing the vector components to be within -1 and 1. In all of the following experiments we run a randomized 3-fold cross validation repeated 5 times with random weight initialization, which allows us to obtain a robust measure of each of the models' performance. We further divide the testing data fold evenly into a test and validation set and perform early stopping with a patience of 50.

4.2 Stock Trend Prediction

For this task, we aim to utilize the financial news headline data with the objective of predicting the stock movement of the S&P500. Specifically, we formulate our prediction problem as a binary classification problem wherein the goal is to determine whether the S&P500 index will increase or decrease in the succeeding time step. For our experiments, we consider time steps of 5 days and define an increasing (decreasing) event to be one in which the average opening stock price of the succeeding time step is greater (less) than the average opening stock price of the current time step.

4.3 Volatility Prediction

As a second task, we examine the problem of stock volatility prediction. As in the stock trend prediction task, we analyze our textual data in time steps of 5 days and pose this as a classification task wherein we use the financial news data of the current time step to predict whether the stock price of the following time step will significantly deviate in mean from the price in the current time step. Formally, if $p_0, ..., p_T$ are the 5-day average stock prices for the period under consideration, we define a volatile event at time step t as $|p_t - p_{t-1}| > \sigma_0$. We then aim to predict the occurrence of such an event only from the financial text at time step $t - 1$. For the following experiments, we take σ_0 to be 1 standard deviation of $\{|p_t - p_{t-1}|\}_{t=1}^{T}$, which amounts to about 28% of the data having a volatile label.

4.4 Results

Improvement in Prediction Accuracy via Semantic Causal Abstraction – Figures 2a and 2b show the results of the stock trend and volatility tasks respectively. When comparing the abstracted models to the baseline ($\epsilon = 0$) models, we observe the largest improvements in the AUROC for the volatility task at the $\epsilon = 0.09$ abstraction level (Fig. 2b) with the WordNet and VerbNet based models achieving relative improvements over the baseline of 8.1% and 7.7% respectively. From Fig. 2a, we observe that the WordNet and VerbNet based models maintain their accuracy for abstractions $\epsilon \leq 0.17$, with only a minor degradation in the VerbNet performance at $\epsilon = 0.17$. Higher levels of abstraction correspond to a more compressed representation of the dataset, so this result indicates that semantic causal abstraction provides a scheme for reducing model size while maintaining model performance. Moreover, the USE-based model peaks in performance at the abstraction level of $\epsilon = 0.17$, which provides further evidence that semantic causal abstraction can be used to improve performance.

Consistency of the Optimal Causal Abstraction – In both the stock trend and volatility tasks, the WordNet and VerbNet based models attain maximal performance around the $\epsilon = 0.09$ abstraction level. This provides evidence that the optimal level of semantic causal abstraction is a relatively task independent

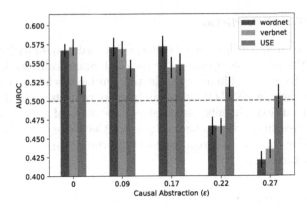

(a) Stock Trend Prediction

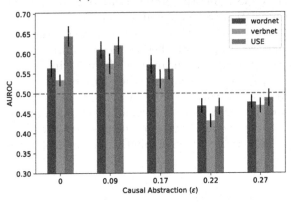

(b) Volatility Prediction

Fig. 2. Results for the stock trend and volatility prediction tasks, across semantic representations and levels of abstraction. We observe an improvement in performance for the causally abstracted models in both the stock trend task and the volatility task with the USE and Verb(Word)net representations respectively. We additionally see similarities in the performance of the WordNet and VerbNet based models across tasks, suggesting that the optimal level of causal abstraction may be task independent.

feature of the original causal graph. However, the results for the USE-based model seem to contradict this claim, since it attains maximal performance with no abstraction for the volatility prediction task, whereas the AUROC peaks at $\epsilon = 0.17$ for the trend task. A possible explanation for this inconsistency is that the USE-based model is underfitting to the trend prediction task. This is evidenced in multiple ways. From Fig. 2a, note that the AUROC score for both the WordNet and VerbNet based models significantly degrades as the abstraction is further increased. This is expected, since at the highest level of abstraction all nodes are grouped into a single node and both causal and predictive information are lost. However, the USE-based model is able to maintain a similar

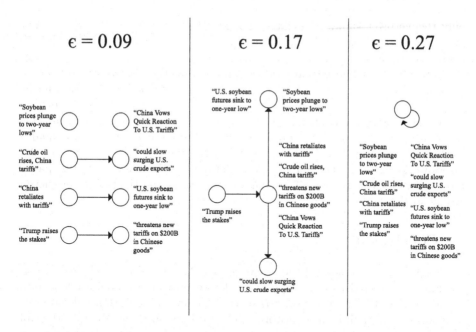

Fig. 3. Depiction of a subset of the news data at multiple levels of abstraction collected in the five day period between 2018-06-15 and 2018-06-20. From this, we can see that a meaningful causal graph, representing the effects of US tariffs imposed on China by Trump, occurs at an intermediate value of ϵ, whereas at the point $\epsilon = 0.09$ we see a disconnected causal structure and at $\epsilon = 0.27$ we see the causal structure collapse to a single node.

performance across abstraction levels, which is comparable to a random classifier (indicated by the dashed red line in the plot). Both the poor performance and flat-trend across abstraction levels provide evidence that the USE-based model is underfitting to this task.

An appropriate causal abstraction is one which preserves meaningful causal relationships between the original and abstracted representations. To demonstrate the appropriateness of semantic causal abstraction and visualize the effect of varying levels of ϵ on the causal graph, an example sub-graph is shown in Fig. 3. Here we can see that a meaningful causal graph, representing the effects of US tariffs imposed on China by Trump, occurs at an intermediate value of ϵ, whereas at $\epsilon = 0.09$ we see a disconnected causal structure and at $\epsilon = 0.27$ we see the causal structure collapse to a single node.

Multiple Locally-Optimal Abstraction Levels – Finally, we note that the performance of all of the semantic representations appear to marginally improve between $\epsilon = 0.22$ and $\epsilon = 0.27$ in the volatility task. This suggests that by varying the abstraction level, one may attain multiple local optima. Empirically, this provides some evidence for the intuitive idea that predictive causal relations exist at multiple levels of granularity. Note that this effect is relatively weak, only the VerbNet shows a deviation outside of the 1 standard deviation error bars, so more supporting evidence would need to be supplied before any definite claims are made.

Note, the overall performance of the proposed method is *not* state-of-the-art for stock trend prediction. Indeed, this paper is not intended to serve as a means for producing an optimal method of stock trend prediction, but instead provide an initial investigation into the benefits of causal language abstraction. The moderate performance of our proposed method is however of a similar quality to that of other fully text-based methods such as those proposed in [5] and [27]. These works use a similar financial news dataset[3] to the one used in this paper, and achieve 58.94% and 59.80% accuracy respectively on the task of predicting 30 day stock trends.

5   Conclusion

In this paper, we introduced the semantic causal abstraction, a scheme for constructing abstracted causal graphs in order to provide a domain-independent approximation to formal causal abstraction for unstructured textual data. We also outlined a graph neural network model for leveraging the abstracted graph for classification tasks. Using the problems of stock price trend and volatility prediction as a case study, we analyzed the joint effects of semantic causal abstraction and semantic representations on our model's prediction performance. Our results suggest that semantic causal abstraction can not only be used to compress the data whilst maintaining model performance, but can also be used to improve model performance. We also show evidence that the impact of abstraction on performance is strongly dependent on the semantic feature representation applied, and that the optimal abstraction level, under a given semantic representation, may remain consistent across prediction tasks.

Future work can be directed towards investigating the consistency of the optimal level of abstraction across different domains and prediction models. This could help provide further understanding of the performance landscape and the extent to which the optimal abstraction for a given text corpus is model and task independent. In the long-term, a theoretical analysis of the relationship between semantic representation, semantic causal abstraction, and formal definitions of causal abstractions could be conducted. If fruitful, this could provide a joint optimization of semantic representation and abstraction level.

References

1. Beckers, S., Halpern, J.Y.: Abstracting causal models. In: Proceedings of the AAAI Conference on Artificial Intelligence, vol. 33, pp. 2678–2685 (2019)
2. Cer, D., et al.: Universal sentence encoder. arXiv preprint arXiv:1803.11175 (2018)
3. Das, S.R., Chen, M.Y.: Yahoo! for Amazon: sentiment extraction from small talk on the web. Manag. Sci. **53**(9), 1375–1388 (2007)

---

[3] This data is no longer publicly available.

4. Devlin, J., Chang, M.W., Lee, K., Toutanova, K.: Bert: Pre-training of deep bidirectional transformers for language understanding. arXiv preprint arXiv:1810.04805 (2018)
5. Ding, X., Zhang, Y., Liu, T., Duan, J.: Using structured events to predict stock price movement: An empirical investigation. In: Proceedings of the 2014 Conference on Empirical Methods in Natural Language Processing (EMNLP), pp. 1415–1425 (2014)
6. Ester, M., et al.: A density-based algorithm for discovering clusters in large spatial databases with noise. In: KDD, vol. 96, pp. 226–231 (1996)
7. Harradon, M., Druce, J., Ruttenberg, B.: Causal learning and explanation of deep neural networks via autoencoded activations. arXiv preprint arXiv:1802.00541 (2018)
8. Hoel, E.P., Albantakis, L., Tononi, G.: Quantifying causal emergence shows that macro can beat micro. Proc. Natl. Acad. Sci. **110**(49), 19790–19795 (2013)
9. Jammalamadaka, A., Warmsley, D., Lu, T.C.: Causal abstraction via emergence for predicting bilateral trade. In: Ma, S., Kummerfeld, E. (eds.) Proceedings of The 2021 Causal Analysis Workshop Series. Proceedings of Machine Learning Research, vol. 160, pp. 39–51. PMLR, 16 July 2021
10. Kao, P.W., Chen, C.C., Huang, H.H., Chen, H.H.: Ntunlpl at fincausal 2020, task 2: Improving causality detection using Viterbi decoder. In: Proceedings of the 1st Joint Workshop on Financial Narrative Processing and MultiLing Financial Summarisation, pp. 69–73 (2020)
11. Kipf, T.N., Welling, M.: Semi-supervised classification with graph convolutional networks. arXiv preprint arXiv:1609.02907 (2016)
12. Liu, C., et al.: Learning causal semantic representation for out-of-distribution prediction. arXiv preprint arXiv:2011.01681 (2020)
13. Madumal, P., Miller, T., Sonenberg, L., Vetere, F.: Explainable reinforcement learning through a causal lens. In: Proceedings of the AAAI Conference on Artificial Intelligence, vol. 34, pp. 2493–2500 (2020)
14. Magliacane, S., van Ommen, T., Claassen, T., Bongers, S., Versteeg, P., Mooij, J.M.: Domain adaptation by using causal inference to predict invariant conditional distributions. arXiv preprint arXiv:1707.06422 (2017)
15. Mariko, D., Abi Akl, H., Labidurie, E., Durfort, S., de Mazancourt, H., El-Haj, M.: The Financial Document Causality Detection Shared Task (FinCausal 2020). In: The 1st Joint Workshop on Financial Narrative Processing and MultiLing Financial Summarisation (FNP-FNS 2020), Barcelona, Spain (2020)
16. Miller, G.A.: Wordnet: a lexical database for English. Commun. ACM **38**(11), 39–41 (1995)
17. Mittal, A., Goel, A.: Stock prediction using twitter sentiment analysis. Standford University, CS229 (2011). http://cs229.stanford.edu/proj2011/GoelMittal-StockMarketPredictionUsingTwitterSentimentAnalysis.pdf, 15 (2012)
18. Perone, C.S., Silveira, R., Paula, T.S.: Evaluation of sentence embeddings in downstream and linguistic probing tasks. arXiv preprint arXiv:1806.06259 (2018)
19. Saedi, C., Branco, A., Rodrigues, J., Silva, J.: Wordnet embeddings. In: Proceedings of the third workshop on representation learning for NLP, pp. 122–131 (2018)
20. Schölkopf, B., et al.: Toward causal representation learning. Proc. IEEE **109**(5), 612–634 (2021)
21. Schuler, K.K.: VerbNet: A broad-coverage, comprehensive verb lexicon. University of Pennsylvania (2005)

22. Skuza, M., Romanowski, A.: Sentiment analysis of twitter data within big data distributed environment for stock prediction. In: 2015 Federated Conference on Computer Science and Information Systems (FedCSIS), pp. 1349–1354. IEEE (2015)

23. Strelnikoff, S., Jammalamadaka, A., Warmsley, D.: Causal maps for multi-document summarization. In: 2020 IEEE International Conference on Big Data (Big Data), pp. 4437–4445. IEEE (2020)

24. Yang, C.H.H., Hung, I., Danny, T., Ouyang, Y., Chen, P.Y.: Causal inference q-network: Toward resilient reinforcement learning. arXiv preprint arXiv:2102.09677 (2021)

25. Zhang, C., Zhang, K., Li, Y.: A causal view on robustness of neural networks. arXiv preprint arXiv:2005.01095 (2020)

26. Zhang, K., Schölkopf, B., Muandet, K., Wang, Z.: Domain adaptation under target and conditional shift. In: International Conference on Machine Learning, pp. 819–827. PMLR (2013)

27. Zhao, S., et al.: Constructing and embedding abstract event causality networks from text snippets. In: Proceedings of the Tenth ACM International Conference on Web Search and Data Mining, pp. 335–344 (2017)

An Evaluation Study of Intrinsic Motivation Techniques Applied to Reinforcement Learning over Hard Exploration Environments

Alain Andres[1,2]([✉]), Esther Villar-Rodriguez[1], and Javier Del Ser[1,2]

[1] TECNALIA, Basque Research and Technology Alliance (BRTA),
48160 Derio, Spain
{alain.andres,esther.villar,javier.delser}@tecnalia.com
[2] University of the Basque Country (UPV/EHU), 48013 Bilbao, Spain

Abstract. In the last few years, the research activity around reinforcement learning tasks formulated over environments with sparse rewards has been especially notable. Among the numerous approaches proposed to deal with these hard exploration problems, intrinsic motivation mechanisms are arguably among the most studied alternatives to date. Advances reported in this area over time have tackled the exploration issue by proposing new algorithmic ideas to generate alternative mechanisms to measure the novelty. However, most efforts in this direction have overlooked the influence of different design choices and parameter settings that have also been introduced to improve the effect of the generated intrinsic bonus, forgetting the application of those choices to other intrinsic motivation techniques that may also benefit of them. Furthermore, some of those intrinsic methods are applied with different base reinforcement algorithms (e.g. PPO, IMPALA) and neural network architectures, being hard to fairly compare the provided results and the actual progress provided by each solution. The goal of this work is to stress on this crucial matter in reinforcement learning over hard exploration environments, exposing the variability and susceptibility of avant-garde intrinsic motivation techniques to diverse design factors. Ultimately, our experiments herein reported underscore the importance of a careful selection of these design aspects coupled with the exploration requirements of the environment and the task in question under the same setup, so that fair comparisons can be guaranteed.

Keywords: Reinforcement learning · Intrinsic motivation ·
Exploration-exploitation · Hard exploration · Sparse rewards

1 Introduction

Over decades, Reinforcement Learning (RL) has been widely acknowledged as a rich and ever-growing research area within Artificial Intelligence aimed to efficiently deal with complex tasks [1,2]. One of the key components of the success

© IFIP International Federation for Information Processing 2022
Published by Springer Nature Switzerland AG 2022
A. Holzinger et al. (Eds.): CD-MAKE 2022, LNCS 13480, pp. 201–220, 2022.
https://doi.org/10.1007/978-3-031-14463-9_13

of RL algorithms is to define a suitable reward function that reflects the objective of the task at hand. However, the design of appropriate reward functions is often difficult – even unfeasible – depending on the peculiarities of the environment and task to be optimized. In this context, the study of environments with so-called *sparse rewards* has gained attention in the last few years. In such scenarios, the RL agent is just positively rewarded when accomplishing the goal, which is representative of manifold problems that arise from real-world applications [3]. Nevertheless, such *hard exploration* RL problems are more complex to address due to sparse informative feedback delivered by the environment, requiring effective means to balance between exploration and exploitation during the agent's learning process.

The aforementioned challenge can be overcome through Intrinsic Motivation (IM, [4]), Imitation Learning [5] and Inverse Reinforcement Learning [6], among other strategies. This work gravitates around the first (IM), which is used to encourage the agent to explore the environment by its inherent satisfaction of curiosity [7]. In practice, the concept of *curiosity* is translated to the RL domain in the form of an intrinsic bonus r_i, which is combined with the extrinsic reward provided by the environment $r = r_e + \beta r_i$ through a weighting factor β. Besides proposing different ways to generate the intrinsic reward r_i, current state-of-the-art algorithms (e.g., RIDE [8], NGU [9], AGAC [10]) also apply novel methods to weight and scale such rewards, being those applicable to prior approaches such as Intrinsic Curiosity Module (ICM, [11]) and Random Network Distillation (RND, [12]). Unfortunately, when different IM-based schemes are compared to each other, those reward scaling techniques are not always in use, making it unclear whether the identified performance gaps are due to the exploration methods themselves or must be attributed to other design choices (i.e., the variation of intrinsic coefficient weights or the architecture of the models inside the RL agent).

Analogously to what is claimed in other performance evaluation works reported recently in [13,14], a fundamental matter in this research area is to discriminate which design criteria impact most on the performance of the RL agent. This is specially relevant in hard exploration environments, since it is known that under such circumstances, the proficiency of the agent is very sensitive regarding the configuration of its compounding modules. For this reason, this manuscript aims to fairly evaluate IM-based solutions present in the state of the art trying to decouple the solver approach from additional weighting and scaling techniques. Under this rationale, this work also incorporates the naive version of IM modules to study their benefit and ascertain the actual advantage of the algorithmic proposal when generating intrinsic rewards. Furthermore, the impact of having different neural network architectures in actor-critic agents and IM modules poses another question that lacks an informed answer in the current literature.

To sum up, this paper investigates the quantitative impact of different design choices when implementing IM-based techniques to understand their relevance when used in agents deployed over sparse reward scenarios. Hence, our contributions are three-fold: (1) we adopt curiosity mechanisms with different implementation choices that impact on how the intrinsic rewards are processed, (2) we conduct a study with multiple current state-of-the-art intrinsic motivation techniques where we compare them fairly in order to evaluate the improvement of generating rewards with different approaches; and (3) we break down experiments, results and conclusions in the interest of providing the reader with independent performance analysis of the set of modules and parameterizations.

The rest of the manuscript is structured as follows: Sect. 2 overviews works related to intrinsic motivation in RL, while Sect. 3 details the factors and the choices to be taken into account when resorting to IM techniques. Next, Sect. 4 presents the experimental setup designed to achieve empirical evidence. Section 5 discusses the obtained results. Finally, Sect. 6 concludes the paper and outlines future research to be developed from this research on.

2   Related Work and Contribution

Before proceeding with the details of this work, we briefly review insights coming from recent research about intrinsic motivation mechanisms to deal with sparse rewards. In the absence of a dense reward function and/or when having hard exploration problems, intrinsic motivation mechanisms have turned up as an effective workaround to overcome poor exploration behaviour. These techniques generate artificial intrinsic rewards based on the novelty of a state[1], which relates to how curious an agent will be when arriving to that state. The less novel a state is, the less curious the agent should be [4]. In this context, several approaches have been proposed up to now to generate such exploration bonuses.

One mechanism to generate the aforementioned intrinsic rewards is by adopting a visitation count strategy, also referred to as *count-based* methods. In this case, intrinsic rewards are assumed to be inversely proportional to the number of counts $N(s)$ that a given state s has been visited, e.g. $r_i^{counts} = 1/\sqrt{N(s)}$. This is a simple, yet effective, solution to quantify the degree to which a state is *unknown* for the agent. However, counts are only applicable when dealing with discrete state spaces. Contrarily, when having more complex domains with continuous state spaces, density models [15], hash functions [16] and also successor features [17] can be applied to extend the concept of counts.

An alternative strategy to produce intrinsic motivation rewards is the use of *prediction-error* methods which, as their name suggests, generate intrinsic rewards based on the error when predicting the consequence of an agent's action in the environment. The aforementioned Intrinsic Curiosity Module (ICM) proposed in [11] belongs to this family of strategies, and operates by learning a state representation that just models the elements that the agent can control and those elements that can affect him. For this purpose, the intrinsic reward is generated based on the prediction error of the next state in a learned latent space:

$$r_i^{ICM} = ||\widehat{\phi}(s_{t+1}) - \phi(s_{t+1})||_2, \tag{1}$$

---

[1] Depending on the task under consideration, the novelty can be associated to the very last performed action and/or the next state visited by the agent in the trajectory.

where $\phi(\cdot)$ denotes the learned latent space mapping; $\widehat{\phi}(s_{t+1})$ is an estimation taking into account $\phi(s_t)$ and the actual action a_t; s_t is the state visited at time t; and $||\cdot||_2$ stands for the L_2 (Euclidean) norm.

Another approach is the use of RND introduced in [12]. Under this strategy, two identical networks are randomly initialized, where one of the networks takes the role of predictor $\widehat{\phi}$ aiming to mimic the output of the other network – namely, the target $\phi(\cdot)$, whose parameters are fixed after initialization. The reward is generated as an MSE loss between the outputs of both networks:

$$r_i^{RND} = ||\widehat{\phi}(s_{t+1}) - \phi(s_{t+1})||^2. \tag{2}$$

Built upon the idea of ICM, a recent work [8] introduced RIDE to use the same mechanism to learn the state embeddings, but they differ on how exploration bonuses are generated. In RIDE, this bonus is given by the difference between two consecutive states in their latent space:

$$r_i^{RIDE} = ||\phi(s_{t+1}) - \phi(s_t)||_2. \tag{3}$$

With this change, RIDE encourages the agent to perform actions that have an impact on the environment. Moreover, by combining *experiment-* and *episode-level* [18] exploration to avoid the agent going back and forth between a sequence of states, the reward is discounted by the episodic state visitation counts:

$$r_i^{RIDE} = \frac{||\phi(s_{t+1}) - \phi(s_t)||_2}{\sqrt{N_{ep}(s_{t+1})}}, \tag{4}$$

where $N_{ep}(s_{t+1})$ denotes the episodic count of visits of state s_{t+1}. Following this idea of combining two levels of exploration (*experiment-* and *episode-*), the Never-Give-Up (NGU) approach in [9] employs different intrinsic weights set in several parallel agents feeding the same network, which parameterizes each agent by making the neural network subject to the intrinsic coefficient used by each of them. A more aggressive strategy is BeBold/NoveID [19], which goes beyond the boundaries of explored regions which only rewards (intrinsically) the first time the agent visits a given state in an episode. The Fast and Slow intrinsic curiosity in [20] combines local and global exploration by generating two different intrinsic rewards, depending on the quality of the reconstruction of two contexts built from the same state. Furthermore, exploration can be enhanced by adversarially forcing an agent to solve a task in different ways [10,21].

Rather than proposing a new intrinsic generation module, the present work offers a study combining different design choices made in recent solutions and fairly compare them under equal experimental conditions. This being said, other benchmarks/studies have been done in recent times: to begin with, [22] evaluates the performance of different exploration bonuses (pseudo-counts, ICM, RND and noisy networks) in the whole Atari 2600 suite with Rainbow [23]. By contrast, [24] carried out a large-scale study based exclusively on prediction error bonuses (ICM) over 54 environments, where they investigated the efficacy of using different feature learning methods with Proximal Policy Optimization (PPO, [25]).

Our work also connects with [13,14,26], a series of evaluation studies aimed to understand what choices among high- and low-level algorithmic options affect the learning process: as such, the studies in [13,14] focus on-policy deep actor-critic methods (examining different policy losses, architectures and advantage estimators), whereas [26] addresses Adversarial Imitation Learning related decisions (multiple reward functions and observation normalization methods).

Contribution: To the best of our knowledge, there is no prior work that exhaustively evaluates different choices for the implementation of intrinsic motivation strategies. Our study takes a step further by analyzing different weight and scale strategies for the combination of intrinsic and extrinsic rewards, as well as the impact of adopting different neural networks architectures and dimensions. The design choices here evaluated are applicable to any intrinsic curiosity generation module, so that conclusions about which ones are the most suitable given a task and an environment with sparse rewards can be drawn.

3   Methodology of the Study

After reviewing different solutions proposed in the literature to cope with hard exploration issues with IM techniques, we now proceed by describing the methodology adopted in this study to gauge the advantages and drawbacks of design choices that are present in some of them, giving an informed hint of their utility when extrapolated to the rest of IM solutions. The methodology is driven by the pursuit of responses to three research questions (RQ):

- RQ1: Does the use of a static, parametric or adaptive decaying intrinsic coefficient weight β affect the agent's training process?
- RQ2: Which is the impact of using episodic counts to scale the intrinsic bonus? Is it better to use episodic counts than to just consider the first time a given state is visited by the agent?
- RQ3: Is the choice of the neural network architecture crucial for the agent's performance and learning efficiency?

Departing from these questions, the following methodology has been devised:

3.1   RQ1: Varying the Intrinsic Reward Coefficient β

In general, it is not advisable to combine raw extrinsic and intrinsic reward signals directly due to their potentially diverging value scales. Moreover, even if taking values from comparable ranges, the agent could need to grant more importance to exploration than to exploitation at specific periods. In fact, in sparse rewards settings, the explorer role of the agent must be strengthen and enlarged in comparison to the exploitative behaviour to guide the agent by an artificial bonus in the absence of knowledge about the target task. This balance between exploration and exploitation is usually controlled by the intrinsic reward coefficient β, whose value is often tuned manually depending on the environment and task to be accomplished. A priory, this value might be fixed and kept unaltered, or dynamically updated, as is further explained in what follows:

Static β: commonly, the β coefficient is stationary along the whole training. In such cases, we refer to this fixed and default value as β_s. On this basis, diverse fixed intrinsic coefficient values can be used to learn a family of policies with different exploration-exploitation balances, so as to concentrate on maximizing the extrinsic reward (a policy with $\beta = 0$) while maintaining a degree of exploration (rest of policies with $\beta > 0$) [9]. Contrarily to the rest of approaches, when using multiple (fixed) intrinsic coefficients training more than one agent is required.

Dynamic β: to focus on the extrinsic signals provided by the environment, it is interesting to modulate the weight given to the intrinsic rewards generated by the agent in a dynamic fashion. Without loss of generality, in our work we consider two different options: parametric decay and adaptive decay. For the *parametric* decay, the value of β decreases by following a modified sigmoid function, which parametrically controls the smoothness of the decay:

$$\beta_t = A + \frac{K - A}{\left(1 + \exp\left(-16B\left(1 - \frac{t}{F}\right)\right)\right)^{20}} \tag{5}$$

where K is a value proven to deliver a good performance and well balanced trade-off between exploration and exploitation (e.g., the fixed value β_s that one could select under a fixed β strategy); A is the final value of β, which can be defined from K (e.g. $A = K/100$) to reflect that at the end of the learning process, the agent should receive hardly any intrinsic signal bonus; and F denotes the number of frames (= sample, steps) we expect the whole train to have. Moreover, B permits to control the *smoothness* of the progression of β throughout the training (Fig. 1). Note that this parametric decay can also be used to sample different β values for each policy learned by means of the approach with multiple static intrinsic coefficients β, by defining F as the number of agents.

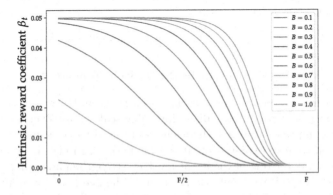

Fig. 1. Example of the parametric decay evolution of β_t for multiple values of the smoothness control parameter B, with $K = 0.05, A = 0.0005$ and $F = 2e7$.

In turn, we can vary the intrinsic coefficient by adopting an `adaptive decay strategy`. Motivated by [27] for concurrent environments, we propose to calculate a decay factor d_i^{τ}[2] based on the ratio between the agent's intrinsic return at the current rollout, G_i^{τ}, and the averaged historical intrinsic return values in past rollouts H^{τ}:

$$\beta_i^{\tau} = \beta_s d_i^{\tau} = \beta_s \min\left[\frac{G_i^{\tau}}{H^{\tau}}, 1\right] = \beta_s \min\left[\frac{G_i^{\tau}}{\frac{1}{K}\sum_{k=0}^{K=i} G_k^{\tau}}, 1\right], \qquad (6)$$

where K is the total number of rollouts the agent collected from the beginning of the training process up to the present rollout i. Consequently, under this rationale the agent is discouraged from exploring those trajectories that are more familiar than the average and means less novelty. Furthermore, the intrinsic return during the training may vary due to the non-stationary nature of the intrinsic reward generation process. Thereby, to stabilize the training, instead of leveraging the whole historical data, we also propose the use of a moving average with a sliding window, H_{ω}^{τ}, which strictly considers just the latest returns and avoids the case of discouraging the exploration due to large initial intrinsic returns that may well bias the decay factor calculation.

3.2   RQ2: Episodic State Counts Versus First-Visit Scaling

As defined in [18], there are different periods in which the exploration mode can be carried out: *step-level, experiment-level, episode-level,* or *intra-episodic.* Over the years the use of *step-level* exploration (i.e. ϵ-greedy) has proven to yield good results in a diversity of simple RL environments. However, advances in learning algorithms have paved the way towards RL problems of higher complexity, in which the exploration is one of the critical parts to be addressed. As has been already argued in the introduction, hard exploration problems can be tackled by letting the agent explore the environment by its inherent satisfaction (intrinsic motivation) rather than being guided by environment provided extrinsic feedback signals. Nevertheless, intrinsic motivation techniques are prone to a quick vanishing of the rewards over the course of the training, reducing attractiveness as the training evolves. This condition is exacerbated when facing long-time horizon problems [20]. Actually, by analyzing the rewards obtained during a concrete episode, few differences in terms of novelty are appreciated between similar/close states, even if one has been already visited and the other remains unexplored. This is due to the persistence of curiosity-related information from past episodes (*experiment-level*), which is propagated forward during the agent's training leaving little novelty difference between similar (even identical) states inside the scope of the same episode. Additionally, in environments where state transitions are reversible, using intrinsic rewards to guide the exploration can lead into an agent bouncing back and forth between sequences of states that are more novel than others in the same episode [8,19].

---

[2] Rollout is denoted as τ, whereas the *i-th* rollout is denoted as τ_i.

As a solution to this issue, recent studies [8, 9, 19, 20, 28] have introduced a visitation count term so that they combine two degrees of novelty rather than just one: local (*episode-level*) and global (*experiment-level*). By virtue of episodic visitation counts, the agent is encouraged to visit as many different states as possible within an episode. However, approaches at the forefront of the state of the art (i.e., ICM, RND) do not implement this idea to scale their rewards. In this context, it is unclear whether new proposed IM modules outperform previous approaches due to state-count regularization or to conceptually new algorithmic schemes. If state-counts regularization contributed to improve the performance, already proposed IM schemes that do not implement it and also future IM methods could adopt this strategy to meliorate their designs. In addition, our experimentation incorporates a more aggressive variation that rewards the agent only when it visits a given state for the first time within the episode [19].

3.3   RQ3: Sensitiveness to the Neural Network Architectures

In the literature related to RL, plenty of network architecture proposals have been used to solve any given problem. As an example, the work in [29] simplified the architectures previously proposed in [8], yet achieving similar results[3]. However, they rely on different base RL algorithms (PPO [25] and IMPALA [30], respectively), thereby hindering a fair comparison, a proper interpretability and attribution of the reported performance results.

To avoid this issue, our specific experimentation evaluates the effect of the network architecture on the performance of the RL agent by considering a fixed RL algorithm and IM module, and by assessing several network configurations. By reporting the dimensions and characteristics of different neural network architectures and the performance of RL agents using them, we can gain intuition about the performance improvement (degradation) incurred when increasing (decreasing) the complexity of the neural architectures in use. Our experiments also measure the required amount of time when using those architectures, so that latency implications can be examined. This third research question is also aligned with practical concerns arising when deciding on which implementation is more suitable for a real-world deployment, specially in resource-constrained scenarios (e.g. embedded robotic devices).

4   Experimental Setup

We answer RQ1, RQ2 and RQ3 over procedurally generated RL tasks from the Minimalistic Gridworld Environment (MiniGrid [31]). This framework allows creating RL tasks of varied levels of difficulty, does not strictly make use of images as observations, and most importantly, runs fast, thereby easing the implementation of massive RL benchmarks.

---

[3] We note that the choice of the neural network architecture is not just for the actor-critic modules, but also for IM approaches that hinge on neural computation.

4.1   Environments

To design a representative benchmark for the study, among all the possible RL environments that can be selected/generated in MiniGrid, we consider 1) those labeled as MultiRoomNXSY (shortened as MNXSY, with X denoting the number of rooms and Y their size), 2) KeyCorridorS3R3 (KS3R3); and 3) ObstructedMaze2D1h (O2D1h). These scenarios belong to hard exploration tasks (i.e., rewards are sparse), in which the agent fails to complete the task without the help of any IM mechanism. Refer to Fig. 2 for further information about each scenario and its associated goal.

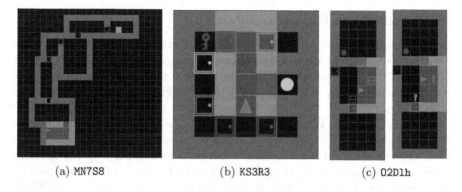

(a) MN7S8   (b) KS3R3   (c) O2D1h

Fig. 2. Examples of MiniGrid scenarios (a) MN7S8: the agent has to open multiple doors to reach the distant goal (green square); (b) KS3R3: the agent has to first collect the blue key in order to open the door of the room leading to the yellow ball that must be picked (c) O2D1h: the agent has to discover keys hidden below some boxes, take the proper key and open the door to the blue ball (target). (Color figure online)

By default, observations in these tasks are essentially egocentric and partially observable views of the environment, where a 7×7 tile set in the direction that the agent is facing composes the observation. Concretely, an observation is featured by a $7 \times 7 \times 3$ matrix, being the 3 features of the last dimension information of interest such as type, colour and status of the object (e.g., doors, keys, balls, or walls) placed in the specific tile. Notice that the agent is incapable to see through walls or doors. 7 basic actions are available to solve all scenarios: **turn left**, **turn right**, **move forward**, **pick up** (an object, for instance keys or balls), **drop** the object (if carried), **toggle** (open doors, interact with objects) and **done**. Nevertheless, some of these actions are only useful at specific locations, whereas others become useless for certain tasks (for instance, **pick/drop** and **done** in MNXSY environments).

Not all the environments require the same amount of steps to be solved. Thus, in MNXSY environments a maximum number of $20 \cdot X$ steps is set to make it dependent on the number of rooms. Consequently, the three considered environments that fall within this set (MN7S4, MN7S8 and MN10S4) are assumed to take at most 140, 140 and 200 steps, respectively. For KS3R3 270 and for O2D1h 576 steps are set as maximum. The rewards are valued according to the number of steps taken. The optimal average extrinsic returns that the agent can achieve are 0.77 (MN7S4), 0.76 (MN10S4), 0.65 (MN7S8), 0.9 (KS3R3), and 0.95 (O2D1h). Actually, since they are procedurally generated environments, each scenario's final reward can slightly change due to the variance on the minimum required steps. In our case, we get these values by taking the median value of an optimal policy (equal to other previous reported optimal results [19]). Moreover, we also refer as suboptimal behavior to those policies that managed to obtain at least a 95% of the optimal score. In terms of complexity, MN7S4 and MN10S4 are the easiest ones to solve, followed by MN7S8 and KS3R3 which are harder. Finally, O2D1h is the most difficult task in our benchmark.

4.2   Baselines and Hyperparameters

All our experiments will employ PPO [25] as the main RL algorithm. On top of it, we will use state-of-the-art IM techniques in order to obtain intrinsic rewards to augment the exploration efficiency, in which a naive PPO model fails [29]: COUNTS[4], RND [12], ICM [11] and RIDE [8]. For PPO we use a discount factor γ equal to 0.99, a clipping factor $\epsilon = 0.2$, 4 epochs per train step and $\lambda = 0.95$ for GAE [32]. We use 16 parallel environments to gather rollouts of size 128. Hence, we set a total horizon of $2,048$ steps between updates. Moreover, a batch size equal to 256 is considered. Unless otherwise specified, the following values - selected from an off-line grid search procedure over MN7S4 - will be used to configure the intrinsic coefficient and entropy: $\beta = 0.05$ and $\varepsilon = 0.0005$ for RND, ICM and RIDE; $\beta = 0.005$ and $\varepsilon = 0.0005$ for COUNTS. In what refers to the dynamic update of β, we select $B = 0.5$ in Expression (5) as it represents a balanced trade-off for the agent to explore in the early stages of the training process, evolving towards a behavior mainly driven by extrinsic signals.

4.3   Network Architectures

Finally, experiments around RQ3 are performed with two different neural network architectural designs, which differ in terms of the type of neural layers (and design) and their number of trainable parameters. Following Fig. 3, on one hand a *lightweight* neural architecture as in RAPID [29] is considered, in which both the actor and the critic are made of 2FC with 64 neurons each. This dual FC-64 architecture also applies to the embedding networks required for RND, ICM and RIDE. Additionally, we include a more sophisticated neural design based on what is proposed in RIDE [8], where both the actor and critic are combined into a two-headed (one for the policy, the other for the critic) shared network with 3 convolutional neural layers (32 3×3 filters, stride equal to 2, and padding 1) and a FC-256 layer. This last architecture will be deemed the *default* architecture to endow the agent with more learning capabilities and to ensure that it is not limited by a restricted network.

---

[4] In this case, we take advantage of the 2D grid (discrete state space) and map each state directly to a dictionary when using COUNTS. Nevertheless, when facing more complex state spaces pseudo-counts [15] can be applied as an alternative as in [22].

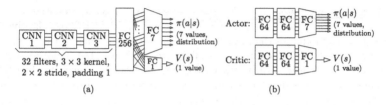

Fig. 3. (a) Sophisticated/default and (b) lightweight network architectures.

5   Results and Analysis

This section is devoted to present experiment results and answer the research questions posed in Sect. 3. Scripts and results have been made available in a public GitHub repository (https://github.com/aklein1995/intrinsic_motivation_techniques_study) to foster reproducibility and follow-up studies. For all the experiments described in this section we provide the mean and standard deviation of the average return computed over the past 100 episodes, performing 3 different runs (each with a different seed) to account for the statistical variability of the results.

RQ1: Does the use of a static, parametric or adaptive decaying intrinsic coefficient weight β affect the agent's training process?

Our first set of results compares the multiple weighting strategies introduced in Sect. 3.1, which differently tune the importance granted to the intrinsic rewards with respect to extrinsic signals coming from the environment. The results are shown in Table 1, where it is straightforward to note that RIDE outperforms COUNTS and RND. At this point we remind the reader that RIDE is configured with episodic count scaling, in accordance with the final solution proposed in [8]. Count-based generated rewards seem to be the best solution when facing easy exploration scenarios (MN7S4 and MN10S4), but its performance degrades when facing scenarios that require more sophisticated exploration strategies. A similar pattern can be observed when analyzing the results of RND, which is unable to solve MN7S8 and O2D1h with any kind of weighting strategy. Contrarily, RIDE manages to solve all the tasks by its basic implementation, although it obtains better results when using more sophisticated weighting exploration strategies.

We now focus the discussion on gaps arising from the use of different weighting strategies. The static (default) weighting strategy (indicated with a suffix _s appended to each approach) is surpassed by any of the other proposed weighting approaches in the majority of the cases. When using multiple static values (_ngu), the only approach that takes advantage of such strategy is RND, yielding worse results for both COUNTS and RIDE in all the cases. This might happen due the slow pace at which the intrinsic rewards values decay in RND in reference to the other strategies. Moreover, the error outputs higher amplitude values than those of RIDE, then being RND a better candidate to get benefit of

Table 1. Results of different IM strategies over several MiniGrid scenarios with static (_s), multiple static (_ngu) (as in NGU [9]), a parametric (_pd) or adaptive decay (_ad) weight β to modulate the importance of the intrinsic bonus in the computation of the reward. Cell values denote the training steps/frames (1e6 scale) at which the optimal average extrinsic return is achieved; in parentheses, steps at which 95% of the optimal average extrinsic return is reached. Best results for every (IM strategy, scenario) combination are highlighted in bold.

	MN7S4	MN10S4	MN7S8	KS3R3	O2D1h
COUNTS_s	**0.93** (0.86)	1.87 (1.78)	> 30	> 30	> 50
COUNTS_ngu	1.17 (1.11)	2.67 (2.35)	> 30	> 30	> 50
COUNTS_pd	0.96 (**0.83**)	2.27 (1.67)	> 30	**22.91 (22.49)**	> 50
COUNTS_ad	1.03 (0.92)	**1.81 (1.65)**	24.23 (24.10)	> 30	> 50
COUNTS_ad1000	1.03 (0.92)	**1.81 (1.65)**	23.63 (23.56)	> 30	> 50
RND_s	3.83 (3.78)	7.84 (7.79)	> 30	10.83 (9.72)	> 50
RND_ngu	2.69 (2.62)	5.78 (5.75)	> 30	8.12 (7.50)	> 50
RND_pd	4.04 (3.94)	6.02 (5.99)	> 30	9.24 (8.07)	> 50
RND_ad	**2.02 (1.39)**	**3.21 (2.65)**	> 30	**6.02 (5.43)**	> 50
RND_ad1000	3.62 (1.42)	3.59 (3.50)	> 30	7.47 (6.66)	> 50
RIDE_s	**2.49** (1.82)	2.27 (2.14)	4.00 (3.68)	6.63 (4.39)	30.88 (25.87)
RIDE_ngu	3.85 (2.40)	2.59 (1.26)	> 30	7.18 (3.91)	36.07 (29.96)
RIDE_pd	5.20 (2.14)	5.01 (1.96)	**3.73 (3.49)**	6.42 (3.87)	29.27 (**20.84**)
RIDE_ad	2.89 (**0.91**)	**1.60 (0.99)**	> 30	5.93 (2.99)	**27.65** (20.91)
RIDE_ad1000	2.54 (**0.91**)	**1.60 (0.99)**	3.88 (3.70)	**4.70 (3.00)**	28.00 (23.01)

applying the _ngu strategy by the use of agents with smaller intrinsic coefficient weights. On the other hand, the use of parametric decay (_pd), which decreases the weight of the intrinsic reward as the train progresses to favor exploration, provides significant gains in almost all simulated scenarios. This approach is similar to _ngu although, instead of using multiple agents with different static intrinsic coefficients, it modulates a single value during the course of training. Hence, when employing _pd strategy, COUNTS is able to get a valid solution in KS3R3, RND improves all its scores and RIDE improves its behaviour in the most challenging scenarios MN7S8, KS3R3 and O2D1h. Nevertheless, _ngu and _pd highly depend on the intrinsic coefficients given to each agent and the evolution of a single intrinsic coefficient during training, respectively. This strongly impacts on the agent's performance for a given scenario and dictates when those approaches might be better.

Finally, the use of adaptive decay (_ad) produces better results in COUNTS and RND when compared to the static case (_s). For RIDE, however, this statement does not strictly hold true, as its performance degrades in MN7S4 and MN7S8 (the agent does not even solve the task in the latter). We hypothesize that this is because the initial intrinsic returns are too high and calculating the historical average intrinsic returns biases the decay factor calculation. As outlined in Sect. 3.2, a workaround to bypass this issue is to calculate returns with a moving average over a window of ω steps/rollouts. We hence include in the benchmark an adaptive decay with a window size of $\omega = 1000$ rollouts (_ad1000). With this modification, RIDE improves its behavior in all the complex scenarios. Nevertheless, _ad1000 performs slightly worse than _ad in RND, but never worse than its static counterpart _s. In general, _ad1000 promotes higher intrinsic coefficient values than _ad, as the calculated average return is better fit to the actual return values. This leads to a lower decay value and a higher intrinsic coefficient, forcing the agent to explore more intensely than with _ad (but less than with _s).

RQ2: Which is the impact of using episodic counts to scale the intrinsic bonus? Is it better to use episodic counts than to just consider the first time a given state is visited by the agent?

Answers to this second question can be inferred from the results of Table 2. A first glance at this table reveals that the use of episodic counts or first-time visitation strategies for scaling the generated intrinsic rewards leads to better results. In the most challenging environments (MNS78, KS3R3 and O2D1h), these differences are even wider, as they require a more intense and efficient exploration by the agent. In fact, when the training stage is extended to cope with the resolution of a more complex task, intrinsic rewards also decrease, inducing a lower explorative behaviour in the agent the more the train is lengthened. What is more, the agent is not encouraged to collect/visit as many different states as possible. Hence, in those scenarios the baseline implementation of intrinsic motivation (_noep) may fail, but with these scaling strategies the problem is resolved (i.e. COUNTS and RND in O2D1h). By contrast, in environments requiring less exploration (MN7S4 and MN10S4), differences are narrower when using *episode-level* exploration and may be counterproductive in some cases (i.e. COUNTS at MN10S4 with _1st).

Table 2. Comparison of different IM strategies when using no scaling (_noep), episodic (_ep) or first-time visit (_1st) to scale the generated intrinsic reward and combine two types of exploration degrees. Interpretation as in Table 1.

	MN7S4	MN10S4	MN7S8	KS3R3	O2D1h
COUNTS_noep	0.93 (0.86)	1.87 (1.78)	> 30	> 30	> 50
COUNTS_ep	**0.76 (0.56)**	**1.55 (1.47)**	2.77 (2.56)	3.99 (2.00)	**33.17 (29.79)**
COUNTS_1st	0.85 **(0.48)**	> 20	**1.64 (1.42)**	1.97 (1.19)	45.26 (37.29)
RND_noep	3.83 (3.78)	7.84 (7.79)	> 30	10.83 (9.72)	> 50
RND_ep	1.41 (0.96)	1.72 (1.34)	3.60 (3.30)	**4.31** (2.63)	**18.54** (14.07)
RND_1st	**1.18 (0.59)**	**1.36 (0.78)**	1.97 (1.72)	4.78 **(2.29)**	21.19 **(9.88)**
RIDE_noep	4.71 (4.54)	5.29 (5.20)	> 30	11.44 (9.63)	39.68 (35.15)
RIDE_ep	**2.49** (1.82)	**2.27 (2.14)**	4.00 (3.68)	6.63 (4.39)	**30.88 (25.87)**
RIDE_1st	3.17 **(1.34)**	3.27 (2.33)	**1.95 (1.83)**	5.13 **(2.26)**	32.14 (28.03)
ICM_noep	2.67 (2.55)	> 20	> 30	8.02 (6.75)	34.04 (26.78)
ICM_ep	3.25 (1.26)	**1.68** (1.59)	> 30	5.32 (3.14)	**19.05** (13.87)
ICM_1st	**1.56 (0.87)**	1.90 **(1.07)**	**2.11 (1.77)**	4.72 (4.23)	20.74 **(10.09)**

To better understand the superiority of RIDE over ICM [8], we also evaluate the performance of both approaches under equal conditions, with (_ep, _1st) and without (_noep) scaling strategies. In this way, we can examine the actual improvement between the two types of exploration bonus strategies. Surprisingly, ICM gives better results in almost all the cases for the analyzed scenarios, yet exhibiting a larger variance in several environments that lead to failure (MN10S4,MN7S8). The reason might lie in how RIDE encourages the agent to perform actions that affect the environment forcing the agent to assess all possible actions, so that the entropy in the policy distribution decays slowly. This hypothesis is buttressed by the results obtained in MN7S4 and MN10S4: we recall that there are 3 useless actions in these scenarios (pick up, drop and done), and RIDE performs clearly worse (except for the _ep case in MN7S4). In complex scenarios, when those actions are relevant for the task, performance gaps between RIDE and ICM become narrower.

Finally, for the sake of completeness in the results exposed in RQ1 and RQ2, Fig. 4 shows the training convergence plots of COUNTS, RND and RIDE for different weighting and scaling strategies. These plotted curves permit to visually analyse the performance during the training process.

RQ3: Is the choice of the neural network architecture crucial for the agent's performance and learning efficiency?

One of the most tedious parts when implementing an algorithm is to determine which network architectures to use. First of all, when using an actor-critic RL framework it is necessary to establish whether a single but two-headed network or two different (and independent) networks will be adopted for the actor and the critic modules. In addition, some IM approaches are based on neural networks to generate the intrinsic rewards. In this work we evaluate two of those solutions: RND and RIDE, evaluating the contribution of different neural network architectures to the overall performance of the agent. We use similar architectures to the ones used in RIDE [8] and RAPID [29][5]: (1) a two-headed shared actor-critic network built upon convolutional and dense layers and (2) two independent MLP networks for the actor and the critic, respectively (check Sect. 3.3 for more information). Moreover, we fix the RL algorithm (PPO) and detail the number of parameters and time taken for the forward and backward passes in each network for an informed comparison.

Table 3 informs about these details of the neural architectures in use for COUNTS, RND and RIDE. The table reports the differences in terms of the number of parameters of each network, and the latency taken by the sum of both forward and backward passes through those IM modules (we note that COUNTS uses a dictionary and not a neural network). In addition, we summarize the total number of parameters depending on the IM module that has been implemented, together with the actor-critic parameters. Referred to the

---

[5] Even with different neural architectures and base RL algorithms, they successfully solve the same tasks in MiniGrid.

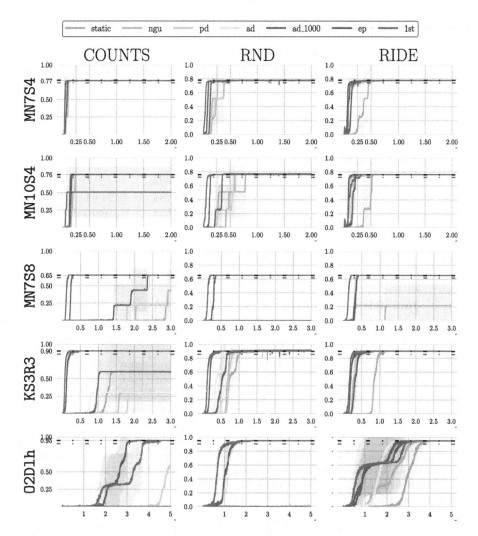

Fig. 4. Convergence plots of the schemes reported in Tables 1 and 2. Each column represents a Intrinsic Motivation type (COUNTS, RND and RIDE from left to right); each row represents the different scenarios (MN7S4, MN10S4, MN7S8, KS3R3 and O2D1h, from top to bottom). All figures depict the average extrinsic return as a function of the number of training steps/frames (in a scale of 1e7). For each scenario, optimal and suboptimal scores are highlighted with horizontal black and brown lines, respectively.

total elapsed time, we report the total amount of time required for a rollout collection. This elapsed time takes into account both the forward and backward passes in the IM modules, and just the forward pass across the actor-critic, among other operations executed when collecting samples. Times are calculated when executing the experiments over an Intel(R) Xeon(R) CPU E3-1505M v6 processor running at 3.00 GHz. The performance of the agent configured with these network configurations is shown in Table 4.

Table 3. Comparison between the network architectures described in Sect. 3.3.

	Lightweight (*lw*)		Default	
	Parameters	Time (ms)	Parameters	Time (ms)
Actor	14,087	-	-	
Critic	13,697	-	-	
Actor+Critic	27,784	-	29,896	-
Dictionary	-	83.66	-	95.11
Total COUNTS	27,784	724.25	29,896	937.37
Embedding	13,632	-	19,392	-
RND	27,264	336.39	38,784	721.64
Total RND	55,048	986.13	68,937	1,408.42
Inverse	12,871	-	18,439	-
Forward	12,928	-	18,464	-
Embedding	13,632	-	19,392	-
RIDE	39,431	388.84	56,295	844.43
Total RIDE	67,215	1,177.75	86,191	1,791.70

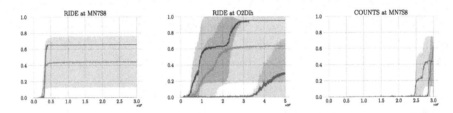

Fig. 5. Convergence plots of COUNTS and RIDE for some scenarios when using the default network (blue), *lw_im* (green) and *lw_tot* (red). All the figures depict the average extrinsic return as a function of the number of training frames. (Color figure online)

Several observations can be drawn by inspecting these tables. On one hand, when reducing the number of parameters in both the actor-critic and the IM modules (*lw_tot*), the agent's behaviour is critically deteriorated. This occurs even with COUNTS (Table 4), where the modification should have had less impact as the generation of intrinsic rewards does not depend on a neural network, but on a dictionary. When inspecting the performance of RIDE, its performance gets worse in all cases except for MN7S4, where the exploration requirements are the lowest among all the analyzed scenarios. Consequently, this modification can be less harmful. As for RND, the lightweight configuration of the networks makes the tasks not solvable by the agent.

The number of parameters to be learned is mostly dependent on the IM networks under consideration, whereas joining the actor and the critic into a single

Table 4. Performance obtained with Counts, RND and RIDE when 1) using the default network configurations, 2) a lightweight architecture for the IM modules and keeping actor-critic with a default configuration (_lw_im), and 3) when both the IM and the actor-critic modules are implemented with the lightweight networks (_lw_tot). Values in the cells represent the training steps/frames (in a scale of 1e6) when the optimal average extrinsic return is achieved. Within brackets, the training steps when a suboptimal behavior is accomplished.

	MN7S4	MN10S4	MN7S8	KS3R3	O2D1h
COUNTS	0.93 (0.86)	1.87 (1.78)	> 30	> 30	> 50
COUNTS_lw_im	0.93 (0.86)	1.87 (1.78)	> 30	> 30	> 50
COUNTS_lw_tot	1.64 (1.48)	2.52 (2.36)	> 30 (29.96)	> 30	> 50
RND	3.86 (3.79)	7.84 (7.79)	> 30	10.84 (9.72)	> 50
RND_lw_im	5.66 (5.44)	6.68 (6.61)	> 30	10.97 (9.45)	> 50
RND_lw_tot	> 20	> 20	> 30	> 30	> 50
RIDE	2.49 (1.82)	2.27 (2.14)	4.01 (3.38)	6.63 (4.39)	30.88 (25.87)
RIDE_lw_im	1.63 (1.31)	1.75 (1.53)	> 30	9.44 (5.08)	> 50
RIDE_lw_tot	1.42 (1.05)	> 20	> 30	8.00 (5.69)	> 50

two-headed network barely increases the dimensionality requirements[6]. Nevertheless, the time required to perform a forward pass increases in approximately 25% when an unique actor-critic network is employed (Table 3). Moreover, by using a single network, part of the parameters of the network are shared between the actor and the critic, which can induce more instabilities but also a faster learning (as the model may share features between the actor and the critic and require less samples to learn a given task). With this in mind, we carry out an additional ablation study considering only the reduction of parameters at IM modules, and maintaining the actor-critic as a single two-head network. Such results are provided in the 2nd row of every group of results in Table 4 (_lw_im).

These results evince that when using RND_lw_im, slightly worse results are achieved with respect to RND with the default network setup. However, its performance does not degrade dramatically down to failure as with RND_lw_tot. Hence, using parameter sharing in a single network yields a faster learning process for this case. Regarding RIDE_lw_im, in some cases (MN7S4 and MN10S4) it attains better results, whereas in MN7S8 and KS3R3 it suffers from a notorious performance decay (MN7S8 is not solved). It can also be observed that the use of the single actor-critic network might be beneficial when reducing the complexity of the IM network (_lw_im), as it mitigates the performance degradation in 3 out of 5 scenarios (yet MN7S8 and O2D1h are not solved) when compared to separated actor-critic networks (_lw_tot), which fail over MN7S8, O2D1h and MN10S4.

Finally, we include Fig. 5 in order to help the reader draw deeper conclusions and gain insights about the behaviour of the learning process. It can be seen

---

[6] We note that the number of parameters is slightly increased, but they also differ in the type of layers that are used in each network (the two-headed network uses CNNs while the independent actor-critic only uses dense layers).

that in the two cases in which RIDE_*lw_im* failed (namely, MN7S8 and O2D1h), in two out of the three experiments that were run (*seeds*) the agent learned to solve the task, which underscores the impact of using different actor-critic architectures. Moreover, with the default actor-critic approach and using the COUNTS approach, the agent is also able to solve the MN7S8 task in two out of the three runs. When using COUNTS_*lw_tot*, the agent reaches suboptimal performance and almost the optimal one within the frame budget.

6   Conclusion

In this work we have studied the actual impact of selecting different design choices when implementing IM solutions. More concretely, we have evaluated multiple weighting strategies to give different importance when combining the intrinsic and extrinsic rewards. Moreover, we have analysed the effect of applying distinct exploration degree levels along with the influence of the complexity of the network architectures on the performance of both actor-critic and IM modules. To conduct the study we have utilized environments belonging to Mini-iGrid as benchmark to test the quality of proposed schemes in a variety of tasks demanding from hard to very hard intensity of exploratory behaviour.

On one hand, we have shown that using a static intrinsic coefficient might not be the best strategy if we focus on sample-efficiency. Adaptive decay strategies have proved to be the most promising ones, although they require a good parameterisation of the sliding window. Parameter decay approach, in turn, have performed competently but it is subject to a decay parametrisation which could be more dependent of the task at hand than the previous scheme, which makes this strategy more sensitive to the environment and the task to be solved (as it happens with ϵ−greedy strategies in Q-learning). The use of multiple agents as in NGU [9], each featuring a different exploration-exploitation balance also suffers from this parametrisation but reports worse results.

On the other hand, the use of *episode-level* exploration along with *experiment-level* strategies seem to be preferable when having environments with hard exploration requirements. It is not a clear winner/preference between episodic counts and first visitation strategies as their performance is not only subject to the environment, but also to the selected IM strategy, although both achieve significant improvement in the performance. Hence, we encourage the implementation of any of these strategies in follow-up IM-related studies.

Last but not least, we have analyzed the impact of modifying the neural network architecture in both the actor-critic and IM modules. The results show that reducing the number of parameters at the IM modules deteriorate the performance of the agent, making it fail in some challenging scenarios which are feasible for the complex neural configuration. What is more, when reducing the IM network dimensions, it is preferable to use a shared two-headed actor-critic as it provides better results, although it is not clear whether those results are due to the use of a single neural network (and the underlying parameter sharing and common feature space for the actor and the critic) or to the adoption of different architectures (e.g. CNNs). Further research is necessary in this direction.

We hope this work can guide readers in the implementation of intrinsic motivation strategies to address tasks with (1) a lack of dense reward functions or (2) at hard exploration scenarios where the classic exploration techniques are insufficient. Aligned with the purposes of academic and industry communities, we make all the experiments available and provide the code to ensure reproducibility [3]. In the future, we will intent to extend these analysis to more environments and algorithms in order to have more representative results.

Acknowledgments. A. Andres and J. Del Ser would like to thank the Basque Government for its funding support through the research group MATHMODE (T1294-19) and the BIKAINTEK PhD support program.

References

1. Silver, D., et al.: Mastering the game of go without human knowledge. Nature **550**(7676), 354–359 (2017)
2. Baker, B., et al.: Emergent tool use from multi-agent autocurricula. arXiv:1909.07528 (2019)
3. Holzinger, A.: Introduction to machine learning & knowledge extraction (make). Mach. Learn. Knowl. Extr. **1**(1), 1–20 (2019)
4. Aubret, A., Matignon, L., Hassas, S.: A survey on intrinsic motivation in reinforcement learning. arXiv:1908.06976 (2019)
5. Ho, J., Ermon, S.: Generative adversarial imitation learning. In: Advances in Neural Information Processing Systems, vol. 29 (2016)
6. Finn, C., Levine, S., Abbeel, P.: Guided cost learning: deep inverse optimal control via policy optimization (2016)
7. Grigorescu, D.: Curiosity, intrinsic motivation and the pleasure of knowledge. J. Educ. Sci. Psychol. **10**(1) (2020)
8. Raileanu, R., Rocktäschel, T.: Ride: rewarding impact-driven exploration for procedurally-generated environments. arXiv:2002.12292 (2020)
9. Badia, A.P., et al.: Never give up: learning directed exploration strategies. arXiv:2002.06038 (2020)
10. Flet-Berliac, Y., Ferret, J., Pietquin, O., Preux, P., Geist, M.: Adversarially guided actor-critic. arXiv:2102.04376 (2021)
11. Pathak, D., Agrawal, P., Efros, A.A., Darrell, T.: Curiosity-driven exploration by self-supervised prediction. In: International Conference on Machine Learning, pp. 2778–2787 (2017)
12. Burda, Y., Edwards, H., Storkey, A., Klimov, O.: Exploration by random network distillation. arXiv:1810.12894 (2018)
13. Andrychowicz, M., et al.: What matters in on-policy reinforcement learning? A large-scale empirical study. arXiv:2006.05990 (2020)
14. Andrychowicz, M., et al.: What matters for on-policy deep actor-critic methods? A large-scale study. In: International Conference on Learning Representations (2020)
15. Bellemare, M., Srinivasan, S., Ostrovski, G., Schaul, T., Saxton, D., Munos, R.: Unifying count-based exploration and intrinsic motivation. In: Advances in Neural Information Processing Systems, vol. 29 (2016)
16. Tang, H., et al.: # exploration: a study of count-based exploration for deep reinforcement learning. In: Advances in Neural Information Processing Systems, pp. 2753–2762 (2017)

17. Machado, M.C., Bellemare, M.G., Bowling, M.: Count-based exploration with the successor representation. In: AAAI Conference on Artificial Intelligence, vol. 34, no. 4, pp. 5125–5133 (2020)
18. Pîslar, M., Szepesvari, D., Ostrovski, G., Borsa, D., Schaul, T.: When should agents explore? arXiv:2108.11811 (2021)
19. Zhang, T., et al.: NovelD: a simple yet effective exploration criterion. In: Advances in Neural Information Processing Systems, vol. 34 (2021)
20. Bougie, N., Ichise, R.: Fast and slow curiosity for high-level exploration in reinforcement learning. Appl. Intell. **51**(2), 1086–1107 (2020). https://doi.org/10.1007/s10489-020-01849-3
21. Campero, A., Raileanu, R., Küttler, H., Tenenbaum, J.B., Rocktäschel, T., Grefenstette, E.: Learning with amigo: adversarially motivated intrinsic goals. arXiv:2006.12122 (2020)
22. Taiga, A.A., Fedus, W., Machado, M.C., Courville, A., Bellemare, M.G.: On bonus-based exploration methods in the arcade learning environment. arXiv:2109.11052 (2021)
23. Hessel, M., et al.: Rainbow: combining improvements in deep reinforcement learning. In: AAAI Conference on Artificial Intelligence (2018)
24. Burda, Y., Edwards, H., Pathak, D., Storkey, A., Darrell, T., Efros, A.A.: Large-scale study of curiosity-driven learning. In: ICLR (2019)
25. Schulman, J., Wolski, F., Dhariwal, P., Radford, A., Klimov, O.: Proximal policy optimization algorithms. arXiv:1707.06347 (2017)
26. Orsini, M., et al.: What matters for adversarial imitation learning? In: Advances in Neural Information Processing Systems, vol. 34 (2021)
27. Jing, X., et al.: Divide and explore: multi-agent separate exploration with shared intrinsic motivations (2022)
28. Seurin, M., Strub, F., Preux, P., Pietquin, O.: Don't do what doesn't matter: intrinsic motivation with action usefulness. arXiv:2105.09992 (2021)
29. Zha, D., Ma, W., Yuan, L., Hu, X., Liu, J.: Rank the episodes: a simple approach for exploration in procedurally-generated environments. arXiv:2101.08152 (2021)
30. Espeholt, L., et al.: IMPALA: scalable distributed deep-RL with importance weighted actor-learner architectures. In: International Conference on Machine Learning, pp. 1407–1416 (2018)
31. Chevalier-Boisvert, M., Willems, L., Pal, S.: Minimalistic gridworld environment for OpenAI gym. http://github.com/maximecb/gym-minigrid (2018)
32. Schulman, J., Moritz, P., Levine, S., Jordan, M., Abbeel, P.: High-dimensional continuous control using generalized advantage estimation. arXiv:1506.02438 (2015)

Towards Generating Financial Reports from Tabular Data Using Transformers

Clayton Leroy Chapman[1], Lars Hillebrand[1,2], Marc Robin Stenzel[2], Tobias Deußer[1,2], David Biesner[1,2(✉)], Christian Bauckhage[1,2], and Rafet Sifa[2]

[1] University of Bonn, Bonn, Germany
claytonleroy.chapman@iais.fraunhofer.de
[2] Fraunhofer IAIS, Sankt Augustin, Germany
{lars.hillebrand,marcrobin.stenzel,tobias.deuer,david.biesner,
christian.bauckhage,rafet.sifa}@iais.fraunhofer.de

Abstract. Financial reports are commonplace in the business world, but are long and tedious to produce. These reports mostly consist of tables with written sections describing these tables. Automating the process of creating these reports, even partially has the potential to save a company time and resources that could be spent on more creative tasks. Some software exists which uses conditional statements and sentence templates to generate the written sections. This solution lacks creativity and innovation when compared to recent advancements in NLP and deep learning. We instead implement a transformer network to solve the task of generating this text. By generating matching pairs between tables and sentences found in financial documents, we created a dataset for our transformer. We were able to achieve promising results, with the final model reaching a BLEU score of 63.3. Generated sentences are natural, grammatically correct and mostly faithful to the information found in the tables.

Keywords: Transformers · Data-to-text · Text mining

1 Introduction

A big part of running a successful business is record keeping. There are many different areas that must be overseen and documented, but arguably one of the most important involves finances. Finance is a broad term, but can be mostly summarized as anything having to do with the creation and management of capital or money. Even a small company can have hundreds or even thousands of records involving finances, so it makes sense that this information should be available in a presentable, easily understandable format. This is formally known

Part of this research is supported by the Competence Center for Machine Learning Rhine Ruhr (ML2R) which is funded by the Federal Ministry of Education and Research of Germany (grant no. 01—S18038B).

A. Holzinger et al. (Eds.): CD-MAKE 2022, LNCS 13480, pp. 221–232, 2022.
https://doi.org/10.1007/978-3-031-14463-9_14

as a financial report. Such reports are necessary when discussing the state of a company, and allow investors and stakeholders to make more informed decisions. These lengthy documents contain financial data for the company from current and previous years, and often include balance sheets, income or revenue, and cash flow.

To make sense of these numbers and tables, the reports also contain written sections. The written sections can contain detailed information about the company, decisions made by management, or relevant changes since the last report. Scattered within the written sections are also summaries and explanations of the various tables and spreadsheets. While these texts are usually similar from year to year, from company to company, they still require a considerable amount of time to be addressed. As they say in business: "time is money", and creating a financial report can be costly for a company. Automating this process, even partially, has the potential to reduce the amount of man hours spent on written sections of these reports each quarter/year, thus saving the company money. There are, of course, a few different ways to go about automating this.

Recently, a new type of model has piqued the interest of many in the Natural Language Processing (NLP) community, known as the Transformer Network. This network is similar to RNNs in that it uses a sort of "memory" and can work with sequential data. It excels at sequence-to-sequence and text generation tasks, among other things. Such a network could be used by transforming table data into an input sequence, and generating a sequence of text as its output, making it an ideal candidate for generating text in financial reports (see a depicted example in Fig. 1). This work will go on to describe a brief history and explanation of Transformer Networks, as well as to determine the legitimacy of this method as a way to generate text for financial documents.

2   Background

2.1   Financial Reports

Financial reports or financial statements are formal records kept by either a business, person or other similar entity that serve to document the financial activities of the company. These records contain pertinent information for existing and potential investors and shareholders, and as such must be organized and easy to understand. For the most part, these documents typically have 4 sections which contain various information from a given time period:

1. The balance sheet, which details the company's assets, liabilities, and equity.
2. The income statement, which lists the company's income, expenses, and profits, as well as losses.
3. The statement of equity, which reports on the changes in equity: the value of shares issued by the company.
4. The cash flow statement, which reports on the cash flow of the company: the amount of money being transferred into or out of the company. This also includes operating, investing, and or financing activities of the company.

2.2   Transformer Networks

The Transformer Network [13] is a deep learning architecture that utilizes multi-headed attention mechanisms, allowing it to process sequential data in parallel

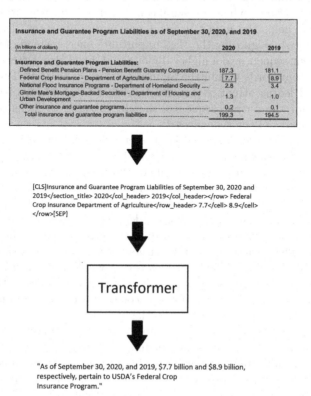

Fig. 1. A visual representation of the working pipeline.

and has theoretically infinitely long windows of memory. The transformer has had great success in the fields of NLP, NLG, and more recently even in CV. It is slowly being recognized as a successor to RNNs and LSTMs, with the speculation that they could even one day succeed the CNN [5]. Since then, several successful transformer architectures have emerged and dominated the field of NLP. BERT from Google is one of the more popular architectures and has achieved state-of-the-art results on several NLP benchmarks [4]. T5, also from Google, has proven itself to be an incredibly capable general-purpose architecture. A single T5 model can be used in place of several different models, where each model is trained on a specific task. T5 sometimes performs almost as well as humans for certain tasks [12].

The basic architecture of a transformer consists of an encoder and a decoder. Each are made up of 6 encoder and 6 decoder layers, respectively, and together they follow the Seq2Seq approach. Seq2Seq models transform an input sequence into an output sequence, which works well for machine translation tasks. The job of the encoder is to map the input sequence into a higher dimensional space, while the decoder transforms this abstract vector into the output sequence. Traditionally, RNN layers were used to encode and decode this information. This means, however, that all features must be sequentially encoded, then decoded. Long sentences and training times remain a problem. To differentiate among the features and determine which are significant within a sequence, attention is used instead, which helps the model to focus on "more important" features within the input data while ignoring the "less important" ones. Instead of transforming the entire input sequence to a higher-dimensional, fixed-size vector, a list of context vectors from words within the input sequence are created. The model then computes and learns attention weights by determining which of these vectors are relevant for the output sequence, thereby learning to focus only on more important features of the sequences [2].

The transformer uses a special kind of encoding called positional encoding. Recall that a transformer is not recurrent but can still work with sequential data. Removing recurrency is what allows the transformer to use stacked layers which can more or less work independently of one another. However, the relative position of data within a sequence must be retained, especially when working with natural languages. Word order can completely change the meaning of the text, for example "The children eat chicken" vs. "The chicken eats children". In order to accomplish this without using recurrency, the transformer uses positional encodings. The basic idea is that when each token is encoded, an additional bit of information explaining its position is encoded with it.

Taking a deeper look inside of the encoder and decoder layers of a transformer, we see that a special type of attention mechanism, known as multi-headed self-attention, is employed. This gives transformers a certain edge in NLP as they are able to extract context and syntactic functions from sentences. Self-attention also receives the advantage over attention because a self-attention layer now acts as the encoder or decoder instead of RNNs in the previous example, allowing for longer sequences and a larger reference window of memory.

$$\text{Attention}(Q, K, V) = \text{softmax}\left(\frac{QK^T}{\sqrt{D_k}}\right) V \qquad (1)$$

3   Related and Previous Work

For many years now, text generation has been a popular use for RNNs [6]. Advancements in the model architecture led to GRUs [3] and LSTMs [7] which helped address the problems that RNN networks had. Recently, a new type of model has emerged that is starting to dominate the field of NLP, known as the transformer network [13]. Transformers were already being used in a few papers to generate text from table datasets. One popular dataset is called ToTTo [10]. The dataset consists of 120,000 samples of tabular data from Wikipedia articles with an accompanying sentence. Each sample has a list of "highlighted" cells from the table, from which the sentence is based on. Another popular dataset is called ROTOWIRE [14]. This dataset contains 4,853 samples of tabular data from basketball games with a written summary of the game. The papers that use the ToTTo and Rotowire datasets have shown that transformers are an appropriate solution to generating text from tables.

In terms of previous work regarding financial text generation, the master thesis of A. Khatipova [8] has been a large motivation for this work. In her work, she attempted to solve the same problem of generating financial documents using tables. Her work began at simple Markov Chains and ended with using RNNs to generate sentences that one might find within a financial report. Her solution was to generate sentences from tables using rule-based generation and then train the network on these generated sentences. The RNN would then only be able to generate text given a starting sequence. Another interesting detail is that all numbers were replaced with <NUM> tokens. <UNK> tokens were also used, representing "unknown" words that appeared less than a certain N number of times, meaning that they were not present in the model's vocabulary. Both of these tokens were a result of the memory constraints of RNNs. The inclusion of these tokens also meant that postprocessing was needed to replace these tokens with reasonable values.

4   Implementation

Originally, we wanted to continue with the work done by [8], which involved generating text from financial documents using RNNs and Markov Chains. Working with this idea and data as a starting point, our goal was to generate text directly from tables found within reports.

Using pre-trained transformers from HuggingFace [1] and datasets similar to this problem (ToTTo and ROTOWIRE), an initial pipeline would be coded that could receive table data and produce text similar to those written by a person. Our dataset is based on ToTTo: The input data would consist of "highlighted" cells, their associated column and row headers, the name of the table and similar metadata, and in some instances the name of the subtable where these cells were located. The output data is a caption or sentence that describes only the highlighted cells from the table.

The documents at hand had already been parsed into a JSON format and contained Blobs of different types. The JSON data is parsed again and the Blobs without the type "text", and "table" are removed. The intention would then be to take these Blobs and try and match each text with a table that contains the same or similar words, as well as numbers, percentages, and dates.

Each text Blob is tokenized and iterated through. For each token, all of the tables are parsed and checked whether the token appears and if so, it is recorded as a match. A counter tracks the matches, and after all tokens are checked, the table with the highest number of matches is paired with the text Blob. Each match represents a highlighted cell. The table is fed to the transformer along with special tokens to separate rows and cells, and mark which cells are highlighted. The transformer being used is Bert2Bert. For each cell, one of the following tokens: </section_title>, </row>, </cell>, </row_header>, or </col_header> is appended to its tokenized value, depending on what type of cell it is. This is to give the transformer a better awareness of how the table was structured in the hope that it can better interpret it.

Text blobs are broken into sentences instead of being left as paragraphs to reduce complexity. The algorithm is adjusted to only check for matches on numbers, specifically currency values, instead of for every token in the sentence. This cuts down on noise and lowers the chance that a sentence will be matched with an incorrect table. Samples are also filtered based on the word count in sentences and the number of unique highlighted cells in tables.

Our final version of this matching algorithm takes further inspiration from the ToTTo paper. Their dataset collection process had 3 different strategies, and their final dataset was randomly sampled from the sets produced from these strategies. Their first strategy was matching tables and sentences with numbers that were at least 3 nonzero digits and not dates. Their second was to match tables and sentences that had more than 3 unique token matches in a single row of a table. Their third was to match sentences and tables that contained respective hyperlinks to each other. We already have a similar strategy to their first one, namely matching currency tokens. Next is to employ a similar strategy to their second one, in order to eliminate as much noise as possible while also looking for the most context-rich pairs.

An upper bound of 30 words is used to preprocess the samples and remove outliers. Next, only unique matches in a row are counted when determining the best table. For the highlighted cell postprocessing, in order for a table-sentence pair to become a sample, the number of unique highlights per row needs to be 2 or more. After preprocessing, the sample size is 181,010, and after postprocessing drops to 35,790.

Most tables in our dataset contain empty cells, which heavily contribute to noise in the input data because they are still being included as </cell> tokens. Other sources of noise present in the table come in the form of irrelevant cells. In the ToTTo paper, they found greater success when limiting the size of the table for their input data. For this reason, only subtables made up of highlighted cells and their column and row headers are used instead of entire tables (Table 1).

5   Results

Table 1. Results after training.

Approach	BLEU	Training time
Bert2Bert (subtables)	63.3	22.7 h
Bert2Bert (full-tables)	29.6	49.0 h
Seq2Seq RNN (512 nodes)	4.7	1.3 h
Seq2Seq RNN (256 nodes)	8.7	1.5 h

The experiments were carried out on a GPU cluster, where enough RAM was available to accommodate a batch size of up to 16. Although our best model

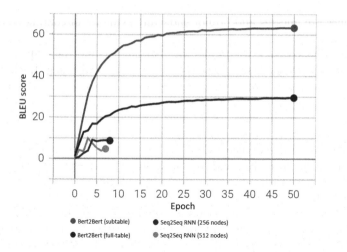

Fig. 2. BLEU scores of validation set during train time.

was trained on subtable samples, another model was trained on full tables for comparison. Our final model was trained on 50 epochs. We decided to use BLEU as our metric, as the BLEU score is a standard metric for text generation and was also used in the ToTTo paper. Final scores are reported as sacrebleu scores in order to ensure consistency across different systems [11] (Fig. 2).

The results from only using subtables were somewhat of a surprise, namely, with the noticeable jump in the model's BLEU score to 63.3 vs. 29.6 when training on full tables. It appeared that using subtables which only included highlighted cells and their headers was enough to produce high-quality sentences. By leaving out the unneeded information, the size of each sample shrunk, and the training time was also noticeably reduced. Sentences generated were of high quality, with correct grammar and sentence structure. There was the occasional hiccup where a word might be incorrectly assumed. An example being "TEUR" instead of "EUR", or a word article will be left out, such as "Ergebnis" instead of "Das Ergebnis".

Some results can be see in Table 2, along with the BLEU score that was achieved. It's clear that certain samples may have been missing important cells in their encoded table when looking at the results with lower BLEU scores. The samples that scored higher had better consistency in terms of their encoded tables. As mentioned, the sentences produced are quite convincing although its clear that a few of them were generated by an algorithm. Impressively, the correct use of articles and grammatical cases are present in most generated sentences. This can be attributed to the pre-training which was carried out on the models, but the way the model retains this information after training is notable. One visible shortcoming is the model's inability to do calculations. In certain samples, the model generates a difference or a sum, however most of these values are guesses and do not show any evidence of coming from any sort of calculation.

It also appeared as if the model was learning the writing style for each company, thanks to the inclusion of the section title special token. Looking through some of the samples, target sentences between companies confirmed there were similarities. Word choice tended to be consistent, and certain sentences followed a similar structure. In some cases, sentences had the same wording except for the year and monetary value. This explains how the model was able to produce diverse and detailed sentences while working with as few as 5 cells. This could also explain why our BLEU score of 63.3 is so high, especially when compared to ToTTo's 44.0. Our task is more focused, with a tighter vocabulary and similarly structured tables. Most tables did not actually vary much, with the majority having only 2 or 3 columns, with years as column headers. In comparison, ToTTo had to deal with a wide range of topics and worked with tables of many different shapes. This is to be expected, as the complexity of our task is comparatively lower than ToTTo.

6   Challenges and Future Work

One of the biggest challenges we faced during this experiment was the quality of our data. We did not have access to the original documents and had to use the already parsed JSONs. Reading through the samples, one can see a mix of partial sentences, words being cut off, as well as numbers and units being incorrectly spaced. While it was possible to filter out or fix some of these issues, it was infeasible to control every single sample. Many samples contained a mix of English and German words, meaning that a spellchecker to remove samples with cut-off words might also remove legitimate samples. This mix of languages also negatively affected our tokenizer. Some words were incorrectly split into multiple tokens, and many of our number values were separated due to commas and decimal points. Fortunately, these tokens still belonged to the same cell and were wrapped by special tokens. Despite this, nearly every date, percent, and number from the generated sentences contained awkward spaces. This can of course easily be remedied by more postprocessing, but we still wonder if this spacing problem has any effect on training.

Inconsistencies within the tables also proved to be quite the challenge for us, and led to some difficulties in our parsing algorithm. There are several different formats used for dates and money within the tables. After trying to manually catch each edge case, we decided to just use a simple regular expression, which assumed that any occurrence of a 4 digit number with the pattern 18XX, 19XX, or 20XX was a date. It would be interesting to see if our results could be improved with a more elegant strategy. We faced a similar problem with currency values. Sometimes the number would have the currency unit inside its cell, other times it would be somewhere else in the table. Luckily, these edge cases were less of an issue to tackle, and the model seemed to do a fine job inferring the difference between "100 Eur" and "100 Mio Eur".

Table 2. Results across various BLEU scores, along with target sentence and encoded table. One can see the hallucination referenced in the results section.

Encoded table	[CLS] RhoenKlinikum-AnnualReport-2015.json </section_title> Zugänge aus Änderung Konsolidierungskreis </row_header> 0,2 </cell> </row> Umbuchungen </row_header> 0,2 </cell> 2,0 </cell> </row> [SEP]
Predicted sentence	2. 2 Mio. EUR (Vj. 2, 2 Mio. EUR)) hat die ML V. 0, 0 Mio. EUR (Vj. 0, 0 Mio. EUR
Target sentence	Auf nicht beherrschende Anteile entfallende Gewinnanteile sind im Vergleich zur Vorjahresperiode transaktionsbedingt um 0, 2 Mio. EUR auf 2, 0 Mio. EUR zurückgegangen
BLEU score	10
Encoded table	[CLS] Commerzbank-QuarterlyReport-2010-Q3.json </section_title> und Sonstige </row_header> 20 </cell> </row> Sonstige </row_header> 9 </cell> 20 </cell> </row> [SEP]
Predicted sentence	Die Risikovorsorge im Privatkundengeschäft lag im dritten Quartal 2010 bei 9 Mio Euro, im Vergleich zu 20 Mio Euro im Vorjahr
Target sentence	Hauptbestandteile des Exposures sind die Teilportfolios Büro (27 Mrd Euro), Handel (20 Mrd Euro) und Wohnimmobilien (9 Mrd Euro)
BLEU score	20
Encoded table	[CLS] FuchsPetrolub-AnnualReport-2013.json </section_title> 2013 </col_header> </row> Europa </row_header> 1.104,2 </cell> 23,5 </cell> </row> [SEP]
Predicted sentence	Europa ist im Jahr 2013 um 23, 5 % bzw. 23, 5 Mio EUR organisch gewachsen und hat sich organisch um 1. 52, 2 Mio EUR bzw
Target sentence	Die europäischen Gesellschaften erwirtschafteten im Berichtsjahr Umsätze von 1. 104, 2 Mio EUR (1. 080, 7), das sind 2, 2 % oder 23, 5 Mio EUR mehr als im Vorjahr
BLEU score	30
Encoded table	[CLS] MTU_Aero_Engines-QuarterlyReport-2014-Q3.json </section_title> 30.9.2014 </col_header> 31.12.2013 </col_header> </row> Summe Eigenkapital und Schulden </row_header> 4.493,1 </cell> 4.410,5 </cell> [SEP]
Predicted sentence	Die Konzernbilanzsumme ist zum 30. September 2014 mit 4. 493, 1 Mio. EUR gegenüber dem 31. Dezember 2013 mit 4. 410, 5 Mio. EUR nahezu unverändert geblieben
Target sentence	Die Konzernbilanzsumme ist um 82, 6 Mio. EUR gestiegen - von 4. 410, 5 Mio. EUR zum 31. Dezember 2013 auf 4. 493, 1 Mio. EUR zum 30. September 2014
BLEU score	70
Encoded table	[CLS] Deutsche_Post-QuarterlyReport-2016-Q1.json </section_title> Q1 2015 </col_header> Q1 2016 </col_header> </row> MEA (MiddleEastandAfrica) </row_header> 254 </cell> 261 </cell> </row> [SEP]
Predicted sentence	In der Region MEA (Middle East and Africa) lag der Umsatz im Berichtszeitraum mit 261 MIO EUR auf dem Niveau des Vorjahres (254 MIO EUR)
Target sentence	In der Region MEA (Middle East and Africa) stieg der Umsatz im Berichtszeitraum um 2, 8 % auf 261 MIO EUR (Vorjahr : 254 MIO EUR)
BLEU score	80
Encoded table	[CLS] STADA-AnnualReport-2011.json </section_title> 2011 </col_header> </row> Italien </row_header> 146,1 </cell> 136,8 </cell> </row> [SEP]
Predicted sentence	In Italien verzeichnete der Konzern im Geschäftsjahr 2011 eine Umsatzerhöhung von 17 % auf 146, 1 Mio. EUR (Vorjahr : 136, 8 Mio. EUR)
Target sentence	In Italien verzeichnete der Konzern im Geschäftsjahr 2011 einen Umsatzanstieg von 7 % auf 146, 1 Mio. EUR (Vorjahr : 136, 8 Mio. EUR)
BLEU score	90

Another observation was model hallucination, which was also mentioned in the ToTTo paper. This problem has also been documented in other table-to-text experiments [9,15]. Similar with the ToTTo dataset, the tables were not unique to the samples, meaning 2 samples could reference the same table. This meant overlapping tables were a possibility. The chance of this happening further increased since we filtered all non-highlighted cells out. We suspect this to be the case, as multiple samples hallucinated words, or context which were not present in the tables' tokens. To try and avoid this from happening, the ToTTo team ran different experiments with and without overlapping tables. Their results showed a nearly 20 point increase in BLEU score when comparing the results from the overlapping table dataset with the non-overlapping table dataset. We assume this would also happen for our experiment if we had taken similar steps to filter out overlapping tables. This observation causes a dilemma. On the one hand, our model generates reasonable results similar to their reference sentences. On the other hand, the results generated are less faithful to their tables. For this reason, perhaps another metric might be more appropriate for this task.

Future work in this area would be greatly benefited from a more consistent dataset. Given more time and resources, target sentences written specifically for samples in this experiment could also be helpful. This would eliminate any shortcomings with the parsing and matching algorithms or possible mismatches between sentences and tables. Most of the samples chosen for the final model only relied on the assumption that "good" samples had multiple unique matches in the same row. It is well within the realm of possibility that many samples which could have been useful were ignored. This also limited much diversity among the samples, because many sentences referencing different rows may not have met the algorithm's criteria for being "good". Having people manually highlight cells and write something insightful for those matches could improve on the robustness of the model. The data could be further improved by having multiple reference sentences for each sample. Additionally, it is common for text generation and translation experiments to have humans rate the quality of generated text alongside automated metrics. This was unfortunately not an option available to us, and BLEU had to instead be solely relied upon. Future work may also include judging the generated sentences by humans. Predicted sentences could be compared directly with reference sentences, and scored based on whether they describe the table's highlighted cells worse, the same, or better than the reference sentence.

7   Conclusion

Financial reports are a necessary part of running a business. As we have seen, a lot of time and effort goes into writing these reports despite their content being somewhat standard. Giving a company the option to automate parts of this process would free up employees' time to work on other essential tasks. There already exists some software which can semi-automate this process but which relies on if-else statements and basic logic. This software must account for all edge cases and outcomes, but the text produced in the end still adheres to the template. There exists a possibility for a more intelligent form of automation.

Previous work using RNNs to try and solve this task improved the variety of sentences generated. This method demonstrated that these deep learning networks could fill this potential niche, but that they also had several shortcomings. RNNs are tedious to train and require a large amount of data, time, and resources. RNNs are plagued by exploding and vanishing gradients. This puts an effective upper bound on the length of the sequences they can process. Being sequential also means that RNNs have limited memory. If a feature must travel through several different cells to where it is needed, the chance that this feature is corrupted or forgotten increases.

Recent research in attention-based models and the emergence of transformer networks has given new hope to many NLP and NLG tasks. The transformer network addresses most of the shortcomings of RNNs. Because of multi-headed self-attention, the transformer can better utilize the GPU. This results in faster training time and more efficient use of resources to train large models. Transformer networks are not affected by exploding or vanishing gradients to the same degree as RNNs. As such, they can work with longer sequences. Through their use of attention, transformer networks have a much larger window of memory when compared to RNNs. Many pre-trained transformer networks are available to download and can be immediately trained on downstream tasks. For these reasons we decided to experiment with using transformers to generate text from tables found in financial reports.

Using the ToTTo paper as a guide, we used the BERT2BERT transformer network. This network uses special tokens to separate sentences within an input sequence, and implementing custom special tokens based on a table was trivial. Our special tokens were: </section_title>, </row>, </cell>, </row_header>, and </col_header>.

Financial reports already parsed as JSON were used for the experiment. Within a JSON, Blobs containing tables and text were collected, and parsing and matching algorithms were used to generate table-text pairs. Each table-text pair kept track of matches between cells and words, and these cells were designated as highlighted. Tables were then broken down into highlighted cells with their respective section titles, and column and row headers. These subtables were tokenized, encoded, and given to the transformer to train. Our final model achieved a BLEU score of 63.3.

The final model demonstrates itself as capable of table to text generation. Generated sentences are mostly grammatically correct and fluent, with word choices appropriate for a financial setting. These sentences also reach reasonable conclusions based on their encoded tables. The model sometimes struggled with hallucinations, an already documented phenomenon from previous table-to-text studies. Such a model could be used in place of a person for partial financial document generation, with additional manual completion and correctness checks.

References

1. Huggingface: The AI community building the future. https://huggingface.co/
2. Bahdanau, D., Cho, K., Bengio, Y.: Neural machine translation by jointly learning to align and translate (2016)
3. Chung, J., Gulcehre, C., Cho, K.H., Bengio, Y.: Empirical evaluation of gated recurrent neural networks on sequence modeling (2014)
4. Devlin, J., Chang, M.-W., Lee, K., Toutanova, K.: Pre-training of deep bidirectional transformers for language understanding, BERT (2019)
5. Dosovitskiy, A., et al.: An image is worth 16×16 words: transformers for image recognition at scale (2021)
6. Graves, A.: Generating sequences with recurrent neural networks (2014)
7. Hochreiter, S., Schmidhuber, J.: Long short-term memory. Neural Comput. **9**(8), 1735–1780 (1997)
8. Khativa, A.: Methods of financial reports generation (2019)
9. Lebret, R., Grangier, D., Auli, M.: Neural text generation from structured data with application to the biography domain (2016)
10. Ankur, P., et al.: A controlled table-to-text generation dataset, Totto (2020)
11. Post, M.: A call for clarity in reporting BLEU scores. In: Proceedings of the Third Conference on Machine Translation: Research Papers, Brussels, Belgium, October 2018, pp. 186–191. Association for Computational Linguistics (2018)
12. Raffel, C., et al.: Exploring the limits of transfer learning with a unified text-to-text transformer (2020)
13. Vaswani, A., et al.: Attention is all you need, Gomez (2017)
14. Wang, H.: Revisiting challenges in data-to-text generation with fact grounding. In: Proceedings of the 12th International Conference on Natural Language Generation (2019)
15. Wiseman, S., Shieber, S.M., Rush, A.M.: Challenges in data-to-document generation (2017)

Evaluating the Performance of SOBEK Text Mining Keyword Extraction Algorithm

Eliseo Reategui[1](✉), Marcio Bigolin[1,2](✉), Michel Carniato[3](✉),
and Rafael Antunes dos Santos[1](✉)

[1] PGIE, UFRGS, Paulo Gama 110, Porto Alegre, RS, Brazil
eliseoreategui@gmail.com, marciobigolin@gmail.com,
rderafa@gmail.com
[2] IFRS, Rua Dra. Maria Zélia Carneiro de Figueiredo, Canoas, RS 870, Brazil
[3] PUCRS, Av. Ipiranga, Porto Alegre, RS 6681, Brazil
michelcarniato@hotmail.com

Abstract. This article presents a validation study of the algorithm implemented in the text mining tool called SOBEK, comparing it with YAKE!', a known unsupervised keyword extraction algorithm. Both algorithms identify keywords from single documents using mainly a statistical method, providing context independent information. The article describes briefly previous uses of SOBEK in the literature, and presents a detailed description of its text mining algorithm. The validation study presented in the paper compares SOBEK with YAKE!. Both systems were used to extract keywords from texts belonging to fourteen public text databases, each containing several documents. In general, their performance was found to be equivalent, with the algorithms outperforming one another in a batch of tests, and reaching similar results in others. Understanding why each algorithm outperformed the other in different circumstances may shed light on the advantages and disadvantages of specific features of keyword extraction methods.

Keywords: Text mining · Keyword extraction · SOBEK · YAKE!

1 Introduction

Text mining is a research field that encompasses different approaches such as information retrieval, natural language processing, information extraction, text summarization, supervised and unsupervised learning, probabilistic methods, text streams and social media mining, opinion mining and sentiment analysis. It aims to retrieve relevant information from semi-structured or nonstructural data (Feldman and Sanger 2006). It's a field in constant development, mostly due to the ever increasing amount of documents found on the Internet. These documents often have to be classified and examined, which can be done through statistical processes and also through the analysis of semantic and syntactic information contained in the texts.

Along the years, text mining applications proliferated to different areas, from patent analyzes (Noh et al. 2015; Tseng et al. 2007), to e-commerce (Pang and Lee 2008;

© IFIP International Federation for Information Processing 2022
Published by Springer Nature Switzerland AG 2022
A. Holzinger et al. (Eds.): CD-MAKE 2022, LNCS 13480, pp. 233–243, 2022.
https://doi.org/10.1007/978-3-031-14463-9_15

Song et al. 2019), to bioinformatics (Krallinger and Valencia 2005; Lamurias and Couto 2019). Other examples include the prediction of stock trading strategies (Sun et al. 2016) and the inspection of diseases and syndromes in neurology reports (Karami et al. 2019).

These applications of text mining often rely on one or more combined processes (Allahyari et al. 2017), such as information retrieval (Marcos-Pablos and García-Peñalvo 2020), natural language processing (Flor and Hao 2021), text summarization (El-Kassaset al. 2021), opinion mining and sentiment analysis (Zvarevashe and Olugbara 2018). In recent years, the introduction of Bidirectional Encoder Representations from Transformers (BERT) (Devlin et al. 2019) also brought significant results related to the learning of word contexts. However, BERT models are based on supervised learning methods, which is out of the scope of the work presented in this paper.

In the field of text mining, a particular task that has been drawing attention is that of unsupervised keyword extraction. Keywords are defined as a sequence of one or more words that provide a compact representation of the content of a document (Rose et al. 2010). They are used in many situations, primarily to allow indexing mechanisms to quickly identify and retrieve documents. Keywords may also improve many natural language, information retrieval and text categorization processes (Hulth and Megyesi 2006). Based on the idea that relevant information is often overlooked by non-proficient readers (Winograd 1984), keyword extraction has also been used to support reading comprehension tasks (Reategui et al. 2020).

Although keywords may be an important element to give people a general idea of text contents and to provide systems with information that may be useful for indexing mechanisms, documents on the web are not always associated with keywords, which sometimes makes it difficult for users and search engines to identify relevant material. Carrying out the task of manually assigning keywords to documents can be not only tedious, but also time-consuming, especially when considering numerous documents.

Although a lot of effort has been made towards the development of unsupervised methods to extract relevant keywords from texts accurately (Hasan and Ng 2014), this is a problem that is still addressed in current research (Firoozeh et al. 2019). The main difficulties related to this problem are language diversity, match restriction, and the high number of candidate keywords that can be generated from a single text (Campos et al. 2020). These are issues that highlight the need for further research to design keyword extraction solutions that work well in different contexts.

In this paper we present a validation study carried out with SOBEK, a text mining tool that extracts keywords from texts using an unsupervised approach. It retrieves recurrent terms and relationships from texts, representing them in the form of a graph. SOBEK has been used in several applications in different research centers and in different contexts, from the extraction of keywords for their use in discourse analysis (Lee et al. 2017) to the analysis of documents (Führ and Bisset Alvarez 2021) and classification of items of a digital library on the History of Science (Bromberg 2018). It has also been used for idea discovery in a pipeline framework designed to unveil the dynamics of the development process of ideas (Lee and Tan 2017). In the field of education, it has been used in the evaluation of students' posts in discussion forums (Azevedo et al. 2014), and the analysis of collaborative writing assignments to support teachers' work (Macedo et al. 2009). Still in the field of education, SOBEK was integrated into other systems to extract terms and relationships from texts and use this information in an intelligent educational system in the area of Industrial Automation (Gonzalez-Gonzalez et al. 2016). Although the tool has been used in several studies in the past, no research to date has shown how SOBEK performs in comparison with other keyword extraction algorithms.

The next section presents SOBEK's text mining algorithm, and Sect. 3 presents a validation study comparing SOBEK's performance with that of YAKE!, an unsupervised keyword extraction algorithm based on statistical methods. We opted to contrast SOBEK's performance with that of YAKE!'s because the latter has been extensively compared to other keyword extraction methods in a recent publication (Campos et al. 2020), in which it showed a superior overall performance. The detailed description of the YAKE key extraction algorithm can also be found in Campos et al. (2020).

2 SOBEK Text Mining

SOBEK is a text mining tool based on an unsupervised algorithm originally designed by Schenker (2003). It identifies keywords from single documents using a statistical method. While other algorithms evaluate the statistically discriminating power of keywords within a given corpus, SOBEK operates on individual documents, providing context independent features.

SOBEK's operation can be divided into three stages. The first one is the identification of keywords. The second step is related to the identification of relationships among those keywords, and the last one concerns the visual representation of the information extracted in the form of a graph.

In the first step in SOBEK's mining algorithm, a text T is split into a set of words W, using spaces and punctuation marks as dividers. The set of words W is then mapped into terms that may consist of a single word (called here a "single term") or many words (called "compound term"). This mapping is a statistical process that considers the frequency τ with which each word is found in the text. When a subset of words $w_n \in W$ is repeated in W (i. e. w_j, w_{j+1} and w_{j+2}} with a frequency equivalent to that of the most frequent single terms, a compound term is formed (e.g. "Global Warming"). Once those compound terms are found, the combination of words that created it are removed from the word list and the words remaining are considered single terms. A word may appear in both a compound term and a single term, as long as SOBEK identifies its appearance in the text individually as well as in the compound term.

In order to identify whether a word should be part of a compound term or if it should figure in the list of single terms, each word $w_n \in W$ is combined with n subsequent words to create a set S of terms:

$$S = \{w_i, w_i \cup w_{i+1}, w_i \cup w_{i+1} \cup w_{i+2}, ..., w_i \cup w_{i+1}... \cup w_{i+n}\}$$

For instance, the sequence of words 'AA BB CC' in a scenario where n = 3 may be used to create the following set of strings: $S = \{$'AA'; 'AA BB'; 'AA BB CC'; 'BB'; 'BB CC'; 'CC'$\}$. Although the value of n could be higher than 3, terms with more than 3 words are not very frequent and the computation required to identify them could not justify the benefits.

Once the process of term identification is finished, the elements in S whose frequency are higher than a minimum value ϕ are selected for further consideration. The frequency ϕ is determined as a threshold considering that the returning set of terms R_c has a minimum size ε. The minimum value accepted for ϕ is 2; otherwise all the words of any text would feature in the resulting graph.

During the process of identification of terms, three functions are used to remove terms and words that do not add information to the graph. The first one is the removal of stop words. These words are mainly articles and prepositions that do not have any specific meaning. The second function is called stemming and it is used to reduce redundancy and remove terms with the same meaning and/or similar spelling. The third function is related to the identification of synonyms by using a thesaurus, which enables SOBEK to further prune terms that have a similar meaning from the resulting graph.

After the identification of terms that have a frequency τ greater than a specified parameter (2 or more), all other words are ignored. Although the number of terms returned from the mining process (ε) can be determined by the user, according to Novak and Cañas (2006), no more than 25 terms should be necessary to identify the central idea of a text. Based on this assumption, SOBEK's default setting is $\varepsilon = 20$. There are several ways to change the value of ε, including selecting arbitrary values for ε or selecting different graph sizes, in which case SOBEK will automatically adjust the value of ε.

SOBEK's second step is to identify relationships between terms. Each term c selected by SOBEK during the mining process belongs to a set C of all terms. A relationship between c_i and c_j implies that there is a connection between them and that they are closely related in T. It could represent different types of relationships, such as cause and consequence, membership, time sequence, or other. A new analysis of the text T relates c_i and c_j when they are no more than z words distant from each other and when there are no full stops between them.

Depending on the size of set C, a term c_i would be related to too many other terms, which would produce a graph in which the connections would have no particular meaning. To reduce the number of connections between terms, a maximum of r links is allowed for any term. However, terms with high frequency do not have the same number of connections as terms with low frequency. Each term may have at most ω connections and this value is proportional to the frequency τ of that term. In this way, more frequent terms have a larger value ω and may be connected to a larger number of terms in the graph. Only the term with the highest τ will have $\omega = r$; the other terms will have at most r times their frequency divided by the frequency of the most frequent term.

$$\omega_{c_i} = \frac{\left(\tau_{c_i} \times r\right)}{max}\{\tau \epsilon C\}$$

In cases when a term could have more relationships than it is allowed to have, the relationships selected to be displayed are those that occur more frequently. There is no lower bound to the number of times a relationship between two terms should occur in the text for it to be considered as a link in the graph. SOBEK uses $z = 5$ and $r = 7$; those specific parameters were defined based on the users' review of the tool. A bigger value for z would link terms that are not always related and a bigger value for r would produce a larger number of connections, which could make their interpretation more difficult.

SOBEK's final step is related to the creation of a graphical representation of terms extracted from the text. In this representation, the terms are represented by nodes and the relationship between them as links between nodes. Figure 1 presents a graphical representation of terms extracted from the Wikipedia text about text mining.

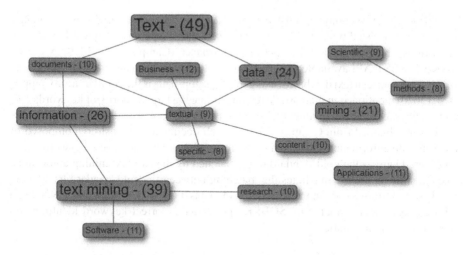

Fig. 1. Graphical representation of terms of the Wikipedia text about text mining.

SOBEK represents terms next to the number of times they appear in the text. More recurrent terms are represented in slightly larger boxes. Relations are represented as lines connecting two terms, denoting a close relationship between them. These relations are identified by tracking the closeness with which terms occur together within the text, an analysis carried out by tracking the occurrence of terms within the same sentence (Allahyari et al. 2017). Although relation extraction is an essential feature, in this article we've focused strictly on the validation of SOBEK keyword extraction algorithm. The next section presents the validation study comparing the performance of SOBEK with that of YAKE!'s by using several text databases, most of them in English and one in Portuguese.

3   Methods

In the field of information retrieval, two metrics are frequently used to evaluate the accuracy of results obtained by retrieval systems: precision and recall. These were the metrics we used to compare SOBEK and YAKE!'s keyword extraction performance.

$$precision = correctK / returnedK$$

$$recall = correctK / totalK$$

where *correctK* is the number of correct keywords returned by the system (SOBEK or YAKE!), *returnedK* is the number of keywords returned, and *totalK* is the total number of keywords considered as correct answers. In addition, the *F-score* measure was computed to combine precision and recall scores, by using the harmonic mean between these values:

$$F\text{-}Score = 2 \times precision \times recall/(precision + recall)$$

The decision to compare SOBEK to YAKE! was because YAKE! was thoroughly tested and contrasted with 10 other unsupervised and 1 supervised keyword extraction algorithms, showing an overall superior performance (Campos et al. 2020). We used YAKE!'s REST API available at DokerHub[1] to test its performance in extracting keywords from fourteen text databases, using the *3-g* setup. This is the configuration Campos et al. (2020) mentioned in their article to concentrate the large majority of keywords correctly identified in the experiments carried out with the tool. The databases used in our work were obtained from Campos (2020), a well documented repository composed of datasets with unformatted texts and their corresponding keywords. These keywords were considered here as the gold standard to be matched by the two text mining algorithms. The example in Table 1 illustrates the matching between the gold standard list of keywords stored for two texts in the Krapivin2009 dataset (files 543016.txt and 609202.txt) with the keywords extracted by SOBEK and YAKE! (correct keyword identification depicted in bold and italic).

Table 1. Comparison of the Gold Standard set of keywords with those extracted by SOBEK and YAKE! from two texts in the Krapivin2009 dataset

Krapivin2009 dataset, file 543016.txt		
GOLD STANDARD	**SOBEK's KEYWORDS**	**YAKE's KEYWORDS**
cue integration, grouping, normalized cut, segmentation, texton, texture	contour, image, lter, pixel. ***segmentation, texton, texture***	computer vision, image, image segmentation, segmentation, ***texture, texture analysis***
Krapivin2009 dataset, file 609202.txt		
GOLD STANDARD	**SOBEK's KEYWORDS**	**YAKE's KEYWORDS**
combining program, data specialization, partial evaluation, program specialization, program transformation	computations, data, ***data specialization,*** program specialized, time	data, ***data specialization,*** program, ***program specialization,*** specialization

---

[1] https://hub.docker.com/r/liaad/yake.

In the first example, SOBEK correctly identified 3 out of the 6 keywords in the gold standard list for text No. 543016, while YAKE! identified only 2. In the second example (text No. 609202), YAKE! identified correctly 2 out of the 5 keywords in the gold standard list, while SOBEK identified only 1.

Our choice of repository has been based on its previous use in testing algorithms for the automatic extraction of keyphrases. SOBEK's web service, configured according to directions presented in its website[2], was used to extract a set of keywords from each text in the selected datasets. The tool was configured to return a number of keywords close to the number of terms of each gold standard list, and to use a tree-tagger in the mining process to prioritize the identification of nouns.

Regarding the accuracy with which the tools identified terms from the gold standard lists, it is important to stress that these gold standard lists had keywords mostly provided by humans to better characterize and index their texts. In this respect, it is somehow understandable that automatic algorithms will not identify exactly the same set of terms, especially when partial matches were not considered in the comparison method.

4 Results

Table 2 shows the computation of the precision, recall and F-score metrics to evaluate the performance of SOBEK and YAKE! in the keyphrase extraction task. Partial matches were not considered when comparing the keyphrases extracted from the tools and the ones originally stored in the datasets. We also used the paired sample *T-test* to check whether the differences between the calculated metrics of SOBEK and YAKE! were significant. In general, it can be said that SOBEK and YAKE! reached a similar level of performance. In some circumstances the methods surpassed one another, in other circumstances they had an equivalent number of correct and incorrect keywords identified. SOBEK surpassed YAKE! in the identification of keywords from 5 datasets (the first 5 lines of the table, ie.: *Pubmed, Citeulike, fao30, Fao780 and Schutz2008*). For these datasets, SOBEK showed a higher F-score measure, attested by the calculation of the paired T-test showing p-values < 0.05 (rightmost column of the table). YAKE!, on the other hand, outperformed SOBEK in the analysis of 5 other datasets, considering the F-score measures and their corresponding T-test results (the following lines of the table, ie.: *Krapivin, Semeval2017, Semeval2010, Inspec, Nguyen2007*). It is interesting to note that among the results in which SOBEK outperformed YAKE, there are only datasets with full articles. YAKE! performed better on both datasets made up of article abstracts, and on 3 other article datasets. It is important to highlight that it is out of the scope of this research to inspect how keywords were obtained for each text dataset, or how accurate they were. As our goal has been to contrast the performance of two keyword extraction algorithms, inaccuracies in the description of keywords in these datasets were understood as part of the challenge to both algorithms.

For the last 4 datasets on the table (i.e. *500N, wiki, Theses100 and PT-110*), SOBEK's and YAKE!'s performance was equivalent. The *F-score* values computed for both tools

---

[2] http://SOBEK.ufrgs.br/#/.

Table 2. Comparison of the performance of YAKE! and SOBEK in the task of extracting keywords from fourteen text datasets

Text dataset	Text type	#docs	SOBEK Recall	YAKE Recall	SOBEK Precision	YAKE Precision	SOBEK F_score	YAKE F_score	T-test Recall	T-test Precision	T-test F_score
Pubmed	Articles	500	0.0763	0.0517	0.0755	0.0517	0.0759	0.0517	0.0000	0.0000	0.0000
Citeulike	Articles	183	0.1655	0.1496	0.1770	0.1473	0.1710	0.1473	0.0229	0.0000	0.0006
fao30	Articles	30	0.1574	0.1195	0.1571	0.1195	0.1572	0.1195	0.0004	0.0004	0.0004
fao780	Articles	779	0.1226	0.1081	0.1222	0.1081	0.1224	0.1081	0.0005	0.0006	0.0005
Schutz2008	Articles	1231	0.3004	0.2341	0.3004	0.2341	0.3004	0.2341	0.0000	0.0000	0.0000
Krapivin	Articles	2304	0.0583	0.0958	0.0608	0.0958	0.0594	0.0958	0.0000	0.0000	0.0000
Semeval2017	Abstracts	493	0.1188	0.2089	0.1502	0.2089	0.1277	0.2089	0.0000	0.0000	0.0000
Semeval2010	Articles	206	0.0871	0.1313	0.0877	0.1313	0.0874	0.1313	0.0000	0.0000	0.0000
Inspec	Abstracts	2000	0.0727	0.1551	0.1087	0.1551	0.0842	0.1551	0.0000	0.0000	0.0000
Nguyen2007	Articles	209	0.1108	0.1772	0.1160	0.1772	0.1132	0.1772	0.0000	0.0000	0.0000
500N	News	500	0.1611	0.2083	0.3762	0.2085	0.2121	0.2084	0.0000	0.0000	0.4525
wiki	Articles	20	0.1036	0.1235	0.1036	0.1183	0.1036	0.1208	0.1208	0.2451	0.1755
Theses100	Theses	100	0.0850	0.0644	0.0850	0.0644	0.0850	0.0644	0.0615	0.0615	0.0615
PT-110	News	100	0.1259	0.1430	0.1267	0.1430	0.1263	0.1430	0.1840	0.2085	0.1950
Average			**0.1247**	**0.1407**	**0.1462**	**0.1402**	**0.1304**	**0.1404**			
T-test			**0.2 278**		**0.7186**		**0.4170**				

were close to one another, and the outcome of the *T-test* for the 4 datasets showed *p-values* $> = 0.05$. These datasets involved 2 text databases with news, one with theses, and one with articles in Portuguese.

5 Discussion and Final Considerations

This article presented an evaluation study comparing SOBEK's and YAKE!'s methods for keyword extraction, based on the analysis of 14 public datasets. The performance of both algorithms were found to be equivalent in a few of the tests, and one outperformed the other in given circumstances. Such results are considered to be positive in the sense that SOBEK reached a similar performance to one of the methods known to be one of the most powerful keyword extraction algorithms (i.e. YAKE). Unlike other keyword extraction systems, SOBEK and YAKE! follow an unsupervised approach that can operate without the need of any annotated text corpora. In addition, they can extract keywords from single documents, which means they can be used in a very simple way in very different applications. However, YAKE! has one important difference which is its ability to identify stop words on the fly. One the one hand, such a feature can be considered as a major advantage, as the tool can operate in any language and no stopword settings have to be done. However, in certain domain-specific applications, having the ability to edit and control stopword lists, as SOBEK does, may be seen as an advantage, as such manipulation may produce more accurate results in terms of keyword identification.

An interesting feature that favors SOBEK current implementation is that it has a free web service that can be used by any external web application to extract keywords and relations from any document. SOBEK has also the ability to identify relations between keywords, a function that is not covered by YAKE! In this respect, SOBEK may be able to benefit from its graph-based mining approach to represent context information and model conceptual knowledge. Graph Neural Networks, for instance, are able to determine multi-modal causability, in which causal links between features are defined through graph connections (Holzinger et al. 2021). A similar use of conceptual knowledge represented in graph connections could serve as a guiding model to improve the performance of the keyword extraction algorithm.

In the study presented in this paper, it was possible to observe that SOBEK had a good performance when it was used to extract keywords from longer articles. In shorter documents, as in the two test datasets of article abstracts, YAKE! outperformed SOBEK. This may be due to the fact that SOBEK is based on a statistical method that will only consider as keywords, terms with a frequency larger than n, a parameter that can be configured. However, for small texts where most terms appear at most twice, the algorithm does not perform well.

These results may interest the text mining community as they provide possible inter-pretations for performance issues related to two current keyword extraction algorithms, a comparison that has not been made in the past. In this sense, however, a more thorough comparison of SOBEK with YAKE! should include an inner look at their algorithms to understand why they perform better with some types of documents and not with others. Such analysis should shed light on how small changes to these algorithms could further improve them. Another future study involving SOBEK should be the analysis of its accu-racy in identifying relations between keywords, comparing it with other unsupervised relation extraction algorithms.

References

Allahyari, M.: A brief survey of text mining: classification, clustering and extraction techniques. In: Proceedings of KDD Bigdas (2017). http://arxiv.org/abs/1707.02919

Azevedo, B.F.T., Reategui, E.B., Behar, P.A.: Analysis of the relevance of posts in asynchronous discussions. Interdisc. J. E-Learning Learn. Objects **10**, 107–121 (2014). https://doi.org/10.28945/2064

Bromberg, C.: History of science: the problem of cataloging, knowledge indexing and information retrieval in the digital space. Circumscribere: Int. J. Hist. Sc. **21**, 41 (2018). https://doi.org/10.23925/1980-7651.2018v21;p41-55

Campos, R.: Datasets of automatic keyphrase extraction (2020). https://github.com/LIAAD/KeywordExtractor-Datasets

Campos, R., Mangaravite, V., Pasquali, A., Jorge, A., Nunes, C., Jatowt, A.: YAKE! Keyword extraction from single documents using multiple local features. Inf. Sci. **509**, 257–289 (2020). https://doi.org/10.1016/J.INS.2019.09.013

Devlin, J., Chang, M.-W., Lee, K., Toutanova, K.: BERT: pre-training of Deep Bidirectional Trans-formers for Language Understanding. Cornell University (2019). https://doi.org/10.48550/arXiv.1810.04805

El-Kassas, W.S., Salama, C.R., Rafea, A.A., Mohamed, H.K.: Automatic text summarization: a comprehensive survey. Expert Syst. Appl. **165**, 113679 (2021). https://doi.org/10.1016/j.eswa.2020.113679

Feldman, R., Sanger, J.: Text Mining Handbook: Advanced Approaches in Analyzing Unstructured Data. Cambridge University Press, Cambridge (2006)

Firoozeh, N., Nazarenko, A., Alizon, F., Daille, B.: Keyword extraction: Issues and methods. Nat. Lang. Eng. **26**(3), 259–291 (2019). https://doi.org/10.1017/S1351324919000457

Flor, M., Hao, J.: Text mining and automated scoring. In: von Davier, A.A., Mislevy, R.J., Hao, J. (eds.) Computational Psychometrics: New Methodologies for a New Generation of Digital Learning and Assessment. Methodology of Educational Measurement and Assessment. Springer, Cham (2021). https://doi.org/10.1007/978-3-030-74394-9_14

Führ, F., Bisset Alvarez, E.: Digital humanities and open science: initial aspects. In: Bisset Álvarez, E. (ed.) DIONE 2021. LNICSSITE, vol. 378, pp. 154–173. Springer, Cham (2021). https://doi.org/10.1007/978-3-030-77417-2_12

Gonzalez-Gonzalez, C.S., Moreno, L., Popescu, B., Lotero, Y., Vargas, R.: Intelligent systems to support the active self-learning in industrial automation. In: IEEE Global Engineering Education Conference, EDUCON, 10–13 April 2016, pp. 1149–1154 (2016). https://doi.org/10.1109/EDUCON.2016.7474700

Hasan, K.S., Ng, V.: Automatic keyphrase extraction: a survey of the state of the art. In: 52nd Annual Meeting of the Association for Computational Linguistics, ACL 2014 - Proceedings of the Conference, vol. 1, pp. 1262–1273 (2014). https://doi.org/10.3115/V1/P14-1119

Holzinger, A., Malle, B., Saranti, A., Pfeifer, B.: Towards a multi-modal causability with graph neural networks enabling information fusion for explainable ai. Inf. Fusion **71**, 28–37 (2021). https://doi.org/10.1016/j.inffus.2021.01.008

Hulth, A., Megyesi, B.B.: A study on automatically extracted keywords in text categorization. In: COLING/ACL 2006 - 21st International Conference on Computational Linguistics and 44th Annual Meeting of the Association for Computational Linguistics, Proceedings of the Conference, vol. 1, pp. 537–544 (2006). https://doi.org/10.3115/1220175.1220243

Karami, A., Ghasemi, M., Sen, S., Moraes, M.F., Shah, V.: Exploring diseases and syndromes in neurology case reports from 1955 to 2017 with text mining. Comput. Biol. Med. **109**(February), 322–332 (2019). https://doi.org/10.1016/j.compbiomed.2019.04.008

Krallinger, M., Valencia, A.: Text-mining and information-retrieval services for molecular biology (2005). https://doi.org/10.1186/gb-2005-6-7-224

Lamurias, A., Couto, F.M.: Text mining for bioinformatics using biomedical literature. In Encyclopedia of Bioinformatics and Computational Biology. Elsevier Ltd. (2019). https://doi.org/10.1016/b978-0-12-809633-8.20409-3

Lee, A.V.Y., Tan, S.C., Lee, A.V.Y., Tan, S.C.: Discovering dynamics of an idea pipeline: understanding idea development within a knowledge building discourse. In: Proceedings of the 25th International Conference on Computers in Education, pp. 119–128 (2017). https://repository.nie.edu.sg//handle/10497/19430

Lee, A.V.Y., Tan, S.C.: Promising ideas for collective advancement of communal knowledge using temporal analytics and cluster analysis. J. Learn. Anal. **4**(3), 76–101 (2017). https://doi.org/10.18608/jla.2017.43.5

Macedo, A.L., Reategui, E., Lorenzatti, A., Behar, P.: Using text-mining to support the evaluation of texts produced collaboratively. In: Proceedings of IFIP World Conference on Computers in Education, Bento Gonçalves, Brazil (2009)

Marcos-Pablos, S., García-Peñalvo, F.J.: Information retrieval methodology for aiding scientific database search. Soft. Comput. **24**(8), 5551–5560 (2018). https://doi.org/10.1007/s00500-018-3568-0

Noh, H., Jo, Y., Lee, S.: Keyword selection and processing strategy for applying text mining to patent analysis. Expert Syst. Appl. **42**(9), 4348–4360 (2015). https://doi.org/10.1016/j.eswa.2015.01.050

Novak, J.D., Cañas, A.J.: The theory underlying concept maps and how to construct them (2008)

Pang, B., Lee, L.: Opinion mining and sentiment analysis. In: Foundations and Trends in Information Retrieval, vol. 2, issue number 2 (2008)

Reategui, E., Epstein, D., Bastiani, E., Carniato, M.: Can text mining support reading comprehension? In: Gennari, R., et al. (eds.) MIS4TEL 2019. AISC, vol. 1007, pp. 37–44. Springer, Cham (2020). https://doi.org/10.1007/978-3-030-23990-9_5

Rose, S., Engel, D., Cramer, N., Cowley, W.: Automatic keyword extraction from individual documents. Text Min. Appl. Theory **1–20** (2010). https://doi.org/10.1002/9780470689646.CH1

Schenker, A.: Graph-Theoretic Techniques for Web Content Mining Graph-Theoretic Techniques for Web Content Mining. University of South Florida (2003). https://scholarcommons.usf.edu/etd

Song, B., Yan, W., Zhang, T.: Cross-border e-commerce commodity risk assessment using text mining and fuzzy rule-based reasoning. Adv. Eng. Inform. **40**(January), 69–80 (2019). https://doi.org/10.1016/j.aei.2019.03.002

Sun, A., Lachanski, M., Fabozzi, F.J.: Trade the tweet: social media text mining and sparse matrix factorization for stock market prediction. Int. Rev. Financ. Anal. **48**, 272–281 (2016). https://doi.org/10.1016/j.irfa.2016.10.009

Tseng, Y.-H., Lin, C.-J., Lin, Y.-I.: Text mining techniques for patent analysis automatic information organization view project Chinese grammatical error diagnosis view project text mining techniques for patent Analysis. Inf. Process. Manage. **43**, 1216–1247 (2007). https://doi.org/10.1016/j.ipm.2006.11.011

Winograd, P.N.: Strategic Difficulties in Summarizing Texts. University of Illinois at Urbana-Champaign, Cambridge (1983)

Zvarevashe, K., Olugbara, O.O.: A framework for sentiment analysis with opinion mining of hotel reviews. In: Proceedings of the Conference on Information Communications Technology and Society (ICTAS), Durban, South Africa, 8–9 March, pp. 1–4 (2018). https://doi.org/10.1109/ICTAS.2018.8368746

Classification of Screenshot Image Captured in Online Meeting System

Minoru Kuribayashi[✉], Kodai Kamakari, and Nobuo Funabiki

Graduate School of Natural Science and Technology, Okayama University,
3-1-1, Tsushima-naka, Kita-ku, Okayama 700-8530, Japan
kminoru@okayama-u.ac.jp

Abstract. Owing to the spread of the COVID-19 virus, the online meeting system has become popular. From the security point of view, the protection against information leakage is important, as confidential documents are often displayed on a screen to share the information with all participants through the screen sharing function. Some participants may capture their screen to store the displayed documents in their local devices. In this study, we focus on the filtering process and lossy compression applied to the video delivered over an online meeting system, and investigate the identification of screenshot images using deep learning techniques to analyze the distortion caused by such operations. In our experimental results for Zoom applications, we can obtain more than 92.5% classification accuracy even if the captured image is intentionally edited to remove the traces of screen capture.

Keywords: Online meeting system · Screen capture · Image classification · Anti-forensics

1 Introduction

Owing to the spread of the COVID-19 virus, our lifestyle has altered significantly. An example is the increase in online meetings via remote connections from various places. Online meetings are often held using applications such as Microsoft Teams[1], Zoom[2], Google Meet[3], etc., that allow multiple people to participate in a conversation simultaneously, and the application's built-in screen-sharing functionality makes documents available to all participants. In several cases, users may be prompted in advance not to capture the screen. However, it is difficult to physically prevent their actions in general. Under these circumstances, there is an increasing threat that confidential materials may be secretly saved as images by screen capture functions and may be leaked via the internet.

Online meeting systems are based on the server-and-client method, in which each participant's PC or smartphone accesses and communicates with a dedicated

---

[1] https://www.microsoft.com/en-us/microsoft-teams/group-chat-software.

[2] https://zoom.us/.

[3] https://apps.google.com/intl/en/meet/.

© IFIP International Federation for Information Processing 2022
Published by Springer Nature Switzerland AG 2022
A. Holzinger et al. (Eds.): CD-MAKE 2022, LNCS 13480, pp. 244–255, 2022.
https://doi.org/10.1007/978-3-031-14463-9_16

server. Video and audio data are captured from participants' web cameras and microphones, compressed (encoded), and sent to other participants. The system receives the contents of these data and restores to its original state (decoding) so that it can be played at the receiver end. To reduce the amount of communication, some unique techniques and configuration of common multimedia codecs which are generally not revealed in public are employed at each application. Therefore, the number of simultaneous connections and quality of video and audio differ depending on the system.

Different from the images created at terminals by capturing their screen, the images captured from an online meeting system involves distortion induced at the encoding and decoding operations at the sender and receiver ends, respectively. Depending on the applied codecs and parameters, the uniqueness of the distortion will be helpful for classifying the normal and screenshot images.

In this study, we investigate a multimedia forensic method that can determine whether an image is a screenshot of the online meeting system or not. The shared screen in online meeting systems must be encoded by a video codec such as H.264, even though the detailed specifications are not revealed. The distortion caused by the compression must provide useful information about the codec and the corresponding online meeting system. We train a binary image classifier using the images captured from both terminal PCs and online meeting systems and evaluate the basic performance. Then, we consider some possible anti-forensics operations, such that an attacker attempts to remove or alter the traces of distortion using some signal processing operations. Using experiments, the robustness against some operations is evaluated and the possibilities of forensic approaches are discussed.

2 Preliminaries

2.1 Encoding of Video Stream

The Moving Picture Experts Group (MPEG) is established as an alliance of the working group for standardizing multimedia format and coding, and includes compression coding of audio and video. Some examples of the standardized format are MPEG-1, MPEG-2, and H.264/MPEG-4 AVC. The basic encoding operation is common among these standards. Each frame of a video stream is divided into fixed-size macro-blocks, and each one of them cover an area of 16×16 samples of the luminance component and 8×8 samples for each of the chrominance components. Each frame in the stream can be encoded in three different ways (I, P, B), depending on the prediction process conducted during the encoding.

There are a few parameters to control the bit-rate and quality of the compressed video stream. The selection of the video codec and its configuration should be considered depending on the required computational resources. Owing to the variety of parameters, the employed techniques in each online meeting system are distinctive generally.

2.2 Multimedia Forensics

For security purposes, the analysis of the authenticity, processing history, and origin of multimedia content is important in the research of multimedia forensics [3,13,16]. It involves the techniques to identify who, when, how, and what has created/edited the content. Owing to the characteristics of hardware devices and software operations, each processing step makes negligible inherent traces which are sometimes difficult to be perceived by human eyes and ears. With the development of machine learning techniques such as support vector machines (SVM) and deep neural networks (DNN), there have been several forensic techniques to identify the traces of editing history and localization of manipulated regions.

Based on the sensor's pattern noise, the types of digital camera utilized to capture the image have been analyzed in [11]. Considering the malicious operations, it also studied how the error rates change with common image processing, such as JPEG compression or gamma correction. The camera response function (CRF) is studied in [7]. The distortion introduced around splicing boundaries helps us to determine a cut-and-paste forgery created from images taken by two different cameras. Such statistical properties are extracted from a given image, and a classifier trained with a large amount of image dataset can identify the camera device and forgery regions.

In case of software-oriented traces, the study of image manipulation in JPEG images has been intensively studied in the past two decades. Because the traces in the hardware-oriented distortion are generally small, it may be removed by JPEG compression. However, the lossy compression process also leaves traces. It is normal to store an image with JPEG format in digital camera devices. Once the JPEG image is edited by a malicious party, the image is compressed again after manipulation. As the quantization tables are not always unique (e.g. different software, quality factor), the traces of double JPEG compression must be involved in the modified image [10]. The classification accuracy for modified images is improved by introducing SVM as the classifier [12]. Even if the quantization table is unique, the double JPEG compression traces can be analyzed by observing the number of discrete cosine transformation (DCT) coefficients whose values alter after subsequent JPEG compression [8].

2.3 Detection of Capturing Traces

The first study to detect physically captured images based on deep learning is the method in [17]. In this method, the Laplacian filter is embedded into the first layer of a CNN to improve the noise signal ratio introduced by capturing operations. In [9], the convolutional operation is introduced as the pre-processing. Features extracted from the trained CNN model are then fed into a recurrent neural network (RNN) to classify the images. To evaluate captured image forensic techniques, a large dataset is prepared for experiments in [1], and some different kinds of Gaussian filtering residuals are also introduced in the first layer to improve the accuracy. As an image's subject is not useful in the classification, those filters attempt to discard such information while the traces induced by the recapturing operation is enhanced.

An interesting branch of image forensics is the classification of computer-generated and natural images. In this case, the computer generated images are assumed to be stored in a lossless compression format such as PNG. In [18], VGG-based architectures are evaluated for computer-generated image detection. It is indicated that their performance could be improved by dropping max-pooling layers. Some common CNN-based architectures such as VGG-19 and ResNet50 are evaluated in [4], and the effects of fine-tuning and transfer learning techniques are measured. More complicated architectures combining CNN and RNN are investigated in [5,9]. Instead of using fixed filters in the pre-processing step, several convolutional operations are employed in [14]. Rahmouni et al. [15] presented a CNN-based system with a statistical features extraction layer which extracted four features: mean, variance, maximum, and minimum.

3   Classification of Captured Images

3.1   CNN-based Binary Image Classifier

"CGvsPhoto"[4] is developed by Rahmouni et al. [15], and the source code is available on GitHub.it. The program implements a classification of computer graphics images and photographs using a CNN with a TensorFlow as the back end.

An input image is divided into small patches of 100×100 pixels, and the classification results of all patches are aggregated to determine the final binary decision. The patch size is chosen to consider the trade-off between computational costs and statistical meaningfulness. The patch-level classifier comprises three operations: filtering, statistical feature extraction and classification. The filtering is the convolution with several 3×3 kernels and rectified linear unit (ReLU) as a non-linear activation function. The statistical feature extraction is regarded as a pooling layer in ordinal CNNs. To extract useful information exhaustively, four statistical quantities (i.e. mean, variance, maximum, and minimum) are calculated from each filtered image with a specified size. The classification operation comprise fully-connected layers with dropout operation.

3.2   Training

For each label, image data is randomly selected to match the number of data pieces actually required for training. The image data is then randomly divided into training, validation, and test data. Each of these three types of data are divided into 100×100 pixel patches from the image data. These patch data are employed for actual training. The batch size and number of epochs are set arbitrarily, and the model is trained and tested on the patch data. The result of each patch is then aggregated by weighted voting on a single image for the final decision.

---

[4] https://github.com/NicoRahm/CGvsPhoto.

The objective of this study is to identify the images captured from the shared screen in an online meeting system. We employ the screenshot image of a PDF file, PowerPoint file, or other document. If the files are displayed at a standalone terminal PC, the captured images are labeled as the "original image", while the images captured from the screen at the online meeting system are labeled as the "captured image".

3.3   Dataset

We collect the original and captured images from the PDF and PowerPoint files at a Full-HD (1920×1080 pixels) size, and divide them into patches of 100×100 pixels as the dataset.

A screenshot is taken by a standard feature of Windows OS pressing the "Windows" and "Print Screen keys" simultaneously, which is saved in the PNG format by default. Owing to the lossless compression format of PNG, there is no distortion by the capturing operation. However, once the screen is shared in an online meeting system, a lossy compression is performed to reduce the amount of communication. In this study, we choose "Zoom" for the experiments.

The original images were created by capturing the PDF or PowerPoint files from the research presentations of students at University. For screenshots, two PCs were connected from inside and outside the University via Zoom, one shared the original image, and the other captured it manually in full screen mode.

4   Experimental Results

4.1   Conditions

The number of original images are 500 and their captured images are 500. These images are randomly separated into 350 for training, 50 for validation, and 100 for test with labels of original and captured, respectively. In this experiment, a batch size of 25 and an epoch number of 25 are adopted for training the binary image classifier.

Each image is divided into non-overlapping patches of 100×100 pixels. As the size of the image is 1920×1080 pixels, 190 patches are created for each image. For training, there are 350 frames with labels of original and captured, respectively, and they are divided into 66500 ($= 350 \times 190$) patches. Similarly, 50 and 100 frames are selected for validation and test from respective folders to avoid using the patches from the same frame at training, validation, and test phase.

In the experiment, the default parameters of the CNN model "CGvsPhoto" are selected. The model is trained on the labeled patches with 100×100 pixels, and outputs the probability that a given patch is determined to be "captured". We use the Adam optimizer with cross entropy as the loss function. Among several possible hyper-parameters, the choice of a binary classifier plays an

Table 1. Computing environment.

CPU	AMD Ryzen 7 3800X
RAM	64 GB (DDR4-3200)
GPU	Nvidia GeForce RTX2080 SUPER
Softwares	TensorFlow-gpu 2.6.0
	TensorBoard 2.6.0
	CUDA ver.11.5
	Python 3.8.8

important role. Some candidates of the classifier include the SVM, Random Forest [6], and Multi-Layer Perceptron (MLP) [2]. Since "CGvsPhoto" implements fully connected layers with dropout operation as the default classifier, we use it for simplicity. This is because the main purpose of this study is to see if it is possible to classify images captured from an online meeting system.

In the test, a given image is divided into patches and the probabilities are calculated by the trained model. If the average of the probabilities exceeds 0.5, the image is considered as "captured". The computing environment for experiments is shown in Table 1.

4.2 Evaluation Metrics

In this study, we evaluate the binary classification results for the original and captured images. There are a total of four patterns in the experimental results, where the data labels are positive or negative, and the predictions are true or false. Each result is denoted as True Positive (TP), True Negative (TN), False Positive (FP), and False Negative (FN).

TP: The number of captured images correctly classified as captured.
TN: The number of normal images correctly classified as normal.
FP: The number of normal images misclassified as captured.
FN: The number of captured images misclassified as normal.

We adopted *Accuracy*, *Precision*, and *Recall* as the evaluation metrics in this experiment. Accuracy is the percentage of true answers out of the total number of answers in the trained model.

$$Accuracy = \frac{TP + TN}{TP + TN + FP + FN} \times 100 \ (\%) \tag{1}$$

Precision, which is the percentage of captured images and predicted data that are actually captured images, is defined by the following equation.

$$Precision = \frac{TP}{TP + FP} \times 100 \ (\%) \tag{2}$$

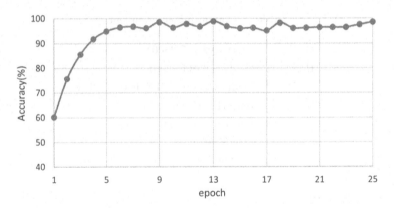

Fig. 1. Classification accuracy in the training phase.

Table 2. Results on classification of original and captured images.

Accuracy	Precision	Recall
97.9	97.3	99.6

Recall is the percentage of captured images that are predicted as captured images out of those that are actually captured images.

$$Recall = \frac{TP}{TP + FN} \times 100 \ (\%) \tag{3}$$

The receiver operating characteristic (ROC) curve is illustrated by plotting the TP rate against the false positive rate at various threshold settings. To measure how accurately the model can distinguish original and captured images, the area under the ROC curve (AUC), which measure the entire two-dimensional area underneath the ROC curve, is also used.

In the binary image classifier adopted in this study, after the model training is completed, the model is evaluated on the patch data for testing. After collecting the binary classification results for divided patches in a given image, the final decision is obtained by the weighted majority voting which calculates the sum of the probabilities of the label "captured".

4.3   Classification

The classification accuracy of original and captured images are evaluated. Figure 1 illustrates the results in a training phase. As observed from the figure, we can obtain high classification accuracy, and the training of binary image classifier rapidly converges in this environment. Table 2 presents the results derived from a test phase. It is not difficult to classify the original and captured images if these images are stored in a lossless format.

Table 3. Results for images after JPEG compression, where a binary image classifier is trained with each QF.

QF	Accuracy	Precision	Recall
50	98.6	95.2	98.8
60	98.6	93.5	98.1
70	96.4	91.9	98.9
80	97.1	90.6	97.8
90	97.7	86.6	97.8

Table 4. Results for images after JPEG compression, where a binary image classifier is trained with various QFs.

Accuracy	Precision	Recall
93.6	89.1	96.1

4.4   Classification After Anti-forensics

Owing to the lossy compression at the communication using the online meeting system, the identification of captured images is possible. From the attacker's point of view, these traces are removed to avoid the identification using signal processing operations, whose operation is called anti-forensics in this study. We evaluate the robustness of the binary image classifier against noise removal operations.

JPEG Compression. JPEG is a standard lossy compression algorithm that can reduce the amount of data in an image with a little sacrifice of quality. For a color image, the RGB color space is converted into YUV space, and the luminance and chrominance are interleaved depending on the setting (e.g. YUV444, YUV422, etc.). The color components are divided into non-overlapping blocks of 8 × 8 pixels, and each block is mapped into frequency components by DCT. After quantizing the DCT components, the compressed file is created by encoding these components.

Similar to JPEG algorithm, the block-wise operation is also a basis of video compression algorithm such as MPEG-1,2,4, and H.264; hence, the JPEG compress may affect the traces of captured images at the online meeting system. In the JPEG compression algorithm, the level of compression ratio can be controlled by the quality factor QF, which changes the quantization table for DCT components. In this experiment, we alter the value $QF = \{50, 60, 70, 80, 90\}$ for evaluation. Table 3 presents the results in a test phase for different QF, where the binary image classifier is trained for the images compressed by each QF. By mixing the images compressed using these five types of QF, the classifier is trained and tested in Table 4. From these results, we observe that the performance of the classifier is not sensitive to the JPEG compression.

It is noteworthy that the performance of the classifier does not always with QF. As the JPEG compression reduces the amount of entropy in an image,

Table 5. Number of images in the training and test datasets for filtering operations.

	Dataset	Original	Captured
Train	I	Non-filtered: 350	Non-filtered: 350
	II	Non-filtered: 350	Non-filtered: 175
			Median filtered ($n = 3$): 175
	III	Non-filtered: 350	Non-filtered: 175
			Median filtered ($n = 5$): 175
	IV	Non-filtered: 350	Non-filtered: 175
			Gaussian filtered: 175
Test	i	Non-filtered: 100	Median filtered ($n = 3$): 100
	ii	Non-filtered: 100	Gaussian filtered: 100

visually non-necessary signals are effectively removed after the lossy compression. However, the traces involved in the captured images remain after the compression, even if they are insignificant. From the results, it is confirmed that the distortion caused by JPEG compression does not correlate well with the traces, and lossy compression can improve classifier performance by suppressing the effects from the image signal.

Filtering Operations. The median filter is a noise removal filter that sorts the order of luminance values in an $n \times n$ pixel area and determines the median value among them. This process is expected to be effective in removing noise from the image because outliers can be removed. By setting a large kernel size, the image luminance values become more uniform than before the filtering process, and the smoothing effect of the image is enhanced.

The Gaussian filter is a filtering process that employs a Gaussian distribution within an $n \times n$ pixel area to weigh pixels according to their distance from the pixel of interest. The closer the pixel is to the pixel of interest, the higher the weight is multiplied. The larger the standard derivation σ, the more uniform the Gaussian distribution becomes, and the greater the smoothing effect of the image. In this experiment we fixed $n = 3$ and $\sigma = 1.3$.

Owing to the smoothing effect of the above filtering operations, the traces of captured images and non-necessary signals involved in the images are expected to be removed. To check the effects, we train the binary image classifier using different datasets and evaluate the performance. The training and test datasets are presented in Table 5, and the performance of the classifier is presented in Table 6. Note that the validation dataset also includes images filtered by such ratios. An example of ROC is calculated for the case of dataset III and i, which result is described in Fig. 2.

In case of training with dataset I, the accuracy is significantly dropped because of the filtering operations. Thus, the median and Gaussian filters act as anti-forensic operations that interfere with the classification of the captured images.

Table 6. Comparison of the performance for different training datasets.

Train	Test	Accuracy	Precision	Recall
I	*i*	70.0	63.3	97.0
	ii	64.0	59.1	95.9
II	*i*	91.6	74.5	**100.0**
	ii	86.0	78.5	**100.0**
III	*i*	**92.5**	**93.1**	**100.0**
	ii	**96.0**	**92.7**	**100.0**
IV	*i*	75.7	87.3	73.1
	ii	90.1	91.3	**100.0**

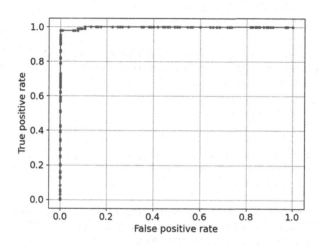

Fig. 2. ROC curve for the results of training dataset III and test dataset *i*, where AUC is 0.998.

By adding the filtered images to the training dataset, we can improve the robustness against such anti-forensics operations from the results of II, III, and IV. Among these three cases, using median filter with $n = 5$ (i.e. training with the dataset III) obtains the best performance. This is because the median filter tends to leave more edges than the Gaussian filter. Increasing the kernel size of the median filter increases the smoothing effect, which works positively in this experiment. The detailed analysis of the effect will be presented in a future work.

5   Concluding Remarks

In this study, we investigated the identification of captured images shared in an online meeting system, and evaluated the classification accuracy using a binary image classifier trained considering the anti-forensics operations. Owing to the characteristics of compression algorithm and its configuration employed in an online meeting system, the traces in the captured image still remained even when the image was compressed by JPEG. In addition, simple noise removal operations such as median filter and Gaussian filter did not alter the traces effectively when the classifier trained with a reasonable size of dataset included some filtered images.

Based on the above experimental results, we can say that the traces generated by the encoding in the online meeting system are useful for the classification of screenshot images. Recall that we used PDF and PowerPoint files containing documents and simple objects drawn in MS-Office. This is because we envision a threat model in which confidential documents are leaked from the shared screen of an online meeting system.

This is an ongoing study, and further experiments with other online meeting systems and anti-forensic operations are required to validate the results obtained in this study. These remain as our future work.

Acknowledgment. This research was supported by JSPS KAKENHI Grant Number 19K22846, JST SICORP Grant Number JPMJSC20C3, and JST CREST Grant Number JPMJCR20D3, Japan.

References

1. Agarwal, S., Fan, W., Farid, H.: A diverse large-scale dataset for evaluating rebroadcast attacks. In: Proceedings of ICASSP 2018, pp. 1997–2001 (2018)
2. Duda, R.O., Hart, P.E., Stork, D.G., et al.: Pattern Classification. Wiley, New York (1973)
3. Farid, H.: Image forgery detection. IEEE Signal Process. Mag. **26**(2), 16–25 (2009)
4. He, M.: Distinguish computer generated and digital images: a CNN solution. Concurr. Comput. Pract. Exp. **31**, e4788 (2018)
5. He, P., Jiang, X., Sun, T., Li, H.: Computer graphics identification combining convolutional and recurrent neural networks. IEEE Signal Process. Lett. **25**, 1369–1373 (2018)
6. Ho, T.K.: Random decision forests. In: Proceedings of 3rd International Conference on Document Analysis and Recognition, vol. 1, pp. 278–282 (1995)
7. Hsu, Y.F., Chang, S.F.: Camera response functions for image forensics: an automatic algorithm for splicing detection. IEEE Trans. Inf. Forensics Secur. **5**(4), 816–825 (2010)
8. Huang, F., Huang, J., Shi, Y.Q.: Detecting double JPEG compression with the same quantization matrix. IEEE Trans. Inf. Forensics Secur. **5**(4), 848–856 (2010)
9. Li, H., Wang, S., Kot, A.C.: Image recapture detection with convolutional and recurrent neural networks. Electron. Imaging Med. Watermark. Secur. Forensics **5**, 87–91 (2017)
10. Lukáš, J., Fridrich, J.: Estimation of primary quantization matrix in double compressed JPEG images. In: Proceedings of Digital Forensic Research Workshop, pp. 5–8 (2003)
11. Lukáš, J., Fridrich, J., Goljan, M.: Digital camera identification from sensor pattern noise. IEEE Trans. Inf. Forensics Secur. **1**(2), 205–214 (2006)
12. Pevný, T., Fridrich, J.: Detection of double-compression in JPEG images for applications in steganography. IEEE Trans. Inf. Forensics Secur. **3**(2), 247–258 (2008)

13. Piva, A.: An overview on image forensics. ISRN Signal Process. **2013**, 1–22 (2013)

14. Quan, W., Wan, K., Yan, D.M., Zhang, X.: Distinguishing between natural and computer-generated images using convolutional neural networks. IEEE Trans. Inf. Forensics Secur. **18**, 2772–2787 (2018)

15. Rahmouni, N., Nozick, V., Yamagishi, J., Echizen, I.: Distinguishing computer graphics from natural images using convolution neural networks. In: Proceedings of WIFS 2017, pp. 1–6 (2017)

16. Stamm, M.C., Wu, M., Liu, K.J.R.: Information forensics: an overview of the first decade. IEEE Access **1**, 167–200 (2013)

17. Yang, P., Ni, R., Zhao, Y.: Recapture image forensics based on Laplacian convolutional neural networks. In: Shi, Y.Q., Kim, H.J., Perez-Gonzalez, F., Liu, F. (eds.) IWDW 2016. LNCS, vol. 10082, pp. 119–128. Springer, Cham (2017). https://doi.org/10.1007/978-3-319-53465-7_9

18. Yu, I.J., Kim, D.G., Park, J.S., Hou, J.U., Choi, S., Lee, H.K.: Identifying photo-realistic computer graphics using convolutional neural networks. In: Proceedings of ICIP 2017, pp. 4093–4097 (2017)

A Survey on the Application of Virtual Reality in Event-Related Potential Research

Vladimir Marochko[1(✉)], Richard Reilly[2], Rachel McDonnell[3], and Luca Longo[1]

[1] The Artificial Intelligence and Cognitive Load Research Lab, The Applied Intelligence Research Center, School of Computer Science, Technological University Dublin, Dublin, Ireland
{vladimir.marochko,luca.longo}@tudublin.ie
[2] School of Medicine, Institute of Neuroscience, Trinity College Dublin, The University of Dublin, Dublin, Ireland
[3] School of Computer Science and Statistics, Trinity College, The University of Dublin, Dublin, Ireland

Abstract. Virtual reality (VR) is getting traction in many contexts, allowing users to have a real-life experience in a virtual world. However, its application in the field of Neuroscience, and above all probing newer activity with the analysis of electroencephalographic (EEG) event-related potentials (ERP) is underexplored. This article reviews the state-of-the-art applications of virtual reality in ERP research, analysing current ways to integrate Head-Mounted Displays (HMD) with EEG head-sets for deploying ecologically valid experiments. It also identifies which ERP components are appropriate in VR settings, along with their paradigms, the technical configurations of the experiments conducted, and the reliability of the findings. Finally, the article synthesises this survey, providing recommendations to practitioners and scholars.

Keywords: Event-related potentials · Virtual reality · Survey

1 Introduction

Event-Related Potentials (ERP) are time-locked voltage fluctuations in EEG-signal connected to physical or mental reactions to some stimuli or action. Their time-locked nature makes it convenient to use them for the investigation of brain functionalities [32]. ERPs are of two types: the early 'sensory' (exogenous) waves (components), peaking within the first 100 milliseconds after the stimulus and depending upon its parameters, and the late 'cognitive' (endogenous) waves, reflecting the ways a human process such a stimulus. ERP components have been found to appear in response to olfactory [39], linguistic [8], visual [2,18,36,44], tactile [15] and auditory [5,26] stimuli. They have found application in the investigation of different neurological conditions [1] making them an efficient diagnostic tool of neural responses. ERPs can also be employed in the non-diagnostic application[13]. An area in which ERP research is getting momentum is Brain-Computer Interfaces (BCI), aimed at mapping, augmenting and supporting humans in sensory-motor functions [15,19,20].

Despite the aforementioned advantages associated with ERP application, their extraction from EEG signals is not a trivial task, because of i) the presence of various types of artifacts contaminating EEG signals, ii) the difficulty of performing experiments in real-world contexts, leading often to lab-based experiment (Fig. 1 left). Although an abundance of research work exists for tackling the first issue, less scientific effort has been devoted towards the resolution of the second one. In fact, it is traditionally difficult to design ERP-based experiments in naturalistic contexts that resemble our everyday environments, because scholars would lose control over many of the factors that might influence them and because of the technical challenges involved in maintaining the electrode-tissue artifact at the EEG electrodes. Some type of environments could be very hard or even dangerous to implement in labs [29]. Participants also tend to act unnaturally in lab environments [35].

A potential solution to the second issue is the use of Virtual Reality (VR) in ERP research. VR is a set of human-digital interaction technologies for the simulation of the real-world experience into an artificial one. Modern VR systems involve visual, auditory and tactile simulation over a 3D visual space that is created by stereoscopic displays covering significant parts of the field of view of a human. VR helps create simply accessible and controllable environments that can be easily manipulated without requiring much professional support. In the context of ERP research, it was demonstrated how the level of human engagement during the execution of experimental tasks for the purpose of ERP component identification could be improved [5,22,23,30]. The spatial presence of participants is reported to be higher when using virtual reality, and this decreases the unnatural behaviour caused by traditional lab-based experimental conditions without it [4]. VR-based environments become more ecological due to higher level of control, as all details are under full control of the VR-environment creator [35]. These environments (Fig. 1 right) can be created by employing multiple displays that are horizontally stacked to create a larger field (usually 180 degrees), or a Head-Mounted Display (HMD), often binocular, that allows the creation of a 365-degree virtual environment, extending the perimeter of possible applications, creating new opportunities for ERP research. Similarly, it supports the creation of Brain-Computer Interfaces (BCI) as well as rehabilitation methods for providing daily life assistance to older people or people with disabilities.

This article is devoted to surveying research work at the intersection of ERP analysis and virtual reality environments. The target audience of this study are researchers working in the field of event-related potentials and brain-machine interfaces who would wish to extend their experimental designs, often conducted in controlled labs, by employing virtual reality. In detail, the structure of this article is as follows: Firstly, the research methods employed to identify relevant articles, and the inclusion criteria, are described in the Sect. 2. Findings of such a literature review follow in a Sect. 3, organised by the methods VR

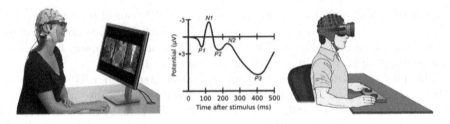

Fig. 1. A traditional lab-based experiment for Event-Related Potential (ERP) (left), and a Virtual Reality-based experiment (right) [1]

has been integrated within EEG-based experiments, by a description of those ERP paradigms that have been adopted in experimental work, and which ERP components received focus. It also presents the technical challenges involved in setting up experiments and the methods VR was used for ERP research, and by discussing the reliability of experimental findings. The article concludes with recommendations for scholars.

2   Research Methods and Inclusion Criteria

A survey was conducted by using Scopus, Google Scholar and the DBLP repositories. Search queries were based on the following terms: 'virtual reality', 'event-related potentials' and 'electroencephalography' applied jointly. In total, 341 articles were retrieved with Google Scholar, 509 with SCOPUS and 2 with DBLP, and these were merged together. They were filtered with a set of inclusion criteria: i) all the pre-prints were discarded, as they did not go through a peer-review process; ii) only empirical research was considered and review articles were discarded; iii) articles strictly adopting a virtual reality headset in conjunction with an EEG headset in the context of ERP research were retained. The final set of research works adhering to the above inclusion criteria included 39 articles.

3   Findings

Out of all the retrieved articles, only 31 employed a virtual reality Head-Mounted Display (HMD), and only 37 articles explicitly mentioned the EEG recording apparatus and its manufacturer. The figurereffig:ManufacturerStats depicts the distribution of the manufacturer of the HDM (left) and the EEG-recording devices adopted in the sampled research works.

3.1   Integration of Virtual Reality and Electroencephalography

The vast majority of researchers have not devised nor discussed any special procedure for integrating an EEG cap with a VR head-mounted display. The articles that tried to measure the interference of the HDM on the EEG data quality found no significant issues [14]. However, the eye, body and participants'

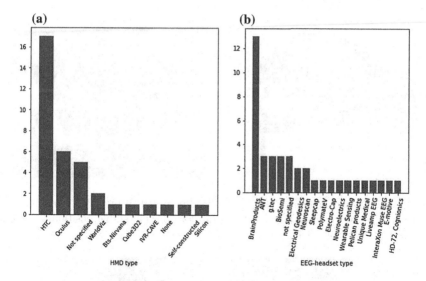

Fig. 2. Distribution of the manufacturers of the a) virtual reality head-mounted displays, b) EEG-recording headsets mentioned in the sample articles.

head movements in a VR-based experiment are more frequent and can lead to various movement artifacts that can increase the noise levels significantly [10, 38]. For this reason, a study implemented a special modification of an EEG cap (Fig. 3, right) in order to decrease physical strain and minimise noise generated by vertical movements. The study also gave advice not to place batteries on the back of the head, but for example, on the belt of a participant. Similarly, they recommended to carefully place the ground electrode right under the central VR strap element, because its misplacement may hamper the whole acquisition of high-quality EEG data [38].

Research shows that scholars tend to use a number of EEG channels between 8 and 32, along with a VR headset. A limited amount of studies used 64, 128 and 256 EEG channels. 64 channels is a more frequent configuration across experiments, except those conducted with the HTC Vive. Out of these configurations, there were no reported difficulties in pairing EEG and head-mounted displays. A correlation analysis was also performed between the year of publication of articles, and the number of EEG channels cited. A Pearson correlation coefficient of 0.1479 between these two variables revealed that there is no relationship between year and number of channels used. It is, therefore, reasonable to expect that with advances in high-density EEG headsets and the decrease in prices, EEG headsets with higher numbers of channels will be more frequently adopted in conjunction with VR-based technologies. The reviewed articles revealed that the average number of participants in the selected experimental studies is 22 with a standard deviation of ±13. The outliers of this distribution refer to those experiments mainly focused on the application of rehabilitation paradigms with patients having rare conditions.

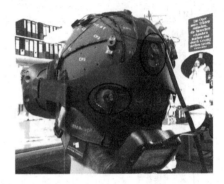

Fig. 3. The original binocular HTC Vive head-mounted display (left) and its modification (right) by creating holes in the strap for allowing stable positioning of EEG electrodes on the scalp [38]

Fig. 4. Integration of head-mounted display (HMD) with cabled EEG system (left) [2] and wireless/bluetooth system (right) [34].

3.2 Experimental ERP Components and Paradigms

As in traditional ERP-based research, most scholars employing VR used baseline correction by considering -200 ms pre-stimuli signal [2]. The vast majority of the experiments conducted in a virtual reality environment were focused on the P3b ERP component, as depicted in the Fig. 5 followed by the N2pc and the N170 ERP components. The P3b response is elicited when a participant is asked to perform an underlying task that demands attention to the deviant stimulus [12]. In this study, the primary task was an Active Visual Oddball Paradigm. Participants were immersed in the environment and when some object (the deviant stimulus) unexpectedly appeared, they had to respond to it, for example, with no physical reaction. Scholars using this paradigm applied virtual reality as a means to create a more naturalistic environment and contribute to the investigation of participants' responses to the deviant stimulus. A study was designed to investigate Spatio-temporal profiles for motion perception using immersive virtual reality [1]. The group that experienced high discomfort revealed higher P3 amplitude mismatch conditions, pointing to a substantial demand for cognitive resources for motion perception in sensory mismatch conditions. Another study investigated the changes in response inhibition following visually induced motion sickness (VIMS) using the two-choice oddball task. Findings revealed a larger deviant-N2 amplitude, a smaller deviant-P3 amplitude, and a delayed deviant-P3 latency after a VR training [43].

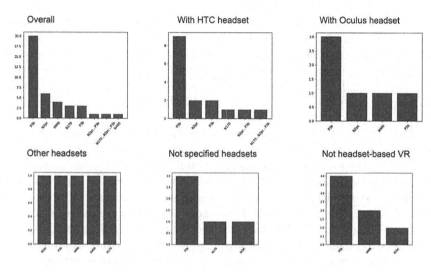

Fig. 5. ERP components analysed across selected articles

The N2pc is linked to selective attention and most of the studies that focused on its analysis aimed at evaluating the level of interaction of participants with an experimental virtual environment, by quantifying error detection [34]. VR can help the creation of virtual environments that are hardly accessible in the real world (such as a space station [43]) or extremely unsafe, such as high buildings [1]). In these environments, the research paradigm was a Simple Visual Search whereby participants observe the environment and report some recognized deviant stimuli without interacting with them, as in the case of the Active Visual Oddball paradigm.

The N170 is an ERP component that reflects the neural processing of faces, familiar objects or words. Visual ERPs of facial expressions (FEs) have been frequently investigated mainly with static stimuli after a baseline, as a nonspecific black screen. However, while studying social events, the ecology of the environment, as well as the stimuli, may be a bias. As a consequence, VR can support the improvement of experimental ecology while keeping stimulus control. For example, full control of the expressions and actions provided by virtual persons could be achieved, as reported in [35]. Here, the primary task for the N170 component is a face perception paradigm where participants have to recognize faces vs non-faces. For example, in [21], the authors have studied how the presented emotional facial expressions in a virtual environment can modulate

ERP components (C1, N170, P2, P3, P4) in conditions with various combinations of passive or active attention, and random or linear presentation sequences. Similarly, in [33] authors have directly compared conventional two-dimensional (2D) and VR-based stimulations by presenting angry and neutral faces to participants. Findings suggest that enhanced immersion with the stimulus materials in the virtual environment can increase the effects of induced brain oscillation to implicit emotion and explicit task effects.

The mismatch negativity (MMN) is an ERP component to an odd stimulus in a sequence of stimuli. Authors use virtual reality to promote better engagement and immersion of participants in activities while researching this component. For example, in [5], participants were requested to play a game or navigate through a virtual city [22] and immersion in these activities, or presence was defied as selective attention to external (non-game related) auditory stimuli, measured as an ERP to an auditory oddball task.

The N400 is an ERP component related to brain response to language (words stimuli) or other meaningful stimuli. In a study, this component was used for investigating the functional differences between bilingual language switching for speakers and for listening to virtual environments. The task for participants was to name a virtual picture in a specific language at the moment it appeared on a virtual screen. The choice of the language depended on the fact that a virtual agent (speaking a specific language) looked it up and gazed at the participant [28]. This experimental paradigm paves the way for the study of bilingual language switching in naturalistic, ecologically valid experimental settings.

3.3   Technical Configurations

Most of the manuscripts reviewed reported no special configurations to the VR headset are accommodated with EEG caps. EEG caps study described a special modification aimed at decreasing the pressure of the head-mounted display on the EEG-cap [38]. An EEG cap study compared wired vs wireless EEG recording. In detail, ERP changes to a simple visual stimulus using wireless and wired setups, in a virtual reality room and in a psychophysiology lab, were investigated [40]. Findings showed that the N1 ERP component, an early component of visual processing, was not affected by the type of signal transmission but, importantly, its latency is different in the two settings. Three of the manuscripts reviewed reported the use of dry electrodes [3,20,44], while four reported the use of wet electrodes [10,11,21,23]. The remaining manuscripts were not explicit on this, thus we could only deduce this information from the model of the EEG recording hardware used. BrainProducts[1], ActiCap[2] and EasyCap[3], the most used EEG-headsets in the sampled articles, uses wet electrodes, whereas g.tec Gamma[4] is based on dry electrodes. The other models can be used with different types of electrodes, so it was not possible to extract whether they were

---

[1] https://www.brainproducts.com/.

[2] https://pressrelease.brainproducts.com/active-electrodes-walkthrough/.

[3] https://pressrelease.brainproducts.com/easycap-cap-overview/.

[4] https://www.gtec.at/product/ggammasys/.

wet or dry. Experience within virtual environments shows that wet electrodes are more accurate, with higher signal robustness and comfortable than dry electrodes which, however, seem to be easier to install and remove than the former type, if the researcher is well-trained and the cap can fit tight to the head. In the case of using an EEG cap with a head-mounted display, pin-based dry electrodes become especially uncomfortable, so it is advisable to use the conventional silver chloride (Ag/AgCl) ring electrodes [6]. Some additional pieces of advice could be found in [6] where a virtual reality-based P300 Human-computer interfaces for games was developed. This suggests that i)using a refresh rate above 70 Hz is optimal, and ii) inviting participants between 20 to 51 years old is advisable to decrease the risk of motion sickness, iii) the camera and the accelerators could be used to remove the muscular artifacts from the EEG signal [6].

3.4   Virtual Reality for ERP Analysis and Application

Half of the selected articles were aimed at investigating the impact of virtual reality on supporting and facilitating ERP analysis. These articles also focused on the new challenges VR creates when employed in conjunction with EEG-recording equipment. These studies investigated how participants' experience differed in VR-based tasks versus non-VR. For example, in [33] it has been found that virtual reality augments human emotions and task effects Similarly, ERP analysis during interaction with virtual objects was presented in [11] whereby ERPs were used to automatically detect visual-haptic mismatches such as those that can cause a loss of the user's immersion. Some studies combined immersive virtual reality and EEG recording for the exploration of whether embodying the errors of an avatar, by watching it from a first-person perspective, could activate an error neural monitoring system of an onlooker [27]. Others focused on the automatic detection of error-related potentials in VR applications, aiming at contributing to the design of adaptive and self-corrective VR applications through the exploitation of information directly from the user [34]. The remaining identified articles employed VR in ERP research mainly to demonstrate its practical application. For example, these focused on changes in ERP results that could demonstrate how virtual reality could improve the quality of learning [7,17,37]. Similarly, some explore specific brain reactions to different stimuli provided by virtual reality and how they appear in different types of tasks [9,24,28,31,41,42], and how virtual reality could improve the game experience [16].

3.5   Reliability of Findings

The quality and robustness of EEG signals recorded in VR environments can be statistically similar to EEG data recorded in typical clinical environments. It is possible to separate articles into two groups. The first contains those studies focused on the comparison of ERP signals in virtual environments under different conditions. The second group includes research focused on the comparison of ERP signals gathered in real-world versus virtual settings. An example within the first group includes an investigation of the early negativity component (prediction error) while detecting visuo-haptic mismatches in VR during an active visual oddball task [11]. Findings were significant ($p < 0.05$) for the specific combination of HTC Vive VR and BrainProducts EEG-headset indicating authors successfully detected haptic conflicts using their technique, demonstrating how ERPs can be used to automatically detect visuo-haptic mismatches in VR, such as those that can lead to a decrease of immersion of participants. In another study, it was found that, through a time-frequency analysis, a significant difference ($p < 0.04$) in the beta frequency band around 350 ms in the temporal lobe exists when processing different facial expressions performed by a virtual avatar over time, in a virtual environment adapted in real-time to the participant's position, in order to increase the feeling of immersion [35]. The goal was to simulate an ecological environment with its natural characteristics while investigating visual ERPs. An example within the second group included the comparison of P300 ERPs responses associated with the completion of drone control tasks both in a virtual environment and in a naturalistic environment with augmented reality [20]. Findings suggested no statistical difference ($p > 0.05$) existed in the averaged ERP responses between the two conditions, demonstrating how the P300 Brain-Computer Interface (BCI) paradigm is relatively consistent across conditions and situations.

Another way to evaluate the quality of the ERPs when HDM are employed in virtual environments is to quantify the decrease in signal-to-noise ratio. This survey found that, overall, the range of rejected trials or rejected channels is between 3 to 28%. The lowest one (3.5% rejected channels) was reported for the joint use of the HTC Vive HDM and EEG headset provided by Electrical Geodesics [33]. 10% of rejected data, in terms of rejected components from an Independent Component Analysis, was reported when the Oculus HDM and the BioSemi EEG headset were jointly used [19,25]. Three studies are all using EEG headsets from BrainProducts and different VR solutions (HTC HDM, WorldViz HDM, and displays). The experiment using the HTC HDM had 10% of noisy epochs removed based on the ranking of the mean absolute amplitude of all channels, standard deviation across all channel mean amplitudes and the Mahalanobis distance of all channels' mean amplitudes [11]. An experiment using the WorldViz HDM had 26.77% trials removed during the averaging process [28]. This experiment employing displays had more than 20% of epochs rejected by an amplitude analysis (more than $100 \mu V$) [40]. In another study, the HDM from HTC Vive was combined with the ANT EEG headset, and 12.5% of epochs, on average, were rejected for all 14 participants [34]. Eventually, in [38], the authors have compared different virtual environments implemented with 2D screens, a 3D dome, and an HTC HDM with and without a special modification of the strap of the HDM (Fig. 3 right). In the case of the 2D screen, 8% of the trials for the static motion were rejected, while 23% for the horizontal motion condition and 28% for the vertical one. In the case of the 3D dome, about 12% of all the trials for all motions were rejected. Regarding the VR condition without modification of the HDM's strap, on the one hand, rejected trials were respectively

29%, 9% and 39% for the static, horizontal and vertical movements. Instead, the condition with the modified strap lead to a rejection rate of 8%, 17% and 15%, respectively for static, horizontal and vertical conditions. Results show that the modification on the HDM's strap had a high impact on noise reduction making the HDM perform better than the screen-based virtual environment, and the unmodified version.

4   Final Remarks and Recommendations

The current survey has shown that Virtual Reality (VR) can bring benefits to ERP research, potentially offering improvements over traditional, lab-based experiments within Neuroscience. However, this field of research is in its infancy, therefore providing a system of recommendations based on current practice is essential for the development of the field.

This survey brought to light that no serious drawbacks were reported in the most used VR headsets, namely the Oculus and HTC products. While using the Oculus head-set in ERP research led to a lower level of rejected trials, there were some studies in which participants reported symptoms of motion sickness. While this did not apply to other VR manufacturers, there is no clear evidence of its superiority in ERP research. Therefore, the choice of which VR headset is ideal for a specific scientific experiment should be based on the usability of the device, the availability of licenses and software, as well as its cost. This explains why head-mounted displays (HMD) such as the HTC Vive and Oculus are often recommended since they can easily fulfil all the above requirements.

Experiments have been conducted with EEG systems that used dry or wet electrodes, with the latter being employed more often than the former because of higher recording quality. Some scholars employed pin-based dry electrodes to avoid using gel-based solutions. However, participants exposed to this configuration reported a higher discomfort while using an HMD. In the case that wet electrodes cannot be used, it has been recommended to choose polymer-based dry electrodes as they cause less discomfort while paired with a VR headset.

The survey did not reveal any significant drop in quality in EEG recording, including the most frequently employed technologies developed by BioSemi, BrainProduct, AntNeuro and G.tec. Although, higher statistically significant levels were reported in studies that employed combinations of the Oculus and g.tec, Oculus and BioSemi and HTC and BrainProducts technologies, no evidence of the high-quality data of these over all the other possible combinations was reported. So the recommendation to scholars on which combination of technologies they can use depends essentially on the type of experiment they need to conduct, whether it is in controlled environments or more ecological contexts, and on the number of EEG channels that are needed as well as the duration of these experiments.

As eye, head and body movements can have a significant impact on recorded EEG signals, it is strongly recommended to greatly focus on artifact minimisation, especially in the case where the stimuli provided to participants appear in a wide field of view (360°), requiring them to actively physically move in space. To support artifact reduction, it is recommended to include accelerometers, gyroscope and other sensors in the experimental configuration, but without precluding the quality of EEG recording. Some EEG recording system has these already embedded, while other allow their addition. Similarly, some HMDs have gyroscopes, accelerometers in the headset and/or the controllers. These additional signals, as used within the wider field of neuroscience, can precisely help scholars locate the temporal dynamics of artifacts, thus providing a means to focus on specific locations of EEG signals and reduce/remove artifacts.

Linked to the problem of artifact reduction, and in order to improve the user experience, some researchers have modified the strap of the head-mounted display, by removing part of the fabric and creating holes in the proximity of EEG electrodes location on the scalp. The purpose was not to have the strap overlapping any electrode thus minimizing their friction, decreasing physical strain, and as a consequence, the generation of artifacts. In this respect, to further improve user experience, some scholars suggested keeping the batteries of the HDM on the waist, in order to minimise the weight placed on the head, therefore minimising fatigue of users.

Eventually, no difference was reported in the quality of EEG recordings when wired or wireless HDM and EEG headsets were used. Most of the experiments used a combination of wired HDM and EEG headsets. The rationale was mainly to have a continuous source of power alimenting the HDM, and cabled EEG electrodes to minimize data loss. Only a smaller amount of scholars used wired HDM and wireless headsets, while no study focused on ERP research fully used both wireless HDM and EEG headsets. In synthesis, the use of a VR-based system in Neuroscience can offer an ecologically valid platform for ERPs research.

Acknowledgement. This work was conducted with the financial support of the Science Foundation Ireland Centre for Research Training in Digitally-Enhanced Reality (D-real) under Grant No. 18/CRT/6224.

References

1. Ahn, M.H., Park, J.H., Jeon, H., Lee, H.J., Kim, H.J., Hong, S.K.: Temporal dynamics of visually induced motion perception and neural evidence of alterations in the motion perception process in an immersive virtual reality environment. Front. Neurosci. **14** (2020)
2. Aksoy, M., Ufodiama, C.E., Bateson, A.D., Martin, S., Asghar, A.U.R.: A comparative experimental study of visual brain event-related potentials to a working memory task: virtual reality head-mounted display versus a desktop computer screen. Exp. Brain Res. **239**(10), 3007–3022, 104107 (2021). https://doi.org/10.1007/s00221-021-06158-w
3. Arake, M., et al.: Measuring task-related brain activity with event-related potentials in dynamic task scenario with immersive virtual reality environment. Front. Behav. Neurosci. **16**, 11 (2022)

4. Baumgartner, T., Valko, L., Esslen, M., Jäncke, L.: Neural correlate of spatial presence in an arousing and noninteractive virtual reality: an EEG and psychophysiology study. CyberPsychol. Behav. **9**(1), 30–45 (2006)
5. Burns, C.G., Fairclough, S.H.: Use of auditory event-related potentials to measure immersion during a computer game. Int. J. Hum Comput Stud. **73**, 107–114 (2015)
6. Cattan, G., Andreev, A., Visinoni, E.: Recommendations for integrating a p300-based brain-computer interface in virtual reality environments for gaming: an update. Computers **9**(4), 92 (2020)
7. Dey, A., Chatburn, A., Billinghurst, M.: Exploration of an EEG-based cognitively adaptive training system in virtual reality. In: 2019 IEEE Conference on Virtual Reality and 3D User Interfaces (VR), pp. 220–226. IEEE (2019)
8. Du, J., Ke, Y., Kong, L., Wang, T., He, F., Ming, D.: 3D stimulus presentation of ERP-speller in virtual reality. In: 2019 9th International IEEE/EMBS Conference on Neural Engineering (NER), pp. 167–170. IEEE (2019)
9. Erdogdu, E., Kurt, E., Duru, A.D., Uslu, A., Başar-Eroğlu, C., Demiralp, T.: Measurement of cognitive dynamics during video watching through event-related potentials (ERPS) and oscillations (EROS). Cogn. Neurodyn. **13**(6), 503–512 (2019)
10. Garduno Luna, C.D.: Feasibility of virtual and augmented reality devices as psychology research tools: a pilot study. Ph.D. thesis, UC Santa Barbara (2020)
11. Gehrke, L., et al.: Detecting visuo-haptic mismatches in virtual reality using the prediction error negativity of event-related brain potentials. In: Proceedings of the 2019 CHI Conference on Human Factors in Computing Systems, pp. 1–11 (2019)
12. Grassini, S., Laumann, K., Thorp, S., Topranin, V.d.M.: Using electrophysiological measures to evaluate the sense of presence in immersive virtual environments: an event-related potential study. Brain Behav. **11**(8), e2269 (2021)
13. Hajcak, G., Klawohn, J., Meyer, A.: The utility of event-related potentials in clinical psychology. Annu. Rev. Clin. Psychol. **15**, 71–95 (2019)
14. Harjunen, V.J., Ahmed, I., Jacucci, G., Ravaja, N., Spapé, M.M.: Manipulating bodily presence affects cross-modal spatial attention: a virtual-reality-based ERP study. Front. Hum. Neurosci. **11**, 79 (2017)
15. Herweg, A., Gutzeit, J., Kleih, S., Kübler, A.: Wheelchair control by elderly participants in a virtual environment with a brain-computer interface (BCI) and tactile stimulation. Biol. Psychol. **121**, 117–124 (2016)
16. Hou, G., Dong, H., Yang, Y.: Developing a virtual reality game user experience test method based on EEG signals. In: 2017 5th International Conference on Enterprise Systems (ES), pp. 227–231. IEEE (2017)
17. Hubbard, R., Sipolins, A., Zhou, L.: Enhancing learning through virtual reality and neurofeedback: a first step. In: Proceedings of the Seventh International Learning Analytics & Knowledge Conference, pp. 398–403 (2017)
18. Hyun, K.Y., Lee, G.H.: Analysis of change of event related potential in escape test using virtual reality technology. Biomed. Sci. Lett. **25**(2), 139–148 (2019)
19. Käthner, I., Kübler, A., Halder, S.: Rapid p300 brain-computer interface communication with a head-mounted display. Front. Neurosci. **9**, 207 (2015)
20. Kim, S., Lee, S., Kang, H., Kim, S., Ahn, M.: P300 brain-computer interface-based drone control in virtual and augmented reality. Sensors **21**(17), 5765 (2021)
21. Kirasirova, L., Zakharov, A., Morozova, M., Kaplan, A.Y., Pyatin, V.: ERP correlates of emotional face processing in virtual reality. Opera Med. Physiol. **8**(3), 12–19 (2021)
22. Kober, S.E., Neuper, C.: Using auditory event-related EEG potentials to assess presence in virtual reality. Int. J. Hum Comput Stud. **70**(9), 577–587 (2012)

23. Li, G., Zhou, S., Kong, Z., Guo, M.: Closed-loop attention restoration theory for virtual reality-based attentional engagement enhancement. Sensors **20**(8), 2208 (2020)
24. Liang, S., Choi, K.S., Qin, J., Pang, W.M., Wang, Q., Heng, P.A.: Improving the discrimination of hand motor imagery via virtual reality based visual guidance. Comput. Methods Programs Biomed. **132**, 63–74 (2016)
25. Lin, C.T., Chung, I.F., Ko, L.W., Chen, Y.C., Liang, S.F., Duann, J.R.: EEG-based assessment of driver cognitive responses in a dynamic virtual-reality driving environment. IEEE Trans. Biomed. Eng. **54**(7), 1349–1352 (2007)
26. Ogawa, R., Kageyama, K., Nakatani, Y., Ono, Y., Murakami, S.: Event-related potentials-based evaluation of attention allocation while watching virtual reality. Adv. Biomed. Eng. **11**, 1–9 (2022)
27. Pavone, E.F., Tieri, G., Rizza, G., Tidoni, E., Grisoni, L., Aglioti, S.M.: Embodying others in immersive virtual reality: electro-cortical signatures of monitoring the errors in the actions of an avatar seen from a first-person perspective. J. Neurosci. **36**(2), 268–279 (2016)
28. Peeters, D.: Bilingual switching between languages and listeners: insights from immersive virtual reality. Cognition **195**, 104107 (2020)
29. Peterson, S.M., Furuichi, E., Ferris, D.P.: Effects of virtual reality high heights exposure during beam-walking on physiological stress and cognitive loading. PLoS ONE **13**(7) (2018)
30. Petras, K., Ten Oever, S., Jansma, B.M.: The effect of distance on moral engagement: Event related potentials and alpha power are sensitive to perspective in a virtual shooting task. Front. Psychol. **6**, 2008 (2016)
31. Pezzetta, R., Nicolardi, V., Tidoni, E., Aglioti, S.M.: Error, rather than its probability, elicits specific electrocortical signatures: a combined EEG-immersive virtual reality study of action observation. J. Neurophysiol. **120**(3), 1107–1118 (2018)
32. Picton, T.W., et al.: Guidelines for using human event-related potentials to study cognition: recording standards and publication criteria. Psychophysiology **37**(2), 127–152 (2000)
33. Schubring, D., Kraus, M., Stolz, C., Weiler, N., Keim, D.A., Schupp, H.: Virtual reality potentiates emotion and task effects of alpha/beta brain oscillations. Brain Sci. **10**(8), 537 (2020)
34. Si-Mohammed, H., et al.: Detecting system errors in virtual reality using EEG through error-related potentials. In: 2020 IEEE Conference on Virtual Reality and 3D User Interfaces (VR), pp. 653–661. IEEE (2020)
35. Simões, M.., Amaral, C.., Carvalho, Paulo, Castelo-Branco, Miguel: Specific EEG/ERP responses to dynamic facial expressions in virtual reality environments. In: Zhang, Yuan-Ting. (ed.) The International Conference on Health Informatics. IP, vol. 42, pp. 331–334. Springer, Cham (2014). https://doi.org/10.1007/978-3-319-03005-0_84
36. Singh, A.K., Chen, H.T., Cheng, Y.F., King, J.T., Ko, L.W., Gramann, K., Lin, C.T.: Visual appearance modulates prediction error in virtual reality. IEEE Access **6**, 24617–24624 (2018)
37. Sun, R., Wu, Y.J., Cai, Q.: The effect of a virtual reality learning environment on learners' spatial ability. Virtual Reality **23**(4), 385–398 (2019)
38. Tauscher, J.P., Schottky, F.W., Grogorick, S., Bittner, P.M., Mustafa, M., Magnor, M.: Immersive EEG: evaluating electroencephalography in virtual reality. In: 2019 Conference on Virtual Reality and 3D User Interfaces, pp. 1794–1800. IEEE (2019)

39. de Tommaso, M., et al.: Pearls and pitfalls in brain functional analysis by event-related potentials: a narrative review by the Italian psychophysiology and cognitive neuroscience society on methodological limits and clinical reliability-Part I. Neurol. Sci. **41**, 3503–3515 (2020)

40. Török, Á., et al.: Comparison between wireless and wired EEG recordings in a virtual reality lab: Case report. In: 2014 5th Conference on Cognitive Infocommunications (CogInfoCom), pp. 599–603. IEEE (2014)

41. Tosoni, A., Altomare, E.C., Brunetti, M., Croce, P., Zappasodi, F., Committeri, G.: Sensory-motor modulations of EEG event-related potentials reflect walking-related macro-affordances. Brain Sci. **11**(11), 1506 (2021)

42. Vass, L.K., et al.: Oscillations go the distance: low-frequency human hippocampal oscillations code spatial distance in the absence of sensory cues during teleportation. Neuron **89**(6), 1180–1186 (2016)

43. Wu, J., Zhou, Q., Li, J., Kong, X., Xiao, Y.: Inhibition-related n2 and p3: Indicators of visually induced motion sickness (vims). Int. J. Ind. Ergon. **78**, 102981 (2020)

44. Yokota, Y., Naruse, Y.: Temporal fluctuation of mood in gaming task modulates feedback negativity: Eeg study with virtual reality. Front. Hum. Neurosci. **15**, 246 (2021)

Visualizing Large Collections of URLs Using the Hilbert Curve

Poornima Belavadi[1] , Johannes Nakayama[1] , and André Calero Valdez[2]([✉])

[1] Human-Computer Interaction Center, RWTH Aachen University,
Campus Boulevard 57, 52076 Aachen, Germany
{belavadi,nakayama}@comm.rwth-aachen.de
[2] Institute for Multimedia and Interactive Systems, University of Lübeck,
Ratzeburger Allee 160, 23562 Lübeck, Germany
calerovaldez@imis.uni-luebeck.de

Abstract. Search engines like Google provide an aggregation mechanism for the web and constitute the main access point to the Internet for a large part of the population. For this reason, biases and personalization schemes of search results may have huge societal implications that require scientific inquiry and monitoring. This work is dedicated to visualizing data such inquiry produces as well as understanding changes and development over time in such data. We argue that the aforementioned data structure is very akin to text corpora, but possesses some distinct characteristics that requires novel visualization methods. The key differences between URLs and other textual data are their lack of internal cohesion, their relatively short lengths, and—most importantly—their semi-structured nature that is attributable to their standardized constituents (protocol, top-level domain, country domain, etc.). We present a technique to spatially represent such data while retaining comparability over time: A corpus of URLs in alphabetical order is evenly distributed onto the so-called Hilbert curve, a space-filling curve which can be used to map one-dimensional spaces into higher dimensions. Rank and other associated meta-data can then be mapped to other visualization primitives. We demonstrate the viability of this technique by applying it to a data set of Google search result lists. The data retains much of its spatial structure (i.e., the closeness between similar URLs) and the spatial stability of the Hilbert curve enables comparisons over time. To make our technique accessible, we provide an R-package compatible with the ggplot2-package.

Keywords: Visualization techniques · Text visualization · URL collections · Computational social science

1 Introduction

In recent years, the world wide web has witnessed an exponential growth of the amount of data. In fact, the total volume of data has increased 30-fold worldwide since 2010 [17]. Generally, visualization methods for structured data are

A. Holzinger et al. (Eds.): CD-MAKE 2022, LNCS 13480, pp. 270–289, 2022.
https://doi.org/10.1007/978-3-031-14463-9_18

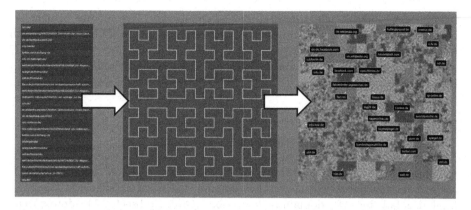

Fig. 1. We visualize large collections of URLs by ordering them alphabetically and taking advantage of the spatial stability of the Hilbert curve. The resulting visualizations provide a spatial mapping of domains and enable comparisons over time.

inherently more mature than for unstructured data and a relatively established repertoire of visualization techniques has been around for some time [21,34]. However, while it is even challenging to process the large amounts of newly generated structured data, it is the unstructured data that adds seemingly untamable complexity. Much of the newly generated data is unstructured textual data from Social Media sites, media outlets, blogs and forums, and many other places online. Visualization is a crucial instrument to reduce complexity, but methods for text visualization remain largely exploratory. As is characteristic for unstructured data, **textual data exhibits high levels of diversity and variability** which makes its visualization an inherently hard task (Fig. 1).

Although textual data can be visualized directly (e.g., word clouds), a more common approach is the visualization of extracted structured meta-constructs like lengths, sentiments, or frequencies (e.g., TF-IDF, bag-of-words models, etc.). For many textual data types and adjacent data domains, the methods for visualization remain largely exploratory. We discovered one of these data domains in our **inquiries into the personalization of web search results** from Google searches. Web search engines like Google provide an access point for information retrieval on the Internet for a large number of users. Personalization of Google search results might thus have a large societal effect in terms of a biased representation of current events. The scientific inquiry into this problem area remains difficult because Google search result lists are not easy to come by on a large scale. In a crowd-sourced data set of Google search result lists, we found that the resulting data can be roughly structured in large collections of URLs that are associated with a user, a search keyword, a time, and a rank.

Our Contribution. This kind of data was inaccessible for us with the current state-of-the-art of data visualization because of its unstructured nature and the specific characteristics of URLs as a data type which we will explain further in this paper. **We present a technique to visualize such data** to make it more readily accessible for scientific inquiry and monitoring. We further present an implementation of this procedure in the programming language R, which makes use of the implementation of the grammar of graphics *ggplot2* [37]. Moreover, we share all study materials and code on the Open Science Framework (link to the repository: https://osf.io/rnkyj).

2   Related Work

The problem that we formulate in this work does not fit easily into existing categories of related work. In the following sections, we review the literature on areas that are adjacent to the problem domain that we are addressing. Visualizing URLs falls into the domain of *text visualization* (Subsect. 2.1). More specifically, the data collections that we target for visualization can be thought of as *corpora* (Subsect. 2.2). However, URLs could also be seen as labels or categories. So it could also be seen as label visualization. Apart from these general considerations, we further review related work on *monitoring of online media streams* (Subsect. 2.3) as the technique we propose in this paper encompasses the aim of making data accessible for scientific inquiry and monitoring. Lastly, our approach falls into the broader scope of *map-like visualizations* which we briefly touch on in Subsect. 2.4.

2.1   Visualizing Textual Data

With the vast amounts of textual data that are available online today, proper visualization techniques are essential to reduce complexity. The literature in the field, especially when it comes to mature, sophisticated methods, is still surprisingly sparse. Kucher and Kerren [23] provide an ongoing visual survey of the field which, at the time of this work, contains 440 publications with **text visualization methods**. It is implemented as a web-based survey browser that lets users filter the works according to a **taxonomy** by the authors (available at: https://textvis.lnu.se/). The categories of this taxonomy are *analytic tasks, visualization tasks, data* (*data source* and *data properties*), *domain*, and *visualization* (*dimensionality, representation*, and *alignment*).

Using this tool, we identified work on text visualization that is similar to what we are aiming to achieve with the technique proposed in this paper. Following is the list of filters applied—*Analytic Tasks:* Event Analysis, Trend/Pattern Analysis, *Visualization Tasks:* Clustering/Classification/Categorization, Overview, Monitoring, *Data Source:* Corpora, *Data Properties:* Time-series, *Domain:* Online Social Media, Other, *Visualization Dimensionality:* 2D, *Visualization Representation:* Pixel/Area/Matrix, *Visualization Alignment:* Other. We chose the *"Other"* option in places where the provided filters did not entirely match our requirements. At the time of this work, this query resulted in only three papers [2,8,30], shedding light on the rich scope for possible research in this area. However, **none of these matched our requirements**.

2.2   Visualization of Large Corpora

In computational linguistics and related fields, texts are often encountered within the context of *corpora*. Text corpora are large structured collections of texts that are rich in information and can be analyzed for a plethora of potential insights. Visualization often plays an important role in this pursuit. A common consideration in corpus visualization is the **distinction between the textual and inter-textual level** which need to be considered in conjunction [1,10,19]. One software solution for corpus analysis that includes both levels is the *DocuScope* software. It provides a set of visualization tools and summary statistics. DocuScope has been used for a wide variety of research purposes including the identification of factors that account for rhetorical variation in canned letters [19] and clarification of collaboration and authorship in the *Federalist Papers* [9]. Building on experiences with DocuScope, Correll, Witmore, and Gleicher [10] present new software for corpus visualization. Their application area is literary scholarship, or—more specifically—the exploration of corpora of tagged text. They introduce the tools CorpusSeparator and TextViewer which are intended to provide insights on the corpus and text level respectively. The software gives literary scholars the tools to quickly identify conspicuous sections of the data in overview and then to zoom into these parts for a more detailed look. **Overview is achieved through visualization of a principal component analysis (PCA)**.

While text corpora can be of a literary nature, the Internet has enabled the collection of large textual data sets in settings of computer-mediated communication (CMC). Abbasi and Chen [1] studied CMC archives and introduced a **technique to perform classification analyses on the textual level**. CMC archives contain rich information about social dynamics and they are often characterized by high numbers of authors, forums, and threads. At the same time, they are inherently hard to navigate. Abbasi and Chen introduce the *Ink Blot* technique which lends itself well to a multitude of visual classification tasks. The technique overlays text with colored ink blots which highlight specific patterns that characterize a particular text within the context of the corpus from which it was derived.

2.3   Monitoring of Online Media Streams

The visualization of corpora is traditionally aimed at analysis, but with the inexorable amount of data that is created constantly, the visualization of online media streams has recently sparked an increasing interest. It has resulted in the development of many media monitoring platforms and tools (www.ecoresearch.net, www.noaa.gov) [32] whose goal is to detect and analyze events and **reveal the different perceptions of the stakeholders and the flow of information**. Such tools often use information extraction algorithms to be able to work with large collections of documents that differ in formatting, style, authorship, and update frequency [32]. Some of the commonly used visualizations by these tools that help in uncovering the complex and hidden relations within the document collection are map-like visualizations, tag clouds, radar charts, and keyword graphs.

Data streams found online can be a rich data source, but their visualization often requires ad-hoc solutions which means that they are cost-intensive. It is thus all the more important to abstract from patterns that occur in online data streams and introduce techniques that apply to a certain problem across contexts. One example is the visualization of news streams, for which Cui et al. [11] introduce a visualization scheme called *TextWheel*. They make use of familiar visual metaphors (Ferris wheel, conveyor belt) to display the development of news streams in one coherent and comprehensible scheme. The technique enables **a comprehensible display of temporal developments** and the authors demonstrate the technique on two example data sets.

2.4   Map-Like Visualization

In visualization research, the term *map-like* can be found describing visualizations that combine the features of cartographic maps to represent abstract data [16]. This representation takes advantage of our ability to recall spatial information or interpret spatial relations between the elements in a map as a similarity measure. Evidence for the **cognitive benefits of using maps** abounds [13]. It is not surprising to see that one of the oldest forms of visualizing spatial data has been in the form of *cartographic maps*.

Investigating ways of leveraging the benefits of maps in visualizing data has been the research focus of the field. In their work on reviewing the state-of-the-art in map-like visualizations, Hograefer et al. [16] have classified the visualizations based on the availability of geographical context into two groups—*schematization* and *imitation*.

Schematization refers to visualizations that are "map-like", where the cartographic **maps are transformed into abstract visualizations** showing emphasis on the thematic data that is spread over a geographical frame of reference. Applying schematization improves the readability by simplifying the map and maintaining the geographic topology, which aids the users to orient themselves in the data space. The visualizations in this technique involve a fundamental trade-off of emphasizing between the visualization of data by applying more schematization and keeping the geographical topology recognizable [4].

Imitation is opposite to schematization, here the abstract visualizations are made to look "map-like" by refining the complex and irregular areas and lines. Imitation depicts **abstract, non-spatial data as visual primitives in a two-dimensional display** by assigning it a position on the plane [16]. To be considered "map-like", however, the positions should achieve a meaningful measure of the distance between visual primitives. Meaningful proximity can be achieved by mapping the dimensions of the data onto the visualization axes, which also helps in understanding the similarity between the data. **Dimensionality reduction methods** like multi-dimensional scaling (MDS), principal component analysis (PCA), and t-distributed stochastic neighbor embedding (t-SNE) **are used to** map n-dimensional data and their distances to 2D [35].

Mapping large data sets into a space suitable for map-like visualizations is both conceptually and computationally hard and requires specialized techniques. Keim [20] proposes the use of space-filling pattern schemes in a new paradigm for visualization that he terms *pixel-oriented techniques*. These techniques use every pixel of a plot panel to maximize the amount of data that can be visualized, making them particularly suitable for the visualization of big data sets. To utilize every pixel space-filling pattern schemes or space-filling curves are used to map the data to a 2D position [18, 25]. These curves have the advantage that **they retain clusters that are present in one dimension** and make them easily discernible in two dimensions. Space-filling curves fall under the *Imitation* technique of visualizing data and are mostly used in combination with a grid. When all the cells that are surrounded by a curve are joined to form a border, the resulting outline forms an area on the "map" [33].

A detailed look into the related literature reinforced the impression that the visualization technique we develop in this paper is novel for visualizing large corpora of URLs and sheds light on the vast possibilities available for research. So far, previous work done in visualizing textual data does not apply to visualizing long lists of URLs because of the structural differences between URLs and normal text data. With the visualization technique applied in this paper, we focus on improving the understanding of personalization. Given the ubiquitous nature of personalization algorithms, this presents a current and critical challenge.

3   Method

Almost a decade ago, Eli Pariser coined the term *filter bubble* [28] as a metaphor for the personalization of web content and the problems that arise because of it. Even though this topic has garnered a lot of public attention since then, data which could be used to address the questions around this phenomenon is still surprisingly sparse. In the following paragraphs, we outline how we addressed the problem of getting an **overview over web search engine personalization** through visualization. Firstly, we describe the requirements of a data set that could be used for such a task.

3.1   Data: Requirements and Description

Procuring the data necessary to address the issues around personalization on web search engines is a challenging task. Krafft, Gamer, and Zweig [22] collected an appropriate data set for the task at hand through "data donation". "Data donors" were asked to install a browser plugin that conducted Google searches in regular time intervals for specific search terms over an extended period of time. In their case, they used the major German political parties and the names of the primary candidates of each party as search terms and **collected Google search results in the months leading up to the German federal election**

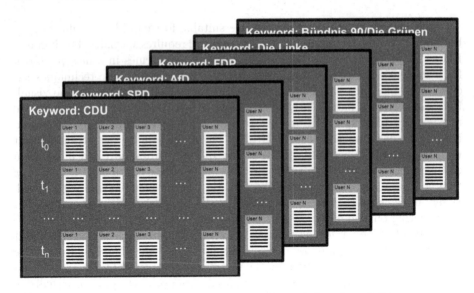

Fig. 2. Structure of a compilation of Google search result lists from different users over an extended period of time.

2017. The resulting data set is a collection of search engine result lists. Each data item is a list of URLs which is associated with a user, a timestamp, and a search keyword (Fig. 2).

Importantly, the result pages that are collected have to be considered as *ranked* lists and not merely as sets. There are strong indications that the rank of a search engine result is strongly related to the attention users pay to the result [24]. While metrics to compare ranked lists exist [36], we advocate for breaking the structure of the data down to make potential results more accessible for interpretation. **We treat the data set as a collection of URLs** each of which is associated with a user, a timestamp, a search keyword, and a rank. The atomic unit of this consideration is thus a singular URL instead of a ranked list of URLs which makes it considerably easier to find a spatial mapping for visualization.

3.2 Problem

URLs are a very particular type of data with specific characteristics that need to be carefully considered. URLs were preceded by a standard access format to documents on the world wide web called Uniform Resource Identifiers (URIs) which followed the same paradigm: object addresses as strings **with a standardized syntax** [5].

URLs were adapted as the standard addressing scheme of web content and with heavy growth rates early on, users were in dire need of aggregation measures and search capabilities. One of the first web search engines was provided by the so-called World Wide Web Worm [26] which cataloged web resources hierarchically and provided keyword search capabilities. Strikingly, it followed a paradigm for web searches that would later become almost universal: curating web search results as a "bulletin board", a list of entries that are relevant to the search specifications, with a hyperlink to the location specified by the associated URL.

While this system provided some much needed complexity reduction at its time, **search engines** were soon confronted with serious scalability issues. These issues were addressed by Sergey Brin and Larry Page who introduced their search engine Google [6] which became a staggering success and the entry point to the world wide web of an unfathomably large number of users. Nowadays, search engines—and first and foremost Google—present the **primary access point to the world wide web** for a large number of users. This means that large portions of a population could get their information on current topics from web searches. For instance, in the 3rd quarter of 2020, the search term "news" is the fifth most popular search term on Google [12]. This presents a potential societal challenge of the following kind: A suspicion concerning search engine results is that they **arrange content differently for different users**, leading to biases and distortions in the users' perceptions. The question is how severe of a problem this actually is and whether the distortion varies over time and between search terms.

3.3 Why Not Map into a Metric Space?

Intuitively, one might think that a viable solution is mapping the data into a (high-dimensional) metric space and to then apply methods like principal component analysis or multi-dimensional scaling for mapping into the 2D plane. However, there are several problems with this solution that, in our view, disqualify it for the problem at hand. First of all, **feature extraction on URLs is not viable** because URLs are short and lack the semantic information content that textual data usually possesses. Second, defining a **string metric** on the set of URLs and treat it as a metric space does not work either. Many **distance measures** that are applicable to strings do not meet the requirement for a metric space in the first place or are otherwise not viable. For instance, the Hamming distance is only defined on strings of the same length and the Cosine distance does not satisfy the triangle inequality. Levenshtein, Damerau-Levenshtein, and Jaccard distance satisfy the triangle inequality, but do not produce sensible results for the problem at hand because of the standardized, semi-structured nature of URLs. String comparison techniques like the aforementioned compare the characters of the whole string. However, character distances in URLs do not equate to meaningful distances in websites found under these URLs (e.g., compare google.com and moodle.com).

3.4 Using Space-Filling Curves

Space-filling curves are **continuous and bijective mappings from one-dimensional into higher-dimensional space**. As the name suggests, the curve passes through every point of an n-dimensional space, which is achieved by folding a one-dimensional line an infinite number of times [3]. Space-filling curves were first discovered by Giuseppe Peano in 1890 [29] and several algorithms mapping one-dimensional lines into higher-dimensional spaces have been discovered since. We will demonstrate that much of the spatial structure in a collection of URLs with the aforementioned meta-data can be retained by ordering the URLs alphabetically and subsequently mapping them into 2D space by means of the Hilbert curve, a space-filling curve that was proposed by the mathematician David Hilbert shortly after Peano's original discovery [15]. The Hilbert curve is a pattern that recursively splits a unit hypercube into quadrants and folds a line into those quadrants according to a set of folding and rotation rules. The simplest and most accessible type of the Hilbert curve is its 2D-instantiation. The first four iterations of this recursive pattern are displayed in Fig. 3. Following, we refer to the 2D-instantiation when we use the term Hilbert curve.

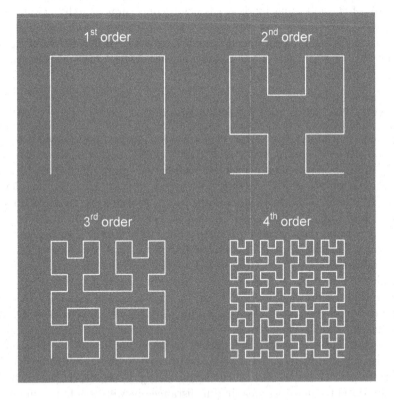

Fig. 3. First to fourth order Hilbert curves.

Hilbert's space-filling curve has fascinated mathematicians for a long time, mostly—however—for its aesthetic appeal. **Useful application areas** have only been proposed relatively recently. Primarily, the Hilbert curve seems to lend itself well to overview and monitoring tasks. Areas in which the Hilbert curve has been applied include the visualization of genomic data [3] and the monitoring of network traffic [31].

Using the Hilbert curve for visualization comes with the benefit that **points that are close in 1D space are also close in the 2D visualization**. This is not necessarily the case in the other direction, where at folding points (e.g., at the centre of the plot), it is possible that data items from distant positions in a vector are close to each other on the 2D plane. However, this effect is weaker for the Hilbert curve than for other space-filling curves [3].

Another benefit is the curve's stability when increasing the resolution of the input data. Increasing the number of iterations yields a longer curve which is more densely folded into the same space, ensuring that **points will always map to a similar position**. This even applies when a small amount of non-uniformly distributed data is introduced to the data.

URLs possess an internal structure that is conducive to a hierarchical treatment of position within a string—which is characteristic for alphabetical orderings. Figure 4 shows a typical URL along with an appraisal of the relative importance of each of its constituents. We suggest that protocol (mostly "https") and the "www." specification do not contain useful information when it comes to evaluating the exposure of users to different web content which is why we excluded these parts in pre-processing. We would further argue that the remaining part of the URL is structured somewhat hierarchically with regard to the importance of informational content. The first elements of the remaining URL are the domain, which specifies the host, and the top-level domain, which can give insight into the location of that host (if it is a country domain) or the kind of institution it represents (".org", ".edu", etc.). Lastly, there is the path to the resource which is usually hierarchically structured from abstract to concrete.

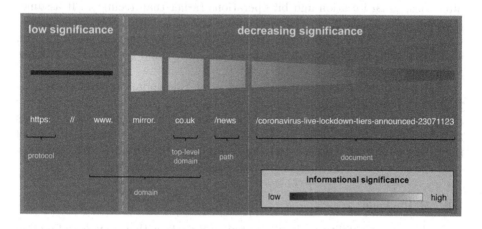

Fig. 4. URL structure and significance estimation.

We propose that for the reasons outlined above, **mapping URLs onto the Hilbert curve in alphabetical order preserves the clustering** essential to the task of visualizing personalization of search engine results. In the following sections, we document an algorithm to achieve a visualization of this kind and present an R package that enables practitioners to use this technique in their own inquiries into the problem domain.

3.5   Implementation

To create a Hilbert curve visualization it is first necessary to understand that n^{th} order Hilbert curves have a number of corners that powers of four. The first order has 4 corners, the second order hast $4^2 = 16$ corners, the third order has $4^3 = 64$ corners, and so on (see Fig. 3). For a map-like visualization that projects individual list entries to a 2D position, we want to **map the entries onto the corners of a Hilbert curve**. It is also possible to map the entries onto the lines connecting the corners, however this would remove the benefit of mapping all entries into the smallest possible 2D representation by creating empty space.

Given the number of corners in a Hilbert curve, it is easiest to also visualize lists of items that have a length that are powers of four. This cannot generally be assumed. To circumvent this problem, we scale the rank numbers of the entries to the next higher power of 4. If we have 10 entries, we use 16 corners. Entry number 8 would be mapped to ceil($8/10 \times 16$) = 13. Each entry is mapped to another unique corner number, creating skipped corners here and there. This creates a **reduced Hilbert curve** (see Fig. 5) which nevertheless retains all of the important properties mentioned above.

From the list of corner numbers we generate x and y coordinates, using an **iterative Hilbert function**. This function "performs the mappings in both directions, using iteration and bit operations rather than recursion. It assumes a square divided into n by n cells, for n a power of 4, with integer coordinates, with $(0,0)$ in the lower left corner, $(n-1, n-1)$ in the upper right corner, and a distance d that starts at 0 in the lower left corner and goes to $n^2 - 1$ in the lower-right corner" [38].

The benefit of creating a visualization using this approach is that it retains relative stability of positions for large data sets (Fig. 6). When individual data points are added into the data, the displacement of existing data tends towards zero for large data sets (Fig. 7). To demonstrate this, we simulated 10 runs of iteratively adding 2200 points of random data into a Hilbert curve. All runs show very little variation regarding root mean squared displacement.

3.6   R Package

To enable reproducibility, we implemented an R package that provides a function to produce a Hilbert visualization for strings. The package is hosted on GitHub (https://github.com/Sumidu/gghilbertstrings). The conversion of ranks into positions, which is the computationally expensive part of the function, was implemented in C++ and made interoperable with Rcpp [14]. The gghilbertstrings

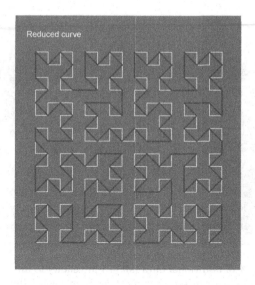

Fig. 5. By scaling the ranks of position we can map an arbitrary number of entries to spatially stable positions. The line in green shows the new reduced curve. (Color figure online)

package enables the creation of graphics like the ones we show in this paper. The core function of the package returns a `ggplot2` object [37]. The major advantage of building on the existing visualization framework provided by `ggplot2` is the flexibility that it grants with regard to customization.

4   Demonstration

For demonstration, we use a **data set collected during the lead-up of the German federal election 2017** [22]. It consists of the Google search results for 16 search terms of over 4000 users over a period of 86 days. The search results were collected daily by a browser plugin that "data donors" could install to contribute to the research project. The keywords were the names of the major German political parties and their top candidates.

Before visualization, the data is pre-processed as described above: First, we remove the low information section in the beginning of the URL, then we collect them in one large list with associated meta-data and sort them alphabetically. For this demonstration, we opted for subsetting the data and only use the result lists for the search keywords "AfD" (right-wing populist party) and "CDU" (christian conservative party). Figure 8 shows the spatial mapping of all URLs that occurred in the respective result lists over the entire time frame covered by the data. **Each colored region in this plot indicates a domain** and the 30 domains that occurred most frequently are indicated with labels pointing to the mean location of URLs in the respective region. Note that larger regions indicate domains that occurred more frequently.

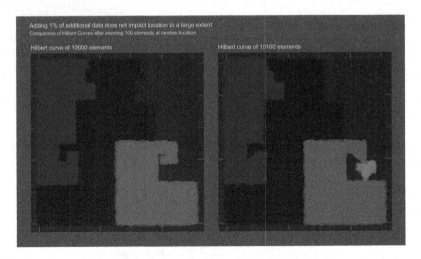

Fig. 6. By adding 1% of additional data (shown in yellow) to an existing Hilbert curve at a random location, we can see that most areas are spatially stable. (Color figure online)

Building on the familiar visual metaphor of topological maps, we chose a color space that is typically used in that domain to encode the regions. As there are 4074 different domains in the data set, we divided the color scale accordingly and randomly assigned each domain one of the generated colors. **The resulting map-like visualization will serve as a reference point** to enable comparisons of different patterns over time.

The landscape for a single point in time can then be displayed by stratifying the data by time. The actual search results for a particular day are mapped onto the reference map as round white markers with high transparency values (akin to viewing clouds from a birds-eye view) to account for overplotting. The size of the markers is contingent of the rank of that particular URL in the search result list that it stems from. Corresponding to their higher relevance, larger markers thus indicate higher-ranked results.

Figure 9 displays the spatial representations of the search result lists of all users for the terms "AfD" and "CDU" on two different days, one of which (September 24, 2017) was the day of the German federal election. The middle-left section of the reference map shows many small domain regions. These account for the websites of local organizations of the CDU party. The **variability in the search results** displayed in the bottom-row plots (CDU) are thus **accounted for by local personalization**. It can be seen that this local personalization is weaker on the day of the election, which is likely due to more media coverage of the CDU on that day, indicated by a higher concentration of solid-looking markers in the bottom right corner of the plot where the regions for *spiegel.de*, *welt.de*, and *zeit.de* (three of the biggest German media outlets) are located. In comparison, there is relatively low variability in the search results for the term

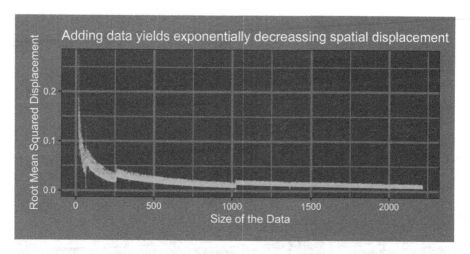

Fig. 7. By consecutively adding data to a Hilbert curve, we see that the amount of displacement continuously shrinks. Occasionally when the order of the curve increases, displacement shortly spikes.

"AfD". A possible explanation for this might be that the party has only existed for a few years and there are not as many local organizations which have websites, resulting in less local personalization of search results. Still, upon closer inspection, one can identify more accentuated clusters of URLs in the regions representing larger media outlets on the day of the election, indicating a higher media coverage than on a reference day two months before the election.

For long-term monitoring and identification of conspicuous patterns in the data, **these graphics can further be animated** with regard to time. Animations of this kind can then be searched for anomalies which in turn could spark further inquiry into that region of the data.

5 Discussion

Text visualization has experienced a surge in research interest since about 2007 [23]. Still, there is a lot of scope for research in this area, particularly because of the wide variety of structural complexities that different types of textual data come with. One such data type that requires dedicated attention is presented by URLs and collections thereof. **URLs differ from conventional text** data in many regards, e.g., in that they cannot be tokenized and often occur in large non-cohesive collections.

We argue that the paradigm of using space-filling curves is promising with regard to creating coherent visualizations of such data. Castro and Burns [7] found that the Hilbert curve method produces an optimal mapping where an arbitrary block of information will be divided a minimum possible number of times in the mapped space [7]. This finding was based on the result of an analytical approach by Mokbel, Aref, and Kamel [27]. With regard to URLs, we showed

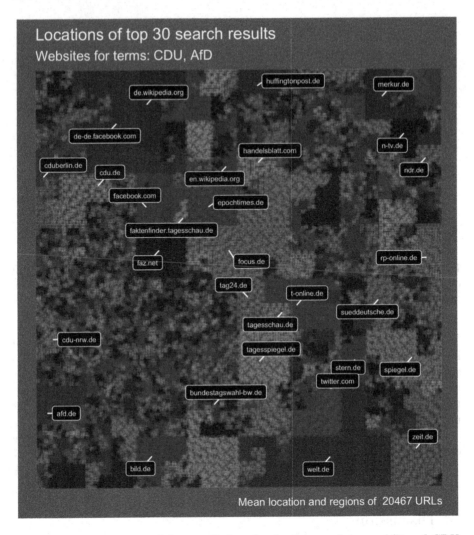

Fig. 8. Relative locations of the top 30 domains for the search terms AfD and CDU. Regions are colored by domain. Larger areas reflect more different URLs returned for the same domain.

that an **alphabetical ordering retains the meaningful similarities** between different URLs (same domain) while still allowing for a spatial mapping that makes different regions distinguishable. Even though the closeness of adjacent regions is not actually indicative of closeness between the domains, the resulting visualization grants a high-level overview of the data that is interpretable and easy to create.

When streaming data is introduced into the visualization, long-term stability over time cannot be guaranteed. If the distribution of domain names in the newly

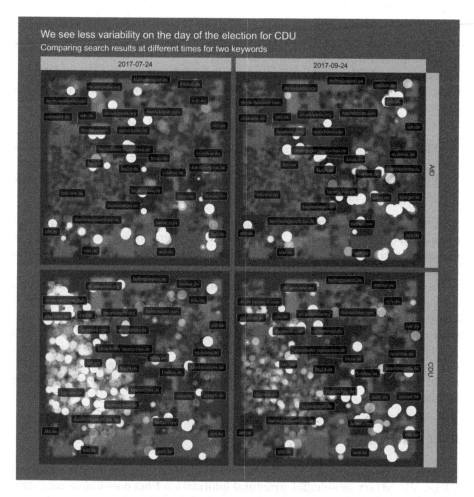

Fig. 9. Comparison of search results of two days. Election day (right column) shows a less diverse set of results for the leading party CDU (bottom row) than two months before the election. Some results stay the same over time. Several of these can we viewed as an animation to see changes over time.

introduced data stays stable, so does the visualization. If this is not the case— e.g., if a large number of URLs with the same domain name is introduced— the positions in the visualization are shifted. However, this would happen in a homogeneous fashion. One idea to address this problem would be to introduce "white noise" into the data: large amounts of equally distributed data points that will not be visualized. Even though this would increase the computational cost as well as **introduce randomness into the visualization**, this adjustment would **improve stability over time**. We also observed that if the input data set contained a large number of links with an identical domain, the resulting visualization would become distorted. In this case, adding noise to the data might not be suitable because the amount of noise would have to be significantly higher than the amount of data. We focus on improving the algorithm to overcome this problem in our future work.

Corpora of URLs might be encountered in different contexts and the technique that we present here may be applicable across different areas of inquiry. However, several caveats have to be taken into consideration. For instance, the naive accumulation of URLs is not necessarily viable in the following cases:

- When link shorteners are used, the lexicographical ordering of URLs loses any meaning.
- When websites do not utilize the URL paths in a meaningful fashion, the only meaningful unit of observation left is the domain.

Still, the technique is generally applicable to text where the lexicographical ordering of entries carries most of the meaning. Other contexts where one might encounter such data include media analyses of hyperlinks used in various news outlets, social media, or any other large scale web-resources.

In general, the application domain lies primarily in the computational social sciences. We thus opted for implementing our approach in the R language, which is a popular choice in this field of research. The package API is interoperable with ggplot2 [37], an implementation of the grammar of graphics [39] which provides future users with the capability to easily customize the generated plots. Our implementation enables the production of visualizations at small computational cost, making it a viable choice in both interactive analysis and real-time monitoring.

6   Conclusion

In this paper, we addressed the complex issue of visualizing multiple long lists of URLs. We extract meaningful structural information from one-dimensional data by using alphabetical ordering as a proxy for the similarity between URLs and use their ranks to map them into 2D space using the Hilbert curve technique.

We demonstrated our technique on a data set containing Google search results of over 4000 users. Practitioners can replicate this method with the R package that we provide on GitHub. In future inquiries, we will address the issue of stability over time for streaming and distorted data.

Acknowledgement. We would like to thank Nils Plettenberg for his help in developing the initial ideas of this project. This research was supported by the Digital Society research program funded by the Ministry of Culture and Science of the German State of North Rhine-Westphalia. We would further like to thank the authors of the packages we have used.

References

1. Abbasi, A., Chen, H.: Categorization and analysis of text in computer mediated communication archives using visualization. In: Proceedings of the 7th ACM/IEEE-CS Joint Conference on Digital Libraries, JCDL 2007, pp. 11–18. Association for Computing Machinery, New York (2007). https://doi.org/10.1145/1255175.1255178
2. Almutairi, B.A.A.: Visualizing patterns of appraisal in texts and corpora. Text & Talk **33**(4–5), 691–723 (2013)
3. Anders, S.: Visualization of genomic data with the Hilbert curve. Bioinformatics **25**(10), 1231–1235 (2009)
4. Barkowsky, T., Latecki, L.J., Richter, K.-F.: Schematizing maps: simplification of geographic shape by discrete curve evolution. In: Freksa, C., Habel, C., Brauer, W., Wender, K.F. (eds.) Spatial Cognition II. LNCS (LNAI), vol. 1849, pp. 41–53. Springer, Heidelberg (2000). https://doi.org/10.1007/3-540-45460-8_4
5. Berners-Lee, T., Cailliau, R., Luotonen, A., Nielsen, H., Secret, A.: The world-wide web. Commun. ACM **37**(8), 76–82 (1994). https://doi.org/10.1145/179606.179671
6. Brin, S., Page, L.: The anatomy of a large-scale hypertextual web search engine. Comput. Netw. **30**, 107–117 (1998). http://www-db.stanford.edu/~backrub/google.html
7. Castro, J., Burns, S.: Online data visualization of multidimensional databases using the Hilbert space–filling curve. In: Lévy, P.P., et al. (eds.) VIEW 2006. LNCS, vol. 4370, pp. 92–109. Springer, Heidelberg (2007). https://doi.org/10.1007/978-3-540-71027-1_9
8. Chi, E.H., Hong, L., Heiser, J., Card, S.K.: ScentIndex: conceptually reorganizing subject indexes for reading. In: 2006 IEEE Symposium on Visual Analytics Science and Technology, pp. 159–166. IEEE (2006)
9. Collins, J., Kaufer, D., Vlachos, P., Butler, B., Ishizaki, S.: Detecting collaborations in text comparing the authors' rhetorical language choices in the Federalist Papers. Comput. Humanit. **38**(1), 15–36 (2004). https://doi.org/10.1023/B:CHUM.0000009291.06947.52
10. Correll, M., Witmore, M., Gleicher, M.: Exploring collections of tagged text for literary scholarship. Comput. Graph. Forum **30**(3), 731–740 (2011)
11. Cui, W., Qu, H., Zhou, H., Zhang, W., Skiena, S.: Watch the story unfold with textwheel: visualization of large-scale news streams. ACM Trans. Intell. Syst. Technol. **3**(2), 1–17 (2012). https://doi.org/10.1145/2089094.2089096
12. DataReportal, We Are Social, Hootsuite: Top Google search queries worldwide during 3rd quarter 2020 (index value) [graph], October 2020. https://www.statista.com/statistics/265825/number-of-searches-worldwide/. Accessed 30 Nov 2020
13. DeLoache, J.S.: Becoming symbol-minded. Trends Cogn. Sci. **8**(2), 66–70 (2004)
14. Eddelbuettel, D., François, R.: Rcpp: seamless R and C++ integration. J. Stat. Softw. **40**(8), 1–18 (2011). https://doi.org/10.18637/jss.v040.i08. http://www.jstatsoft.org/v40/i08/
15. Hilbert, D.: über die stetige abbildung einer linie auf ein flächenstück. Math. Ann. **38**, 459–460 (1891)
16. Hogräfer, M., Heitzler, M., Schulz, H.J.: The state of the art in map-like visualization. In: Computer Graphics Forum, vol. 39, pp. 647–674. Wiley Online Library (2020)
17. IDC, Statista: Volume of data/information worldwide from 2010 to 2024 (in zettabytes) [graph], May 2020. https://www.statista.com/statistics/871513/worldwide-data-created/. Accessed 19 Nov 2020

18. Irwin, B., Pilkington, N.: High level internet scale traffic visualization using Hilbert curve mapping. In: Goodall, J.R., Conti, G., Ma, K.L. (eds.) VizSEC 2007. MATH-VISUAL, pp. 147–158. Springer, Heidelberg (2008). https://doi.org/10.1007/978-3-540-78243-8_10

19. Kaufer, D., Ishizaki, S.: A corpus study of canned letters: mining the latent rhetorical proficiencies marketed to writers-in-a-hurry and non-writers. IEEE Trans. Prof. Commun. **49**(3), 254–266 (2006). https://doi.org/10.1109/TPC.2006.880743

20. Keim, D.A.: Pixel-oriented visualization techniques for exploring very large data bases. J. Comput. Graph. Stat. **5**(1), 58–77 (1996)

21. Keim, D.A.: Information visualization and visual data mining. IEEE Trans. Vis. Comput. Graph. **8**(1), 1–8 (2002)

22. Krafft, T.D., Gamer, M., Zweig, K.A.: What did you see? A study to measure personalization in Google's search engine. EPJ Data Sci. **8**(1), 38 (2019)

23. Kucher, K., Kerren, A.: Text visualization techniques: taxonomy, visual survey, and community insights. In: 2015 IEEE Pacific Visualization Symposium (PacificVis), pp. 117–121. IEEE (2015)

24. Lorigo, L., et al.: Eye tracking and online search: lessons learned and challenges ahead. J. Am. Soc. Inf. Sci. Techno. **59**(7), 1041–1052. https://doi.org/10.1002/asi.20794. https://onlinelibrary.wiley.com/doi/abs/10.1002/asi.20794

25. Markowsky, L., Markowsky, G.: Scanning for vulnerable devices in the Internet of Things. In: 2015 IEEE 8th International Conference on Intelligent Data Acquisition and Advanced Computing Systems: Technology and Applications (IDAACS), vol. 1, pp. 463–467. IEEE (2015)

26. McBryan, O.A.: GENVL and WWWW: tools for taming the web. In: Proceedings of the First International World Wide Web Conference, pp. 79–90 (1994)

27. Mokbel, M.F., Aref, W.G., Kamel, I.: Performance of multi-dimensional space-filling curves. In: Proceedings of the 10th ACM International Symposium on Advances in Geographic Information Systems, pp. 149–154 (2002)

28. Pariser, E.: The Filter Bubble: What the Internet is Hiding from You. Penguin UK (2011)

29. Peano, G.: Sur une courbe, qui remplit toute une aire plane. Math. Ann. **36**(1), 157–160 (1890)

30. Rohrer, R.M., Ebert, D.S., Sibert, J.L.: The shape of Shakespeare: visualizing text using implicit surfaces. In: Proceedings IEEE Symposium on Information Visualization (Cat. No. 98TB100258), pp. 121–129. IEEE (1998)

31. Samak, T., Ghanem, S., Ismail, M.A.: On the efficiency of using space-filling curves in network traffic representation. In: IEEE INFOCOM Workshops 2008, pp. 1–6. IEEE (2008)

32. Scharl, A., Hubmann-Haidvogel, A., Weichselbraun, A., Wohlgenannt, G., Lang, H.P., Sabou, M.: Extraction and interactive exploration of knowledge from aggregated news and social media content. In: Proceedings of the 4th ACM SIGCHI Symposium on Engineering Interactive Computing Systems, pp. 163–168 (2012)

33. Schulz, C., Nocaj, A., Goertler, J., Deussen, O., Brandes, U., Weiskopf, D.: Probabilistic graph layout for uncertain network visualization. IEEE Trans. Vis. Comput. Graph. **23**(1), 531–540 (2016)

34. Shneiderman, B.: The eyes have it: a task by data type taxonomy for information visualizations. In: Proceedings 1996 IEEE Symposium on Visual Languages, pp. 336–343. IEEE (1996)

35. Skupin, A., Fabrikant, S.I.: Spatialization methods: a cartographic research agenda for non-geographic information visualization. Cartogr. Geogr. Inf. Sci. **30**(2), 99–119 (2003)

36. Webber, W., Moffat, A., Zobel, J.: A similarity measure for indefinite rankings. ACM Trans. Inf. Syst. (TOIS) **28**(4), 1–38 (2010)
37. Wickham, H.: ggplot2: Elegant Graphics for Data Analysis. Springer, New York (2016). https://doi.org/10.1007/978-0-387-98141-3. https://ggplot2.tidyverse.org
38. Wikipedia contributors: Hilbert curve – Wikipedia, the free encyclopedia (2020). https://en.wikipedia.org/w/index.php?title=Hilbert_curve&oldid=990914971. Accessed 3 Dec 2020
39. Wilkinson, L.: The grammar of graphics. In: Gentle, J., Härdle, W., Mori, Y. (eds.) Handbook of Computational Statistics. SHCS, pp. 375–414. Springer, Heidelberg (2012). https://doi.org/10.1007/978-3-642-21551-3_13

How to Reduce the Time Necessary for Evaluation of Tree-Based Models

Viera Anderková[✉] and František Babič

Department of Cybernetics and Artificial Intelligence, Faculty of Electrical Engineering and Informatics, Technical University of Košice, Letná 1/9, 042 00 Košice-Sever, Slovakia
{viera.anderkova,frantisek.babic}@tuke.sk

Abstract. The paper focuses on a medical diagnostic procedure supported by decision models generated by suitable tree-based machine learning algorithms like C4.5. The typical result in this situation is represented by a set of trees that should be evaluated by the medical expert. This step is often lengthy because the models may be too detailed and extensive, or the expert is not always 100% available, several experts differ in their opinion. Based on our experience with this type of tasks like diagnostics of Metabolic Syndrome, Mild Cognitive Impairment, or cardiovascular diseases, we have designed and implemented a prototype of a Clinical Decision Support System to improve the tree-based model with selected interpretability methods like LIME, SHAP, and SunBurst interactive visualization. Next, we designed a mechanism containing selected methods from Multiple-Criteria Decision Making (MCDM) and evaluation metrics like functional correctness, usability, stability, and others. We primarily focused on metrics used to evaluate the quality of software products like functional suitability, performance efficiency, usability, etc. Presented proof of concept is further developed into a functional prototype which will be experimentally verified in the form of a pilot study.

Keywords: Interpretability · Decision-making · Evaluation metrics · Model quality

1 Introduction and Motivation

In recent years, there has been a growing interest in topics related to the explanation, interpretation, and understanding of machine learning (ML) models. This problem is becoming more and more attractive and has been addressed in several studies. In psychology, Lombrozo defines *explanations* as a central element of our sense of understanding and how we can exchange views [1]. According to Ribeiro et al., the primary criterion for the explanations is that they must be interpretable and therefore must provide a good understanding between the input variables and the output [2]. The authors emphasize that every ML model may or may not be interpretable. For example, if hundreds or thousands of attributes contribute significantly to a prediction, it is not appropriate to expect every user to understand why the prediction was made.

© IFIP International Federation for Information Processing 2022
Published by Springer Nature Switzerland AG 2022
A. Holzinger et al. (Eds.): CD-MAKE 2022, LNCS 13480, pp. 290–305, 2022.
https://doi.org/10.1007/978-3-031-14463-9_19

Interpretability itself is a comprehensive and not sufficiently defined concept. Velez and Kim define it as explaining or understandably presenting the result [3]. In ML systems, interpretability is the ability to explain or understandably present a result to humans. Miller developed a non-mathematical definition: "The interpretability is the degree to which a human can understand the cause a decision [4]." Kim et al. define interpretability as: "a method are interpretable if the user can correctly and effectively predict the method's results [5]." They also examine the role of transparency and interpretability in ML methods in human decision-making, arguing that interpretability is particularly important in areas where decisions may have significant consequences.

In medical applications, achieving high prediction accuracy is often as crucial as understanding prediction [6]. Gilpin et al. describe the primary goal of interpretability as the ability to explain the ins and outs of the system in an understandable way to end-users [7]. The interpretability provides confidence in end-users' ML systems development process [8]. The interpretation of ML models is often driven by different motivations, such as interpretability requirements, the impact of high-risk decision-making, societal concerns, ML requirements, regulations, and others. Ahmad et al. describe the model as interpretable if it can be evaluated by the end-user, explaining the rationale for the prediction that gives end-users reasons to accept or reject predictions and recommendations [9].

In general, there is no agreement on the definition of interpretability in the scientific community [3, 10]. The concept of interpretability often depends on the domain of application [11] and the target explanatory [8], i.e., the person for whom the interpretations and explanations are intended (end-user/domain expert), so a universal definition may be unnecessary. As synonymous with interpretability in the ML literature, we can consider *intelligibility* [12–14] and *understandability* [10] that are often interchangeable.

Another term prevalent in the literature is *explainability*, which leads to explainable Artificial Intelligence (AI) [15]. This concept is closely related to interpretability, and many authors do not distinguish between these two concepts [4, 8, 14, 16, 17]. Rudin points out the apparent differences between interpretable and explainable ML [11]. Explainable ML usually focuses on deep learning and neural networks, presenting what a node represents and its importance to performance models. In contrast, interpretability is the ability to determine the cause and effect of a ML model by domain experts in practice. In this work, we will stick to the difference proposed by Rudin [11] and the definition of interpretability proposed by Ahmad et al. [9].

A key component of the Decision Support Systems (DSS) and ML models is explaining the decisions, recommendations, predictions, or steps he has made and the process through which the to the decision [18]. Each explanation is related to interpretability, but on the other hand, several other terms or metrics are related to interpretability. The traditional way of evaluating models is as follows:

1. The data analyst shall generate classification or prediction models.
2. Subsequently, the total number of generated models is presented to the domain expert.
3. The domain expert goes model-by-model through the results and evaluates them.
4. He will choose the best for him in terms of results and the easiest to understand.

This process is often lengthy because the expert must evaluate each model, experts are not always 100% available, some cannot understand and correctly evaluate many models, and domain experts often differ in opinion. Such experience motivated us to faster and more effective understanding of decision tree models generated from the medical (healthcare) data on the side of end-users using existing expert knowledge.

1.1 Related Works

Most ML models have focused more often on predicting accuracy and rarely explain their predictions in a meaningful way [9, 19]. It is particularly problematic in several areas of research. Several papers already exist pointing to different areas of research on interpretable ML. In psychology, general concepts of interpretability and explanation have been studied more abstractly [1, 20]. The authors have developed tools for interpretation within adult income to help stakeholders better understand how ML models work [21]. The areas of Fake News, Fake News Detection, and Final Interpretation of the Proposed Models have been the subject of studies [22–25]. Another group of authors focused on using interpretable ML in the financial segment [26, 27] or research on the judicial environment [28, 29]. Interpretation and understanding of the resulting models play the most significant role in the field of medicine and healthcare, for example, in the diagnosis of Alzheimer's disease [30], biomedicine and health care [9, 31–34], in predicting the risk of pneumonia [13], the early prediction of sepsis [35] or COVID-19 [36]. Arik and Lantovics have designed a new, comprehensive hybrid medical system called IntHybMediSys, combining human and computing systems' benefits to solve complex medical diagnostic problems [37]. It is mainly a solution to problems with a serious medical diagnosis. IntHybMediSys can process information that arises when solving medical problems, which allows to accurately determine the most influential contributor (doctor or artificial agent) for each contribution to the problem. Within the system, it is possible to solve problems that a doctor or medical system cannot individually.

The paper is organized as follows: the first section introduces the topic, our motivation, and several selected existing works presenting the actual state-of-the-art in the investigated domain. The second section continues with presenting methods improving the ML model's interpretability. The third section presents already achieved results and the conclusion summarizes the paper and points out the future work.

2   Methods Improving the ML Models' Interpretability

If we talk about interpretable explanations of individual models, several methods exist to interpret the result in a comprehensible and trustworthy way [8, 9]. They point out why the model decided this way and what led it to do so. It would not be easy to understand these decisions without robust interpretability techniques. In our work, we considered the following methods: Local Interpretable Model-Agnostic Explanations (LIME) [2], SHapley Additive ExPlanations (SHAP) [38], and interactive tree-based visualisation called SunBurst [39].

We categorize interpretability approaches into model-specific or model-agnostic and local or global interpretability (see Table 1). Model-specific interpretation methods are derived by explaining the parameters of the internal model and are limited to specific models [40]. Model-agnostic methods can be applied to any ML model and are usually applied post hoc [41], where the parameters of the internal model are not checked because the model is treated as a black-box [40], thereby differs from model-specific methods where this is the case.

The interpretability of the global model helps to understand the distribution of the target result based on attributes. The interpretability of the global model is challenging to achieve in practice. Locally, the prediction can only depend linearly or monotonically on some attributes instead of having a complex dependence on them. Local explanations may be more accurate than global explanations.

Table 1. Types of interpretable methods.

	Local	Global
Model-specific	Set of rules Decision tree k-Nearest Neighbor Visualization techniques	Decision tree Linear regression model Logistic regression model Naïve Bayes classifier
Model-agnostic	LIME, Ribeiro et al. [2] SHAP Lundber and Lee [38] Anchors (Ribeiro et al. [42]) Model-Agnostic Interpretable Rule Extraction (MAIRE, Sharma et al. [43])	SHAP Lundber and Lee [38] Model understanding through subspace explanations (MUSE, Lakkaraju et al. [44]) Learning to Explain (L2X, Chen et al. [45])

LIME provides a patient-specific explanation for a given classification, thus enhancing the possibility for any complex classifier to serve as a safety aid within a clinical setting [46]. The authors used various black-box classifiers such as Adaboost, Random Forest, and Support Vector Machine. The LIME implementation was verified by comparison with the explanations provided by the physicians, their trust and reliance. The results show an encouraging degree of physician consensus and satisfaction with the LIME's visualization combined with the observed cautiously positive degree of physician trust and reliance on it. Meske and Bunde implemented LIME with a Convolutional Neural Network (CNN) and a Multi-Layer Perceptron (MLP) to detect malaria based on the cell images dataset [47]. LIME helped the authors to quickly identify the relevant areas of a predicted class and compare them with their interpretations. Cruz et al. focused on an acute kidney injury as a common complication of patients who undergo cardiac surgery. They applied decision trees and gradient-boosted decision trees combined with LIME providing insight into which feature dimensions are most relevant to the results [48].

SHAP is based on Shapley's values from game theory [49]. Molnar describes that SHAP provides global and local explanations of ML models [41]. The SHAP method comes with both local and global visualizations based on the aggregation of Shapley values. Shapley value is a concept derived from a game theory describing how a "payout" can be fairly distributed among individual "players." Based on its transformation to the ML domain, it is a question of how the individual attributes contribute to the prediction of the ML model. The attribute's contribution to the model's predictions is obtained as a weighted average marginal contribution of the attribute calculated based on all possible combinations of attributes. As a result, as the number of attributes increases, the time required to calculate Shapley values increases. Lundberg and Lee present SHAP values as a method offering a high level of model interpretability based on the extension of the Shapley value [38]. The original method of Shapley values is described by three main properties of additive feature attribution methods: Local accuracy, Missingness, and Consistency.

Tree-based models are, in most cases, too wide, and it is not easy to visualize them so that the whole tree is visible at once. It is necessary to look at the tree gradually through the individual branches. However, some methods like Sunburst or Treemap eliminate this problem. The Sunburst diagram visualizes hierarchical data radially [50]. It is a visualization of DT, where we look at the tree from above, not from the side. Sunburst may resemble nested pie charts, with the top of the hierarchy, the root of the DT at the centre, and the deeper levels tied to that centre. The Sunburst diagram consists of rings that represent the individual levels of the DT. These rings are composed of arcs, each characterizing a specific attribute that further branches the DT. The arcs, therefore, represent the nodes of the original DT. The length of the individual arcs corresponds to the percentage of the dataset. It is given based on the attribute percentage depending on the total number of examples in the dataset.

3   Evaluation Metrics of Model Interpretability and Quality

We can use some of typically used evaluation metrics including [51–54] (Fig. 1):

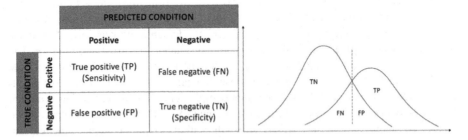

Fig. 1. The general structure of confusion matrix

- Accuracy – is defined as the ratio of the number of all correctly classified examples to all examples (TP + TN)/(TP + TN + FP + FN).
- Precision – is defined as the ratio of the number of positively classified examples to all positively classified examples TP/(TP + FP).

- Specificity/recall – is defined as the ratio of the number of positively correctly classified examples to all classified ones suffering from a given disease TP/(TP + FN).
- Sensitivity – is defined as the ratio of the number of negative correctly classified examples to all classified, which are healthy TN/(TN + FP).
- F1-score – is the harmonic average of recall and precision (2*Precision*Recall)/(Precision + Recall).

However, newer metrics are also used to evaluate the interpretability of decision model explanations. These metrics focus on comprehensibility, and simplicity of explanation and have become increasingly used in recent years.

Understanding the reasons of predictions, causes and recommendations are essential for several reasons. One of the most important reasons is the *trustworthiness* of the result [55]. Lipton argues that trust is not only about "how often the model is right" but also "for which examples it is right [10]." For example, if a model tends to make mistakes only in those inputs where people make mistakes, and on the other hand, it is usually accurate whenever people are accurate, they can trust the model because there is no doubt of mistrust. However, the trustworthiness and applicability of these explanations depend to a large extent on their accuracy and stability [56]. Justification can be used in response to requests for explanations to build human confidence in the accuracy and reasonableness of the system [7].

The System usability scale (SUS), created by John Brooke in 1986, consists of 10 questions, with each item having five answers for end-users [57]. SUS has proven to be very effective and necessary to quickly determine the newly designed user interface's usability. SUS measures how usable the system's user interface is.

Holzinger et al. 2019 [58] propose the notion of *causability* as the degree to which an explanation to a human expert reaches a specified level of causal understanding with effectiveness, efficiency, and satisfaction in a specified context of use. In other words, causability is a person's ability to understand the explanations of ML models. Subsequently, Holzinger et al. 2020 [59] are the first to focus on measuring the quality of individual explanations or interpretable models and propose a *system causability scale (SCS)* metric combined with concepts adapted from a generally accepted usability scale.

Stability states that cases belonging to the same class must have comparable explanations. Alvarez et al. define stability as stable to local input errors, or in other words, similar inputs should not lead to significantly different explanations [38]. Hancox-Li proposes additional stability definition [37]. He argues that the stable explanations reflecting real patterns in the world are those that remain the same compared to a set of equally well-performing models. The measurement of the stability metric is performed by grouping the explanations of all instances in the test data set using the K-means grouping algorithm so that the number of groupings is equal to the number of designations of the data set. For each instance in the test data set, we compare the cluster designation assigned to its clustered explanation with the predicted instance class designation, and if they match, that explanation meets the stability metric.

Also, the interpretable explanations should be explained in *understandable* terms to a human [3]. Interpretable models are also related to *usability* or *user experience*, which is traditionally associated with software products. ISO 9241-11 defines usability as "the extent to which a product can be used by specified users to achieve specified goals with effectiveness, efficiency and satisfaction in a specified context of use." Efficiency, effectiveness, and user satisfaction need to be considered when measuring interface usability [59].

Efficiency can be calculated using the task time (in seconds or minutes) that the participant needs to complete the task successfully. Next, the time required to complete the task can be calculated by simply subtracting the start time from the end time, as shown in the equation below:[1]

$$\text{A.} \quad \textit{Time Based Efficiency} = \frac{\sum_{j=1}^{R} \sum_{i=1}^{N} \frac{n_{ij}}{t_{ij}}}{NR}$$

where N is the total number of tasks (goals), R represents the number of users n_{ij} is the result of task i by user j; if the user successfully completes the task, then N_{ij} is 1, if not, then N_{ij} is 0, t_{ij} represents the time spent by user j to complete task i. If the task is not successfully completed, then time is measured till the moment the user finish the task.

Effectiveness - can be calculated using the task completion rate. The completion rate is the primary usability metric is calculated by assigning a binary value of "0" if the test participant fails to complete the task and "1" if he/she fails to complete the task (see footnote 1).

$$\text{B.} \quad \textit{Effectiveness} = \frac{\textit{Number of tasks completed successfully}}{\textit{Total number of tasks undertaken}} \times 100\%$$

Currently, only a few authors focus on applying of software product quality metrics to ML decision models. The ISO/IEC 25010 standard defines "software and system quality models, which consists of features and ancillary features that determine the quality of a software product and the quality of the software used". The product quality model consists of eight characteristics; each contains a list of relevant metrics.

Functional suitability - represents the extent to which the system provides the fulfilment of expected functions. Three sub-characteristics further describe this property:

a) Functional completeness is the degree to which the set of functions covers all specified tasks and goals of the user.
b) Functional correctness is represented as the degree to which a system provides correct results with the required degree of precision.
c) Functional appropriateness is the extent to which functions facilitate the fulfilment of specified tasks and objectives.

Performance efficiency - is defined by two sub-characteristics:

a) Time behaviour - defines whether the system's response time, processing time, and throughput meet the requirements when performing its functions.
b) Capacity - the degree to which the maximum limits of the system meet the requirements.

---

[1] Available on http://ui-designer.net/usability/efficiency.html.

Compatibility - represents the extent to which the system can exchange information with other systems and perform the required functions when sharing the same hardware or software environment. It consists of two sub-properties:

a) Co-existence - defines the extent to which the system can effectively perform the required functions and at the same time share a common environment and resources with other products without harmful impact.
b) Interoperability - the extent to which two or more systems can exchange information and use the information exchanged.

Usability - defines the extent to which specified users can use the system to achieve specified objectives with efficiency, effectiveness, and satisfaction in the specified context of use. This property consists of the following sub-properties:

a) Appropriateness recognizability - the degree to which users can identify whether a system is appropriate for their needs.
b) Learnability - the extent to which specified users can use a product or system to achieve specific goals.
c) Operability - the degree to which a system has features that facilitate its operation and control.
d) User error protection - the extent to which the system protects users from error.
e) User interface aesthetics - the degree to which a user interface allows for a pleasant and satisfying interaction for the user.
f) Accessibility - the degree to which a system can be used by people with the broadest range of characteristics and abilities to achieve a set goal in a specified context of use.

Reliability - represents the extent to which the system performs specified functions under specified conditions during a specified period. This property consists of the following sub-properties:

a) Maturity - the degree to which a system meets the requirements for reliability in regular operation.
b) Availability - the degree to which a system is operational and accessible when needed.
c) Fault tolerance - the degree to which a system functions as it should, despite the presence of hardware or software errors.
d) Recoverability - the degree to which a system can recover directly affected data in the event of an interruption or failure and restore the desired system state.

Security - is defined as the extent to which the system protects information and data so that persons or systems have access to the data appropriate to their type and levels of authorization. This property consists of the following sub-properties:

a) Confidentiality - the degree to which the system ensures that data is accessible only to those with access rights.

b) Integrity - represents the degree to which the system prevents unauthorized access to or modification of computer programs or data.
c) Non-repudiation - is the extent to which it can be proved that actions or events have taken place so that events or actions cannot be rejected later.
d) Accountability - present the degree to which the entity's actions can be clearly traced to the entity. e) Authenticity - the degree to which an object or resource can be proved.

Maintainability - this characteristic defines the degree of efficiency and effectiveness with which the system can be modified to improve, edit, or adjust to changes in the environment and requirements. This property consists of the following sub-properties:

a) Modularity is the degree to which a system or computer program is composed of discrete components so that changing one component has minimal effect on the other components.
b) Reusability - is the degree to which an asset can be used in more than one system or the construction of other assets.
c) Analysability - represents a measure of efficiency and effectiveness with which it is possible to assess the impact of an intended change of one or more parts on a system, diagnose a product for defects or causes of failures, or identify parts to be modified.
d) Modifiability - the degree to which a system can be effectively and efficiently modified without introducing errors or reducing the quality of an existing product.
e) Testability - The degree of effectiveness and efficiency with which test criteria can be established for a system and tests completed to determine whether those criteria have been met.

Portability - represents the degree of efficiency and effectiveness with which the system can be transferred from one hardware, software, or other operating or user environment. This property consists of the following sub-properties:

a) Adaptability – is defined as the measure to which a system can be efficiently and effectively adapted to different or developing hardware, software, or other operating or user environments.
b) Installability – represents the extent of efficiency and effectiveness with which a system can be successfully installed or uninstalled in a specified environment.
c) Replaceability – defines the extent to which a product can substitute another selected software product for the same purpose in the same environment.

When evaluating the quality and interpretability of the decision models, some metrics overlap, and not all metrics are suitable for proposed approach.

4 Proposed Approach

At first, we have designed and implemented a prototype of a Clinical Decision Support System to improve the tree-based model with selected interpretability methods like LIME, SHAP, and SunBurst interactive visualization. We experimentally evaluated this system with publicly available data about heart diseases[2] and the dataset about Metabolic Syndrome collected by the cooperated medical expert.

Next, we designed a mechanism containing selected metrics and methods from Multiple-Criteria Decision Making (MCDM), see Fig. 2.

This mechanism works in the following steps:

1. The data analyst shall generate interpretable representations of models in the graphical form in the implemented clinical DSS.
2. Subsequently, it is checked whether similar models are already stored in the database (DB of results).
3. If the models are not stored yet, a domain expert (in our case, a doctor) enters the process.
4. The doctor will determine the values of the metrics.
5. In the MCDM process, the doctor then determines the metrics preferences of which metric is more important to the other.
6. The calculation is performed, the Saaty matrix is filled in, and the weights for each metric are calculated using the normalized geometric diameter of this matrix. The higher the weight, the more critical the metric.
7. The values of metrics and metric weights shall be used for calculations performed in the TOPSIS method. Steps in the TOPSIS method:
   a) Compilation of the criterion matrix
   b) Adjust all metrics to the same extreme type.
   c) Normalization of the adjusted criterion matrix.
   d) Creating a weighted criterion matrix.
   e) From the weighted criterion matrix, we determine the ideal and basic models.
   f) Finally, we calculate the distance of the model from the ideal model and the basic model, and using these values, we calculate the relative indicator of the distance of the model from the basic model.
8. We arrange the individual models in descending order.
9. The results database will store individual models, model order, values, weights, and preference metrics.
10. The doctor will obtain a list of recommended models.

If the database contains the same model, its rand will be the same as the one stored. In this way, we would save the doctor's time and make the process more effective because the expert will check only the results filtered by the collected expert knowledge.

4.1 Use Case

Using a simple use case, we want to show how the proposed approach would work in practice:

- The data analysts (DA) process and analyse existing medical dataset within available DSS. He or she generates interpretable models in selected graphical format like typical tree-based, LIME or SHAP.

---

[2] https://archive.ics.uci.edu/ml/datasets/heart+disease.

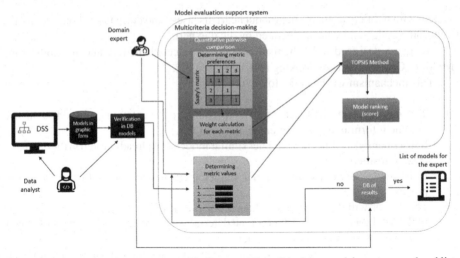

Fig. 2. Step-by-step scheme starting with the generation of decision models up to an ordered list of these models according to the expert's preferences.

- DA imports the generated models into the evaluation system.
- The physician opens the system and displays step by step models needed for evaluation. He or she determines relevant metric values for each model. These values are stored in the database.
- Next, the physician could specify the reference metrics, i.e., which metric is more important to the other.
- After that, the system automatically fills the Saaty matrix, using the normalized geometric diameter of this matrix calculates the weights for each metric, and stores them in the database. The higher the weight, the more critical the metric.
- If all values and metric weights are already in the database, then system calculates the TOPSIS method in the background.
- After this calculation, the physician obtains the descending order of the interpretable models, and the individual model order is stored in the database.
- If the database contains the same models, the physician obtains a list of those models according to his previous preferences and what scores these models have.

4.2 Selected Metrics

Some metrics take values automatically, but some need to be evaluated by a domain expert. In the first phase, we suggest using the following metrics for evaluation:

- Functional suitability:

  - Functional completeness - this metric will be assigned yes or no value after all models have been evaluated. If the doctor can evaluate all the metrics assigned, the value is yes, otherwise the value is no.

– Functional correctness - we will evaluate this metric using accuracy, sensitivity, specificity, and Area under the ROC curve (AUC).

- Usability and user experience:

  – Appropriateness recognizability – the doctor will evaluate this metric using the Likert scale.
  – User interface aesthetics – the doctor will evaluate this metric using the Likert scale.
  – Efficiency – we will use a mathematical formula that is essential to involve a physician in evaluating this metric.
  – Effectiveness - Effectiveness - we will use a mathematical formula that is essential to involve a physician in evaluating this metric.

- Stability - this metric will be evaluated by k-Means algorithm.
- Accuracy - we evaluate this metric as the difference between the accuracy of the local explanation (average of all local explanations) and the accuracy of the original model. The smaller the difference, the more appropriate the explanation.
- We're also considering modifying the SUS and SCS metrics, selecting some of the questions from these metrics, or coming up with five custom questions.

4.3 Multiple-Criteria Decision Making

The task of multicriteria evaluation of alternatives is describing the objective reality of selection using standard procedures, formalizing the decision-making problem and translating it into a mathematical model of a multicriteria decision-making situation [60]. Multicriteria decision-making is the modelling of decision-making situations in which we have defined alternatives and criteria to evaluate alternatives.

The metrics match the criteria, and the alternatives match the models in our system. In the phase of multicriteria decision making, we will analyse two problems:

- modelling of preferences between metrics (criteria), how important (weight) individual metrics are for the user,
- modelling of preferences between models (alternatives) in terms of individual metrics and their aggregation to express the overall preference.

To determine preferences between metrics and the subsequent calculation of the weights of individual metrics, we decided to use the method of quantitative pairwise comparison using the Saaty matrix [61]. It is a method based on comparing two criteria and adding importance to the criterion with the help of a domain expert. Then we calculated the weights using the normalized geometric diameter of the Saaty matrix.

To fully align the models, we minimize the distance from the ideal model (alternative), requiring the domain expert to provide important information (weight and values of metrics). The most used method is the TOPSIS method [62], and this method is based on selecting the model closest to the ideal (best) model and farthest from the basal (worst) model. This method gives us a complete arrangement of all models (alternatives), so it is determined to choose the best model.

5   Conclusion and Future Work

This paper focuses on an interesting topic coming with a continuously increasing amount of data in medicine and healthcare. Today, doctors are overwhelmed with new information knowledge and constantly care about their time to take care of patients. Therefore, based on our experience, the proposed approach should reduce time, reduce the cognitive burden on doctors, increase decision accuracy, and improve treatment and patient care processes when evaluating decision models and recommending doctors only preferred models that are effective, understandable, and trustworthy, easy to manage timesaving. Based on our experience in collaboration with medical professionals, we dare to say that it is necessary to communicate and subsequently produce many results that meet the expectations and requirements of end-users.

In our future work, we will focus on implementing the proposed mechanism and preparing case studies in the Slovak and international conditions, which will aim to fill the proposed system with data and experimentally verify the proposed approach.

Acknowledgements. The work was supported by The Slovak Research and Development Agency under grant no. APVV-20-0232 and The Scientific Grant Agency of the Ministry of Education, Science, Research and Sport of the Slovak Republic under grant no. VEGA 1/0685/2.

References

1. Lombrozo, T.: The structure and function of explanations. Trends Cogn. Sci. **10**(10), 464–470 (2006). https://doi.org/10.1016/j.tics.2006.08.004
2. Ribeiro, M.T., Singh, S., Guestrin, C.: 'Why should I trust you?' Explaining the predictions of any classifier. In: Proceedings of the ACM SIGKDD International Conference on Knowledge Discovery and Data Mining, 13–17 August 2016, pp. 1135–1144 (2016). https://doi.org/10.1145/2939672.2939778
3. Doshi-Velez, F., Kim, B.: Towards A Rigorous Science of Interpretable Machine Learning, no. Ml, pp. 1–13 (2017). https://arxiv.org/pdf/1702.08608.pdf
4. Miller, T.: Explanation in artificial intelligence: insights from the social sciences. Artif. Intell. **267**, 1–38 (2019). https://doi.org/10.1016/j.artint.2018.07.007
5. Kim, B., Khanna, R., Koyejo, O.O.: Examples are not enough, learn to criticize! criticism for interpretability. In: Advances in neural Information Processing Systems, vol. 29 (2016)
6. Stiglic, G., Kocbek, P., Fijacko, N., Zitnik, M., Verbert, K., Cilar, L.: Interpretability of machine learning-based prediction models in healthcare. Wiley Interdiscip. Rev. Data Min. Knowl. Discov. **10**(5), 1–13 (2020). https://doi.org/10.1002/widm.1379
7. Gilpin, L.H., Bau, D., Yuan, B.Z., Bajwa, A., Specter, M., Kagal, L.: Explaining explanations: an overview of interpretability of machine learning. In: 2018 IEEE 5th International Conference on data science and advanced analytics (DSAA), pp. 80–89 (2018)
8. Carvalho, D.V., Pereira, E.M., Cardoso, J.S.: Machine learning interpretability: a survey on methods and metrics. Electronics **8**(8), 1–34 (2019). https://doi.org/10.3390/electronics8080832
9. McKelvey, T., Ahmad, M., Teredesai, A., Eckert, C.: Interpretable machine learning in healthcare. In: Proceedings of the 2018 ACM International Conference on Bioinformatics, Computational Biology, and Health Informatics, vol. 19, no. 1 p. 447 (2018)

10. Lipton, Z.C.: The mythos of model interpretability. Commun. ACM **61**(10), 35–43 (2018). https://doi.org/10.1145/3233231
11. Rudin, C.: Stop explaining black box machine learning models for high stakes decisions and use interpretable models instead. Nat. Mach. Intell. **1**(5), 206–215 (2019). https://doi.org/10.1038/s42256-019-0048-x
12. Lou, Y., Caruana, R., Gehrke, J.: Intelligible models for classification and regression. In: Proceedings of the 18th ACM SIGKDD International Conference on Knowledge Discovery and Data Mining, pp. 150–158 (2012)
13. Caruana, R., Lou, Y., Gehrke, J., Koch, P., Sturm, M., Elhadad, N.: Intelligible models for healthcare: predicting pneumonia risk and hospital 30-day readmission. In: Proceedings of the 21th ACM SIGKDD International Conference on Knowledge Discovery and Data Mining, pp. 1721–1730 (2015)
14. Murdoch, W.J., Singh, C., Kumbier, K., Abbasi-Asl, R., Yu, B.: Definitions, methods, and applications in interpretable machine learning. Proc. Natl. Acad. Sci. U. S. A. **116**(44), 22071–22080 (2019). https://doi.org/10.1073/pnas.1900654116
15. Došilović, F.K., Brčić, M., Hlupić, N.: Explainable artificial intelligence: a survey. In: 2018 41st International Convention on Information and Communication Technology, Electronics and Microelectronics (MIPRO), pp. 210–215 (2018). https://doi.org/10.23919/MIPRO.2018.8400040
16. Dyatlov, I.T.: Manifestation of nonuniversality of lepton interactions in spontaneously violated mirror symmetry. Phys. At. Nucl. **81**(2), 236–243 (2018). https://doi.org/10.1134/S1063778818020060
17. Vellido, A.: The importance of interpretability and visualization in machine learning for applications in medicine and health care. Neural Comput. Appl. **32**(24), 18069–18083 (2019). https://doi.org/10.1007/s00521-019-04051-w
18. Biran, O., Cotton, C.: Explanation and justification in machine learning: a survey. In: IJCAI-17 Workshop on Explainable AI, pp. 8–13 (2017). http://www.cs.columbia.edu/~orb/papers/xai_survey_paper_2017.pdf
19. Elshawi, R., Al-Mallah, M.H., Sakr, S.: On the interpretability of machine learning-based model for predicting hypertension. BMC Med. Inform. Decis. Mak. **19**(1), 146 (2019). https://doi.org/10.1186/s12911-019-0874-0
20. Keil, F.C.: Explanation and understanding. Annu. Rev. Psychol. **57**, 227–254 (2006). https://doi.org/10.1146/annurev.psych.57.102904.190100
21. Kaur, H., Nori, H., Jenkins, S., Caruana, R., Wallach, H., Wortman Vaughan, J.: Interpreting interpretability: understanding data scientists' use of interpretability tools for machine learning. In: Proceedings of the 2020 CHI Conference on Human Factors in Computing Systems, pp. 1–14 (2020)
22. Mohseni, S., Ragan, E.: Combating Fake News with Interpretable News Feed Algorithms, no. Swartout 1983 (2018). http://arxiv.org/abs/1811.12349
23. Mohseni, S., Ragan, E., Hu, X.: Open Issues in Combating Fake News: Interpretability as an Opportunity (2019). http://arxiv.org/abs/1904.03016
24. Malolan, B., Parekh, A., Kazi, F.: Explainable deep-fake detection using visual interpretability methods. In: 2020 3rd International Conference on Information and Computer Technologies (ICICT), pp. 289–293 (2020). https://doi.org/10.1109/ICICT50521.2020.00051
25. Trinh, L., Tsang, M., Rambhatla, S., Liu, Y.: Interpretable and trustworthy deepfake detection via dynamic prototypes. In: Proceedings of the IEEE/CVF Winter Conference on Applications of Computer Vision (WACV), pp. 1973–1983 (2021)
26. Chen, C., Lin, K., Rudin, C., Shaposhnik, Y., Wang, S., Wang, T.: An Interpretable Model with Globally Consistent Explanations for Credit Risk, pp. 1–10 (2018). http://arxiv.org/abs/1811.12615

27. Hajek, P.: Interpretable fuzzy rule-based systems for detecting financial statement fraud. In: MacIntyre, J., Maglogiannis, I., Iliadis, L., Pimenidis, E. (eds.) AIAI 2019. IAICT, vol. 559, pp. 425–436. Springer, Cham (2019). https://doi.org/10.1007/978-3-030-19823-7_36

28. Tan, S., Caruana, R., Hooker, G., Lou, Y.: Distill-and-compare: auditing black-box models using transparent model distillation. In: AIES 2018 - Proceedings of 2018 AAAI/ACM Conference AI, Ethics, Society, pp. 303–310 (2018). https://doi.org/10.1145/3278721.327 8725

29. Soundarajan, S., Clausen, D.L.: Equal Protection Under the Algorithm: A Legal-Inspired Framework for Identifying Discrimination in Machine Learning (2018)

30. Das, D., Ito, J., Kadowaki, T., Tsuda, K.: An interpretable machine learning model for diagnosis of Alzheimer's disease. PeerJ **7**, e6543 (2019)

31. Miotto, R., Li, L., Kidd, B.A., Dudley, J.T.: Deep patient: an unsupervised representation to predict the future of patients from the electronic health records. Sci. Rep. **6**(1), 26094 (2016). https://doi.org/10.1038/srep26094

32. Mamoshina, P., Vieira, A., Putin, E., Zhavoronkov, A.: Applications of deep learning in biomedicine. Mol. Pharm. **13**(5), 1445–1454 (2016). https://doi.org/10.1021/acs.molpharma ceut.5b00982

33. Jackups, R., Jr.: Deep learning makes its way to the clinical laboratory. Clin. Chem. **63**(12), 1790–1791 (2017). https://doi.org/10.1373/clinchem.2017.280768

34. Nori, H., Jenkins, S., Koch, P., Caruana, R.: InterpretML: A Unified Framework for Machine Learning Interpretability, pp. 1–8 (2019). http://arxiv.org/abs/1909.09223

35. Nemati, S., Holder, A., Razmi, F., Stanley, M.D., Clifford, G.D., Buchman, T.G.: An interpretable machine learning model for accurate prediction of sepsis in the ICU. Crit. Care Med. **46**(4), 547–553 (2018). https://doi.org/10.1097/CCM.0000000000002936

36. Wu, H., et al.: Interpretable machine learning for covid-19: an empirical study on severity prediction task. IEEE Trans. Artif. Intell. (2021)

37. Arik, S., Iantovics, L.B.: Next generation hybrid intelligent medical diagnosis systems. In: Liu, D., Xie, S., Li, Y., Zhao, D., El-Alfy, E.S. (eds.) Neural Information Processing, pp. 903–912. Springer, Cham (2017). https://doi.org/10.1007/978-3-319-70090-8_92

38. Lundberg, S.M., Lee, S.I.: A unified approach to interpreting model predictions. In: Advances in Neural Information Processing Systems, vol. 2017-December, no. Section 2, pp. 4766–4775 (2017). https://arxiv.org/pdf/1705.07874.pdf

39. Stasko, J., Catrambone, R., Guzdial, M., McDonald, K.: An evaluation of space-filling information visualizations for depicting hierarchical structures. Int. J. Hum. Comput. Stud. **53**(5), 663–694 (2000). https://doi.org/10.1006/ijhc.2000.0420

40. Du, M., Liu, N., Hu, X.: Techniques for interpretable machine learning. Commun. ACM **63**(1), 68–77 (2019)

41. Molnar, C.: Interpretable Machine Learning. A Guide for Making Black Box Models Explainable. Book, p. 247 (2019). https://christophm.github.io/interpretable-ml-book

42. Ribeiro, M.T., Singh, S., Guestrin, C.: Anchors: high-precision model-agnostic explanations. In: Proceedings of the AAAI Conference on Artificial Intelligence, vol. 32, no. 1 (2018)

43. Sharma, R., Reddy, N., Kamakshi, V., Krishnan, N.C., Jain, S.: MAIRE - a model-agnostic interpretable rule extraction procedure for explaining classifiers. In: Holzinger, A., Kieseberg, P., Tjoa, A.M., Weippl, E. (eds.) CD-MAKE 2021. LNCS, vol. 12844, pp. 329–349. Springer, Cham (2021). https://doi.org/10.1007/978-3-030-84060-0_21

44. Lakkaraju, H., Kamar, E., Caruana, R., Leskovec, J.: Faithful and customizable explanations of black box models. In: Proceedings of the 2019 AAAI/ACM Conference on AI, Ethics, and Society, pp. 131–138 (2019). https://doi.org/10.1145/3306618.3314229

45. Chen, J., Song, L., Wainwright, M.J., Jordan, M.I.: Learning to explain: an information-theoretic perspective on model interpretation. In: 35th International Conference on Machine Learning, ICML 2018, vol. 2, pp. 1386–1418 (2018). https://arxiv.org/pdf/1802.07814.pdf

46. Kumarakulasinghe, N.B., Blomberg, T., Liu, J., Leao, A.S., Papapetrou, P.: Evaluating local interpretable model-agnostic explanations on clinical machine learning classification models. In: 2020 IEEE 33rd International Symposium on Computer-Based Medical Systems, pp. 7–12 (2020)

47. Meske, C., Bunde, E.: Transparency and trust in human-AI-interaction: the role of model-agnostic explanations in computer vision-based decision support. In: Degen, H., Reinerman-Jones, L. (eds.) HCII 2020. LNCS, vol. 12217, pp. 54–69. Springer, Cham (2020). https://doi.org/10.1007/978-3-030-50334-5_4

48. Da Cruz, H.F., Schneider, F., Schapranow, M.-P.: Prediction of Acute Kidney Injury in Cardiac Surgery Patients: Interpretation using Local Interpretable Model-agnostic Explanations (2019)

49. Thomson, W., Roth, A.E.: The Shapley Value: Essays in Honor of Lloyd S. Shapley, vol. 58, no. 229 (1991)

50. Altarawneh, R., Humayoun, S.R.: Visualizing software structures through enhanced interactive sunburst layout. In: Proceedings of the International Working Conference on Advanced Visual Interfaces (2016)

51. Pourhomayoun, M., Shakibi, M.: Predicting mortality risk in patients with COVID-19 using machine learning to help medical decision-making. Smart Heal. **20**, 100178 (2021). https://doi.org/10.1016/j.smhl.2020.100178

52. Xu, W., Zhang, J., Zhang, Q., Wei, X.: Risk prediction of type II diabetes based on random forest model. In: 2017 Third International Conference on Advances in Electrical, Electronics, Information, Communication and Bio-Informatics (AEEICB), pp. 382–386 (2017). https://doi.org/10.1109/AEEICB.2017.7972337

53. Kumar, S., Sahoo, G.: A random forest classifier based on genetic algorithm for cardiovascular diseases diagnosis (research note). Int. J. Eng. **30**(11), 1723–1729 (2017)

54. Khalilia, M., Chakraborty, S., Popescu, M.: Predicting disease risks from highly imbalanced data using random forest. BMC Med. Inform. Decis. Mak. **11**(1), 51 (2011). https://doi.org/10.1186/1472-6947-11-51

55. Yasodhara, A., Asgarian, A., Huang, D., Sobhani, P.: On the trustworthiness of tree ensemble explainability methods. In: Holzinger, A., Kieseberg, P., Tjoa, A.M., Weippl, E. (eds.) CD-MAKE 2021. LNCS, vol. 12844, pp. 293–308. Springer, Cham (2021). https://doi.org/10.1007/978-3-030-84060-0_19

56. Hancox-Li, L.: Robustness in Machine Learning Explanations: Does It Matter? (2020)

57. Brooke, J.: SUS-A quick and dirty usability scale. Usability Eval. Ind. **189**(194), 4–7 (1996)

58. Holzinger, A., Langs, G., Denk, H., Zatloukal, K., Müller, H.: Causability and explainability of artificial intelligence in medicine. WIREs Data Min. Knowl. Discov. **9**(4), e1312 (2019). https://doi.org/10.1002/widm.1312

59. Holzinger, A., Carrington, A., Müller, H.: Measuring the quality of explanations: the system causability scale (SCS). KI - Künstliche Intelligenz **34**(2), 193–198 (2020). https://doi.org/10.1007/s13218-020-00636-z

60. Fiala, P., Jablonský, J., Maňas, M.: Vícekriteriální rozhodování. Vysoká škola ekonomická v Praze (1994)

61. Saaty, T.L.: The Analytic Hierarchy Process: Planning, Priority Setting, Resource Allocation. McGraw-Hill International Book Company (1980)

62. Hwang, C.L., Yoon, K.: Multiple Attribute Decision Making: Methods and Applications A State-of-the-Art Survey. Springer, Heidelberg (1981). https://doi.org/10.1007/978-3-642-483 18-9

An Empirical Analysis
of Synthetic-Data-Based Anomaly
Detection

Majlinda Llugiqi[1,2] and Rudolf Mayer[1,2]

[1] Vienna University of Technology, Vienna, Austria
[2] SBA Research, Vienna, Austria
rmayer@sba-research.org

Abstract. Data is increasingly collected on practically every area of human life, e.g. from health care to financial or work aspects, and from many different sources. As the amount of data gathered grows, efforts to leverage it have intensified. Many organizations are interested to analyse or share the data they collect, as it may be used to provide critical services and support much-needed research. However, this often conflicts with data protection regulations. Thus sharing, analyzing and working with those sensitive data while preserving the privacy of the individuals represented by the data is needed. Synthetic data generation is one method increasingly used for achieving this goal. Using synthetic data would useful also for anomaly detection tasks, which often contains highly sensitive data.

While synthetic data generation aims at capturing the most relevant statistical properties of a dataset to create a dataset with similar characteristics, it is less explored if this method is capable of capturing also the properties of anomalous data, which is generally a minority class with potentially very few samples, and can thus reproduce meaningful anomaly instances. In this paper, we perform an extensive study on several anomaly detection techniques (supervised, unsupervised and semi-supervised) on credit card fraud and medical (annthyroid) data, and evaluate the utility of corresponding, synthetically generated datasets, obtained by various different synthetisation methods. Moreover, for supervised methods, we have also investigated various sampling methods; sampling in average improves the results, and we show that this transfers also to detectors learned on synthetic data. Overall, our evaluation shows that models trained on synthetic data can achieve a performance that renders them a viable alternative to real data, sometimes even outperforming them. Based on the evaluation, we provide guidelines on which synthesizer method to use for which anomaly detection setting.

Keywords: Anomaly detection · Synthetic data · Privacy preserving · Machine learning

© IFIP International Federation for Information Processing 2022
Published by Springer Nature Switzerland AG 2022
A. Holzinger et al. (Eds.): CD-MAKE 2022, LNCS 13480, pp. 306–327, 2022.
https://doi.org/10.1007/978-3-031-14463-9_20

1   Introduction

With increased data collection, also data analysis becomes more wide-spread. One important data analysis task is anomaly detection [5], which aims at finding unusual behavior in datasets. Anomalies may be caused by variations in machine behavior, fraudulent behavior, mechanical defects, human error, instrument error and natural deviations in populations [15]. Anomalies in data lead to important actionable information in a broad range of application domains, including cyber-security intrusion detection, defect detection of safety-critical devices, credit card fraud and health-care [5].

From the above mentioned examples it becomes clear that many detection tasks will operate on sensitive and personal data, and therefore techniques to use those data while preserving privacy are needed. Thus, data privacy has become a concern also among the anomaly detection research community. Different approaches for data privacy were proposed. One of those is K-anonymity which protects against a single record linking threat [34]. Different releases, however, might be connected together to compromise k-anonymity. Extension of the original concept, such as l-diversity, protect against further risks, such as attribute disclosure. However, these have shown to destroy the data utility too much [3].

Another approach is generating synthetic data. The fundamental concept behind synthetic data is to sample from suitable probability distributions to replace some or all of the original data, while preserving their important statistical features [26]. In this paper, we explore in depth the problem of anomaly detection models learned from synthetic data. We analyse different synthetic data generation methods and different anomaly detection methods, to answer whether it can be used as a suitable surrogate for utilising the original data – which might not be a viable option due to data sharing or usage limitations.

We use three different data synthesizers, and multiple different supervised, semi-supervised and unsupervised techniques to detect anomalies, on two frequently used, benchmark data sets with sensitive data (credit card fraud and annthyroid) in the anomaly detection domain. Moreover, since the imbalanced data problem is a core issue in anomaly detection, for the supervised setting, we balance the data using three separate sampling approaches: oversampling, undersampling, and the Synthetic Minority Oversampling Technique (SMOTE) [6], and compare that effect also on the synthetic data.

The remained of this paper is organised as follows. Section 2 discusses related work and state of the art results. We describe our evaluation setting in Sect. 3, and discuss results in Sect. 4. We then conclude in Sect. 5 and discuss directions for future work.

2   Related Work

Privacy preserving of sensitive data has been the subject of extensive research, and several different approaches have been considered. K-anonymity [34] has been a traditional solution to privacy concerns. However, concerns over residual risks, and lack of utility [3] have given rise to other techniques. One recently widely studied approach is synthetic data generation. Synthetic data is created by building a model based on real-world source data, from which samples are drawn to form a surrogate dataset. While the data is (close to) statistically indistinguishable from the real dataset, it no longer has a link to real individuals, and thus can be used, exchanged and transferred with less restrictions.

Rubin et al. [32] were among the first to propose synthetic data generation for disclosure control, namely repeated perturbation of the original data as a replacement for the original data. Ping et al. use a Bayesian network-based data synthesis approach, the *Data Synthesizer* [28]. They further also provide an independent attribute mode, which generates data for each attribute independently of the others. The *Synthetic Data Vault* (SDV) offers among other approaches one based on a multivariate version of the Gaussian Copula to model the covariances between the columns in addition to the distributions [27]. Nowok et al. propose a technique based on classification and regression trees (CART) in their *Synthpop* tool [26]. The synthetic values for the attributes are created progressively from their conditional distributions. Acs et al. [2] utilise generative neural networks. They first cluster the initial datasets into k clusters, and build synthesizer models for each cluster.

A high utility of the generated synthetic data is vital to successfully substitute original data. Several earlier works have specifically analysed and measured the utility of synthetic data for certain data analysis tasks, termed the application fidelity [7], e.g. for classification [12] or regression [13], or more generically supervised learning tasks [31], or for specific data types, e.g. microbiome data [14]. An earlier work addressing specific anomaly detection in synthetic data is provided by [23], focusing on a single dataset. We extend their work by considering more datasets, more detection techniques, and the incorporation of sampling methods, which are shows improvements for several of the supervised setting. Further, we re-create state-of-the-art results on the original dataset to find a more viable baseline for adequately assessing the comparative performance of the synthetic data based detectors, and thus achieve higher scores than [23] in the baseline.

Anomaly detection is a form of unbalanced data problems [20]. In anomaly detection, the majority of samples are "normal" data, whereas the minority samples are anomaly data. Fraud detection [39], disease detection [18], intrusion detection [19], identification systems [16], and fault diagnostics [29] are some example application domains. There are several ways to categorize anomaly detection methods. Goldstein et al. [11] distinguished three settings based on the availability of the data as illustrated in Fig. 1: **Supervised** anomaly detection refers to a setting in which the data consists of training and test data sets labeled with normal and anomaly instances. **Semi-supervised** anomaly detection also employ training and test datasets, with training data consisting only of normal data, but no anomalies. A model is learnt on the normal class, and then anomalies may be found if they are deviating from that model. **Unsupervised** anomaly detection does not require any labels and no differentiation is made between a training and a test dataset.

A further distinction can be on the learning approach. **Classification-based** techniques learn a machine learning model from a set of labeled examples, and

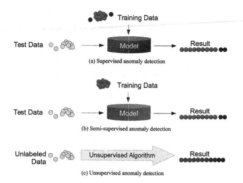

Fig. 1. Different anomaly detection modes depending on the availability of labels [11]

then classify a test instance into one of the classes using that model [5]. These techniques are used in supervised settings.

Statistical anomaly detection methods fit a statistical model (typically for normal behavior) to the available data, and then use a statistical inference test to check if an unseen instance fits the distribution [5]. Anomalies are assumed to occur in the low probability areas of the model. Statistical techniques can be used in an unsupervised context, without the necessity for labeled training data, if the distribution estimate phase is robust to data anomalies.

Clustering-based can be one of three categories. The first category is based on the premise that normal behavior data are grouped into clusters whereas anomalies are those samples not belonging to any of the clusters. The second category is based on the idea that anomalies are far away from their nearest cluster centroid, whereas normal data instances are close. One issue is that if data anomalies create clusters on their own, these methods will fail to detect them. To address this problem, a third category is based on the premise that anomalies belong to small or sparse clusters. Clustering is primarily an unsupervised technique. However, approaches from the second category can work in a semi-supervised mode.

Information theoretic approaches uses different information theoretic measures, such as Kolmogorov or entropy, to examine the information content of a data collection [5]. Anomalies are detected based on the assumption that anomalies in data cause inconsistencies in the data set's information content. Information theory techniques can be used in an unsupervised setting.

The specific techniques and algorithms that we have used for anomaly detection are described in Sect. 3.4.

3 Experiment Setting

In our experiment we used two datasets, *credit card fraud*[1] and *annthyroid*[2]. An overview of the experiment process can be seen in Fig. 2. It starts with the

---

[1] https://www.kaggle.com/mlg-ulb/creditcardfraud.
[2] https://archive.ics.uci.edu/ml/datasets/Thyroid+Disease.

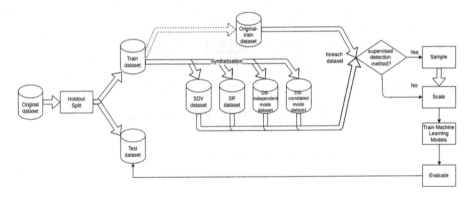

Fig. 2. Experiment flow diagram

original datasets; to be able to evaluate the synthesizers, both datasets are split into train-validation splits (holdout method). From the train splits of the datasets we generate synthetic data with each synthesizer. After the synthetization step, depending if the target variable is available, the original and synthesized train splits are sampled using one of the sampling techniques mentioned in Sect. 3.3. As part of the training process, to improve the model performance, the hyperparameters are tuned using a grid search, for which the values of the hyperparameters are shown in Table 1. In the end the performance (effectiveness) is evaluated on the validation set with the best performing model from the train phase using the F2 score, accuracy and precision, which are described below.

In a two-class problem, we try to tell the difference between anomalies and normal behaviour data. If the anomaly class is considered as "positive", and the non-anomalies as "negative" class, we distinguish four outcomes: "true positives" for correctly predicted anomalies, "false positive" for normal samples incorrectly predicted anomalies, "true negative" for correctly predicted normal samples as such, and "false negative" for incorrectly predicting an anomaly as normal sample. Based on this, we use the following evaluation metrics: **Precision**, which is the ability of a classification model to identify only the relevant data points, and can be calculated as the number of true positives divided by the number of true positives plus the number of false positives.

Recall, which is the ability of a model to find all the relevant cases within a dataset and can be calculated as the number of true positives divided by the number of true positives plus the number of false negatives.

Precision and recall are not representative on their own, as it is trivial to improve one at the cost of the other, but difficult to have both with high values at the same time. Therefore, we also use the **F1 and F2 scores**, which combines precision and recall. In contrast to the balanced F1, F2 puts an emphasis on recall, which is suitable for anomaly detection, where it is more critical to identify the majority of anomalies, and a certain amount of false positives may be allowed.

3.1 Datasets

The credit card dataset (see Footnote 1) includes credit card transactions made by European cardholders with two days in September 2013. It contains 492 frauds out of 284,807 transactions, and is thus highly unbalanced, with the positive class (frauds) accounting for just 0.172% of all transactions.

The dataset has 31 features, all of them are numerical. Features V1,...,V28 are the principal components obtained with a Principal Component Analysis (PCA). Further, one attribute holds the amount of the transaction, and the attribute 'Time' represents the time since the first transaction in the dataset.

The Annthyroid dataset (see Footnote 2) includes patients records from Garavan Institute. This dataset comprises of 7,200 records and has 22 features, all numerical. The target class contains "hyperfunction", "subnormal functioning" and "normal" (not hypothyroid), respectively. As common in literature [10,21], we grouped hyperfunction and subnormal functioning as one group that represents anomalies in this dataset. Thus, 534 entries are anomalies, i.e. 7.4%.

3.2 Dataset Synthetization

For generating synthetic data we used SDV, Synthpop and two modes of Data-Synthesizer, independent attribute and correlated attribute mode. We thus generate four synthetic datasets for each datasets.

3.3 Dataset Pre-procesing

As the datasets are highly imbalanced, we employ three well known sampling techniques, namely Random Undersampling, Random Oversampling and Synthetic Minority Oversampling Technique (SMOTE) [6], to potentially improve the performance of supervised techniques (sampling can only be applied if we have labels for both anomalies and normal data). We used the implementation from the python package "imblearn"[3].

3.4 Anomaly Detection Methods

We use different supervised, semi-supervised and unsupervised machine learning techniques from the Python machine-learning library sk-learn[4].

The supervised methods include: (i) Ada Boost, (ii) XGB (extrem gradient boosting), (iii) Gaussian Naive Bayes, (iv) Linear SVC, (v) k-nearest Neighbors, (vi) Random Forest, and (vii) Logistic Regression.

This selection of supervised machine learning techniques includes a wide range of different approaches, including probabilistic, linear, and rule-based classifiers, and ensemble techniques.

For semi-supervised techniques, we use: (i) AutoEncoder, and (ii) Gaussian Mixture.

---

[3] https://imbalanced-learn.org/.
[4] https://scikit-learn.org/stable/modules/outlier_detection.html.

The selected unsupervised approaches are: (i) Isolation Forest, (ii) Local Outlier Factor, (iii) One Class SVM.

Table 1. Parameter grid for supervised methods

Method	Parameter	(Grid) values
GaussianNB	var_smoothing	$[1.0e^{-02}, 1.0e^{-04}, 1.0e^{-09}]$
k-NN	n_neighbors	[5, 10, 15]
Random Forest	n_estimators	[100, 200, 300]
Logistic Regression	C	np.logspace $(-4, 4, 3)$
Linear SVC	C	[1, 100, 1000]
AdaBoost	n_estimators	[100, 200, 300]
XGB	n_estimators	[100, 200, 300]
IsolationForest	n_estimators	[100, 200, 300]
LocalOutlierFactor	n_neighbors	[5, 10, 15]
OneClassSVM	gamma	$[1.0e^{-03}, 1.0e^{-05}, 1.0e^{-08}]$
GaussianMixture	n_components; n_init	1; 5
AutoEncoder	epochs; batch-size	10; 128

To improve the results for the supervised methods, we executed a grid search on the training set through a number of parameters and values, shown in Table 1. Each of the parameter values are taken through the full pipeline and evaluated on a five fold cross-validation inside the training set. For evaluation we used an unseen validation set, consisting of 20% of the original data.

4   Results

In this section, we present the results from our evaluation, based on the setup described in Sect. 3. We primarily discuss the F2 scores, but provide further results details, namely precision and recall for supervised, semi-supervised and unsupervised methods, for both datasets, in Sect. A.

4.1   Credit Card Dataset

From Table 2, we can observe that for the credit card dataset, in most cases, sampling methods can not improve the scores, with a few exceptions (such as random oversampling or SMOTE for the already very successful XGB on real data). In average over all classifiers, sampling decreases the performance. For better visual comparison, Fig. 3 shows F2 scores for the different synthesizers and supervised methods when no sampling method was used.

We can further see that the performance drops when learning anomaly detection on real versus the synthetic data, implying that in general, it is slightly more difficult to learn an anomaly representation after synthesizing the data. However, in many settings, the drop is relatively small (within a few percent), and there are even a few cases where the best performance on real data is increased after synthetization, e.g. for the Gaussian Naive Bayes (from an albeit low base) and Logistic Regression without sampling.

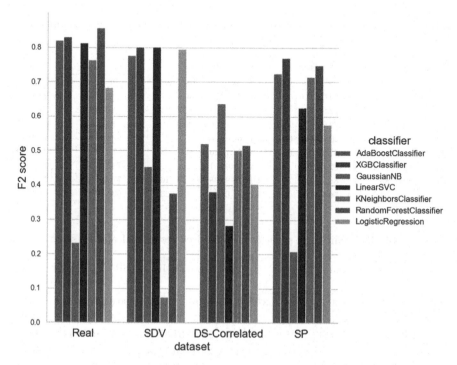

Fig. 3. Credit card: F2 scores for supervised techniques on datasets generated using no sampling

Moreover from Tables 2, 12 and 13 when analyzing the synthetic datasets, we can see that the best performing synthesizers are the Synthetic Data Vault (SDV) and Synthpop (SP); the latter is the best choice for k-NN and Random Forest. Overall, it is difficult to generally recommend which of these synthesizers is the better choice.

Regarding individual classifier performance, from the average columns in Tables 2, 12 and 13 we can observe that XGB achieves a relatively low recall of 77.5% compared to the other supervised methods, however, it has the highest F2 score and precision of 58.3% and 57.5% respectively, making it the most suitable for outlier detection tasks on the credit card dataset. On the other hand, we can see that Gaussian Naive Bayes is not useful for this task, achieving a good recall of 80%, but a very low precision and F2 score, 8.4% and 25.1% respectively, which is 49.1 and 33.2% points less than the best performing classifier on the table, respectively.

When looking only at the sampling methods, it can be observed that on average none of the sampling methods provide a performance increase. There are some exceptions for specific cases where the sampling methods helped such as when synthesizing with DataSynthesizer in correlated mode, using XGB classifier and SMOTE sampling method, where the F2 score is increased by relative 65.1%.

Table 2. Credit card: supervised results, F2 score (ROS: random oversampling, RUS: random undersampling, SM: SMOTE sampling)

Dataset	Real				SDV				DS-Corr.				SP				Avg
Sampl.	No	ROS	RUS	SM	No	ROS	RUS	SM	No	ROS	RUS	SM	No	ROS	RUS	SM	
AB	82.0	65.5	12.3	59.0	77.6	59.9	17.3	74.7	52.0	46.8	17.4	16.8	72.6	37.1	10.2	29.2	45.7
GNB	23.2	21.1	16.0	22.8	45.5	1.2	4.5	0.1	63.7	43.9	44.7	42.6	21.0	19.0	13.3	18.6	25.1
KNN	76.4	79.4	24.8	72.2	7.5	18.5	3.5	3.1	50.2	18.8	5.9	10.2	71.4	56.3	21.1	50.8	35.6
LSVC	81.3	26.7	7.5	30.9	80.2	63.7	53.3	64.8	28.3	27.5	31.4	29.7	62.6	15.9	14.4	15.9	39.6
LR	68.2	24.2	14.6	42.3	79.5	62.2	53.5	64.1	40.4	24.7	28.4	27.6	57.7	14.1	12.7	22.6	39.8
RF	85.6	81.4	18.7	82.7	37.7	0.0	6.5	1.0	51.7	3.8	33.7	28.3	74.8	75.7	11.4	76.3	41.8
XGB	83.0	83.9	14.3	86.4	80.2	78.6	9.0	78.9	38.1	57.9	16.8	62.9	77.1	78.3	12.1	76.1	58.3
Avg	71.4	54.6	15.4	56.6	58.3	40.6	21.1	41.0	46.4	31.9	25.5	31.2	62.5	42.3	13.6	41.4	

Table 3. Credit card: semi- & unsupervised results, F1 and F2 score

Dataset method	Real		SDV		DS-Corr.		DS-Ind.		SP		Avg	
	F1	F2	F1	F2	F1	F2	F1	F2	F1	F2	F1	F2
AutoEncoder	30.4	43.2	24.4	37.9	37.2	48.6	46.3	55.2	30.6	44.1	33.8	45.8
GMM	23.4	34.1	21.2	34.4	27.6	42.5	43.1	52.4	26.5	36.6	28.4	40.0
Isol.Forest	7.4	16.2	8.0	17.3	15.3	29.9	34.3	31.3	6.7	15.0	14.3	22.0
LOF	0.8	1.8	4.7	10.8	17.5	26.9	20.5	37.1	1.5	3.2	9.0	16.0
1-ClassSVM	0.6	1.6	0.8	1.9	1.8	4.2	2.4	5.6	0.6	1.7	1.23	3.0
Avg	12.5	19.4	11.8	20.5	19.9	30.4	29.3	36.3	13.2	20.1		

Figure 4 and Table 3 show the results for semi- and unsupervised settings. Mind again that for these technique, we can not apply sampling strategies, as there is no knowledge about the different types of samples (anomaly or not) for the unsupervised, and no information about relative sizes of the two classes for the semi-supervised case, as in this case, we only have samples from the "normal" class. Therefore, the corresponding tables and figures only show results without any sampling.

The first observation is that unsupervised and semi-supervised methods on average perform worse than the supervised counterparts, the major impact is on precision, where many methods struggle to obtain high values. However, this is expected, as this setting is more difficult, due to less information that can be exploited. In line with this, the semi-supervised methods (AutoEncoder and GMM) work better than the unsupervised ones.

Another noteworthy finding is that, in contrast to the supervised techniques, for unsupervised and semi-supervised techniques synthesizing the data increases the precision, recall and F2 score on average. This improvement in F2 score is 27.6% for semi-supervised and unsupervised approaches.

When comparing synthetic datasets, from Tables 3, 8 and 9 we can see that in semi-supervised and unsupervised setting the best performing synthesizer is DataSynthesizer in independent mode with an average F2 score of 36.3%. This is especially intriguing when compared to the supervised techniques, where there was a significant loss of quality (not depicted) when the data was synthesized with DataSynthesizer in this mode. On the other hand, the synthetic dataset generated by SDV is the worst performing synthesizer for semi-supervised methods with an average F2 of 36.15%, whereas for unsupervised methods the worst performing synthesizer is Synthpop with an average F2 score of 6.3%, due to a very low precision of only 1.5%.

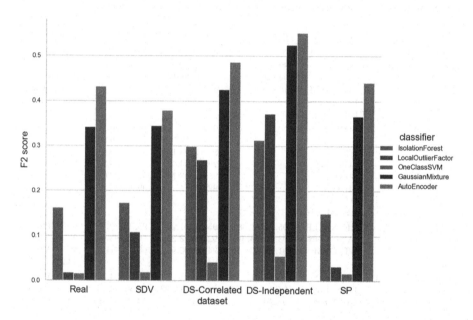

Fig. 4. Credit card: F2 scores for semi-supervised and unsupervised techniques

For the unsupervised methods, Isolation Forest obtains the best precision values, as well as a high recall value, resulting in the highest F2 score of 22%, whereas One Class SVM obtains the lowest F2 score of 3%, due to 0.6% precision. A significantly higher performance is achieved by semi-supervised methods. On average across the original and synthetic datasets, Auto Encoder outperforms Gaussian Mixture Model with an average precision of 23.8%, which is 4.3 higher than the Gaussian Mixture Model's precision and with an average recall of 61.4%, which is 4.5 higher than Gaussian Mixture Model's recall. Thus, also in regards to the F2 score, Auto Encoder outperforms the Gaussian Mixture Model, with an average F2 score of 45.8%, which is 5.8 higher than the Gaussian Mixture Model's average F2 score.

In Table 4 we show the results that are achieved in the literature using several different machine learning methods. If we compare these results with our results for the credit card dataset from Tables 2 and 13, we can observe that we achieved better results for Logistic Regression, Random Forest and XGB compared to [22], with 20.8, 26.8 and 21.6 higher precision. Mittal et al. [25] has the best precision for these three classifiers, achieving a precision of 99% for all three classifiers, which is 15.9, 4.7 and 5 higher than our precision for these three methods. On the other hand, Dornadula et al. [9] have a significantly better performance for Random Forest and Logistic Regression when using a sampling

Table 4. Credit card: benchmark algorithms and results (SM: SMOTE sampling, US: undersampling; '*' indicates scores we calculated from other scores)

Authors	Methods	Recall	Precision	Accuracy	F2-score
Yann-Ael Le Borgne[a]	LR, RF, XGB	N\A	62.3, 67.8, 69.4	N\A	N\A
Mittal et al. [25]	NB, RF, KNN, LR, XGB, SVM, IF, LOF, K-Means	82, 99, 0, 99, 99, 93, 100, 100, 0	6, 99, 0, 99, 99, 0, 99, 99, 99	N\A	23.2*, 99*, 0*, 99*, 99*, 0*, 99.8*, 99.8*, 0*
Dornadula et al. [9]	LOF, IF, SVM, LR, DT, RF, LOF-SM, IF-SM, LR-SM, DT-SM, RF-SM	N\A	0.38, 1.47, 76.81, 87.5, 88.54, 93.10, 29.41, 94.47, 98.31, 98.14, 99.96	89.90, 90.11, 99.87, 99.90, 99.94, 99.94, 45.82, 58.83, 97.18, 97.08, 99.98	N\A
Trivedi et al. [37]	RF, NB, LR, SVM, KNN, DT, GBM	95.12, 91.98, 93.11, 93, 92, 91.99, 93	95.98, 91.20, 92.89, 93.23, 94.59, 90.99, 94	95, 91.89, 90.45, 93.96, 95, 91, 94	95.29*, 91.8* 93.1*, 93.05*, 92.5*, 91.8*, 93.2*
Dhankhad et al. [8]	SC, RF, XGB, KNN, LR, GB, MLP, SVM, DT, NB	95, 95, 95, 91, 94, 94, 93, 93, 91, 91	95, 95, 95, 91, 94, 94, 93, 93, 91, 91	95.27, 94.59, 94.59, 94.25, 93.92, 93.58, 93.24, 93.24, 90.88, 90.54	95*, 95*, 95*, 91*, 94*, 94*, 93*, 93*, 91*, 91*
Bachmann[b]	LR, KNN, SVM, SVC, LR-US, LR-SM	94, 93, 93, 93, N\A, N\A	94, 93, 94, 93, N\A,N\A	94, 93, 93, 93, 94.21, 98.70	94*, 93*, 93.2*, 93*, N\A, N\A
Mayer et al. [23]	NB, SVM, KNN, RF, LR, IF, LOF, 1CSVM, GMM, AE	76.8, 70.5, 77.7, 71.4, 76.8, 76.8, 31.3, 83, 71.4, 55.4	6.6, 88.8, 94.6, 97.6, 86, 5.4, 0.6, 2.8, 87, 19.3	N\A	24.5, 73.6, 80.6, 75.5, 78.5, 21, 2.7, 12.4, 74.1, 40.3

[a]https://github.com/Fraud-Detection-Handbook/fraud-detection-handbook
[b]https://www.kaggle.com/janiobachmann/credit-fraud-dealing-with-imbalanced-datasets

method to balance the dataset, with around 17 and 84 higher precision, respectively compared to our results. Overall, our results are however well in line with what is achievable in literature.

When comparing our unsupervised approaches to those in the literature, from Tables 4, 8 and 9 we can observe that our results for Isolation Forest and LOF slightly outperform the results from [9], achieving 2.43 and 0.02 better precision, respectively. On the other hand, for these two methods Mittal et al. [25] achieved an almost perfect results, with 100% recall and 99% precision for both methods, which is highly suspect due to the lack of similar results on any state-of-the-art paper; also, the paper is not clear on which data they evaluate on, i.e., whether they use an independent validation set, or not – hence, these results should be taken with a grain of salt.

4.2 Annthyroid Dataset

When comparing synthesizers with each other, we can observe from Tables 5, 14 and 15 that SDV achieves by far the lowest precision, recall and F2 scores. Data generated by Synthpop, on the other hand, achieves the best performance, with achieving even better precision and F2 score than the original dataset. On average over all sampling methods, the dataset generated by Synthpop achieves the best F2 score of 77.05%, due to the highest precision and recall of 68.5% and 86.95% respectively.

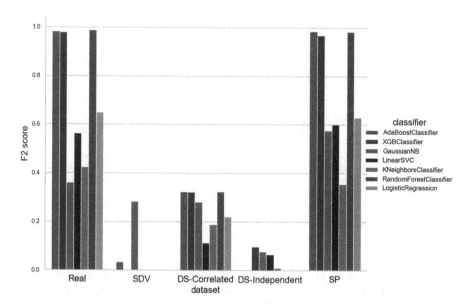

Fig. 5. Annthyroid: F2 scores for supervised techniques on datasets generated using no sampling

Table 5. Annthyroid: supervised results, F2 score (ROS: random oversampling, RUS: random undersampling, SM: SMOTE sampling)

Dataset	Real				SDV				DS-Corr.				SP				Avg
Sampl.	No	ROS	RUS	SM	No	ROS	RUS	SM	No	ROS	RUS	SM	No	ROS	RUS	SM	
AB	98.3	98.3	98.2	97.2	3.5	30.7	46.7	57.0	32.3	44.2	39.7	48.4	98.3	98.0	96.1	97.6	67.8
GNB	36.1	33.8	32.8	34.3	28.4	28.5	40.2	28.8	28.1	31.8	34.6	32.1	57.6	35.8	35.8	32.5	34.4
KNN	42.5	55.2	51.1	55.7	0.0	9.7	36.2	25.3	18.9	29.1	37.2	34.1	35.5	45.8	45.3	50.8	35.8
LSVC	56.2	80.3	64.9	68.2	0.0	41.2	54.4	45.6	11.4	55.7	52.5	53.8	60.0	88.5	80.8	74.6	55.5
LR	65.0	94.7	87.5	95.0	0.0	44.6	53.5	44.5	22.1	55.6	52.9	53.7	63.0	93.1	92.7	93.9	63.2
RF	98.7	98.5	97.3	98.5	0.0	0.0	62.2	23.5	32.4	36.2	44.4	47.8	98.3	98.3	95.9	99.1	64.4
XGB	97.9	98.5	97.1	98.5	0.0	5.8	55.5	23.6	32.1	34.9	40.3	46.9	96.7	97.4	96.7	99.1	63.8
Avg	70.7	79.9	75.5	78.2	4.5	22.9	49.8	35.4	25.3	41.1	43.1	45.3	72.8	79.6	77.6	78.2	

Table 6. Annthyroid: semi- & unsupervised results, F1 and F2 score

Dataset method	Real		SDV		DS-Corr.		DS-Ind.		SP		Avg	
	F1	F2	F1	F2	F1	F2	F1	F2	F1	F2	F1	F2
AutoEncoder	23.1	37.7	18.6	33.5	18.0	31.0	18.6	34.5	23.0	35.3	20.3	34.4
GMM	13.8	28.6	13.8	28.6	13.8	28.6	13.8	28.6	13.8	28.6	13.8	28.6
Isol.Forest	10.9	9.3	16.2	18.5	6.5	5.3	5.4	4.3	10.2	8.4	9.8	9.1
LOF	18.0	17.8	16.8	22.1	7.5	6.9	11.1	8.6	21.0	20.7	14.9	15.2
1-ClassSVM	15.8	28.1	13.4	19.6	15.4	27.5	14.4	21.6	15.6	28.1	14.9	25.0
Avg	16.3	24.3	15.8	24.5	12.2	19.9	12.7	19.5	16.7	24.2		

Table 7. Algorithms used in literature and their results for annthyroid dataset ('*' indicates scores we calculated from other scores in the table)

Authors	Methods	Recall	Precision	Accuracy	F2-score
Salman et al. [33]	DT, SVM, RF, NB, LR, KNN	N\A	N\A	90.13, 92.53, 91.2, 90.67, 91.73, 91.47	N\A
Sidiq et al. [35]	KNN, SVM, DT, NB	N\A	N\A	91.82, 96.52, 98.89, 91.57	N\A
Chandel et al. [4]	KNN, NB	N\A	N\A	93.44, 22.56	N\A
Sinhya et al. [36]	NB, RF	N\A	N\A	95, 99.3	N\A
Ionita et al. [17]	NB, DT, MLP	N\A	N\A	91.63, 96.91, 95.15	N\A
Maysanjaya et al. [24]	RBF, LVQ, MLP, BPA, AIRS	95.3, 93.5, 96.7, 69.8, 93.5	95.3, 94, 96.8, 48.7, 93.5	95.35, 93.5, 96.74, 69.77, 93.5	95.3*, 93.6*, 96.72*, 65.23*, 93.5*
Tyagi et al. [38]	KNN, SVM, DT	N\A	N\A	98.62, 99.63, 75.76	N\A
Raisinghani et al. [30]	SVM, DT, LR, RF	96, 99, 97, 99	96, 99, 97, 99	96.25, 99.46, 97.5, 99.3	96*, 99*, 97*, 99*
Rehman et al. [1]	KNN, DT, NB, SVM, LR	90, 67, 100, 70, 88	N\A	91.39, 74.19, 100, 80.46, 90.32	N\A

Another interesting observation can be seen in the 'avg' columns, where we can see that Gaussian Naive Bayes, even though it achieves the best recall on average that ranges around 81.9%, has a very low precision of 13.8%, making it not suitable for our task. The other algorithms have a significantly higher precision, with Adaboost being the best with an average F2 score of 67.8%, due to good precision and recall of 68.7% and 74% respectively. Random Forest, XGB and Logistic Regression are the next best algorithms, and achieve an average F2 score of 64.4%, 63.8% and 63.2% respectively.

When comparing the sampling methods used to balance the datasets, we can see that, contrary to the credit card dataset's results, in almost all of the cases balancing helped to achieve better results. This is especially true for SDV and DataSynthesizer in correlated mode, though from a rather low base. When comparing the sampling methods between each other, we see that the best performing sampling method is Random Undersampling, followed by SMOTE with a slightly lower performance, whereas supervised methods performed worst on the datasets sampled with Random Oversampling.

Figure 5 depicts an average comparison of F2 scores for the different synthesizers and supervised methods when no sampling method was used.

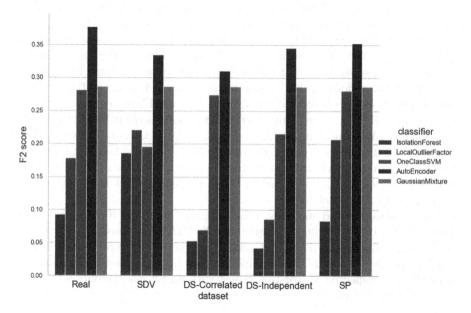

Fig. 6. Annthyroid: F2 scores for semi-supervised and unsupervised techniques

Table 7 shows the results reported in literature. When we compare these with our results in Table 15, we can see that for Random Forest we achieve slightly better recall compared to [30], but 1.8 lower precision. For Logistic Regression, Raisinghani et al. [30] and Rehman et al. [1] achieved better recall than we do, with 36.3 and 27.3 higher recall. Overall, our results are well embedded into the state-of-the-art results.

Tables 6, 10 and 11 show F2 scores, precisions and recalls for semi-supervised and unsupervised methods on original and synthetic datasets. Another visual representation of F2 scores is shown in Fig. 6 for semi-supervised and unsupervised methods. As with the credit card dataset, and as expected, supervised approaches outperform unsupervised and semi-supervised ones. Moreover, similar to the supervised techniques, synthesizing the data decreases the F2 score on average by 9.5%, from 24.3% to 22%, this decrease on F2 score is due to precision and recall, with 11.7% and 5.8% decrease respectively. Thus, for the annthyroid dataset, synthesizing the dataset marginally decreases the performance of semi-supervised and unsupervised methods. An exception is the Gaussian Mixture Model, which achieves same F2 score, precision and recall, before and after synthesizing the dataset with any synthesizer that we have considered in our experiment.

Moreover, from Tables 6, 10 and 11 we can observe that datasets generated with Synthpop and Synthetic Data Vault outperform the datasets generated with DataSynthesizer in both correlated and independent modes. For semi-supervised methods dataset generated with Synthpop outperforms the other synthesizers, with a precision of 11%, which is on average around 17.8% better than other synthesizers, and with a F2 score of 31.95%, which is 3.6% better than other synthesizers. On the other hand, semi-supervised methods perform worst on data generated by DataSynthesizer in correlated mode. On the other hand, when we compare the unsupervised methods to each other, data generated by Synthetic Data Vault outperforms other synthesizers, whereas data generated by DataSynthesizer in independent mode results in the worst F2 scores.

As expected, in all cases semi-supervised methods perform better than unsupervised ones. We note that Auto Encoder outperforms Gaussian Mixture Model in terms of F2 score and precision achieving on average 34.4% F2 score, and 12.0% precision over original and synthetic datasets. On the other hand GMM achieves the best recall compared to other semi-supervised and unsupervised methods, with a value of 100%.

As for unsupervised methods, even though One Class SVM shows a low precision of only 9%, having the highest recall of 47.3% compared to other unsupervised methods makes it the best performing method with a F2 score of 25%, which is around 10 and 16% points higher than Local Outlier Factors and Isolation Forest, respectively. Isolation Forest with an F2 score of only 9.1% is not well suited for this task.

4.3   Observations

Based on the results shown in the Sects. 4.1 and 4.2 we can conclude that the quality of anomaly detection in both credit card and annthyroid datasets for supervised methods is slightly reduced when using synthetic data, but they achieve results that are very competitive to the real data. As expected, unsupervised and semi-supervised approaches perform worse than supervised ones, which is expected, considering that having an actual label for the data provides more exploitable information than the other approaches have available.

In both datasets, XGB performed very well in the supervised techniques, while Gaussian Naive Bayes yields the worst results. When looking at semi-supervised methods for both datasets AutoEncoder outperforms Gaussian Mixture Model. Furthermore, the best synthesizer for the credit card dataset and annthyroid dataset in almost all of the cases is Synthpop. SDV performed well on the credit card dataset, but not so well on annthyroid. DataSynthesizer was overall the worst of the synthesizers.

One of the most noteworthy distinctions is the effect of sampling methods on the results in the two cases for supervised methods. For the annthyroid dataset, sampling significantly improved the performance of supervised methods in most cases. On the other hand, in most of the cases it has bad effect on the quality of anomaly detection in credit card dataset. We see that both effects also transferred to synthetic data, that is, for annthyroid, also synthetic data profits from choosing a fitting sampling methods.

When looking at non-sampled datasets for supervised techniques, another significant difference is the influence synthetization has on the data for the two datasets. After synthesizing the credit card dataset using the Synthetic Data Vault, the performance dropped only marginally, whereas the same synthesizer on the annthyroid dataset incurs a substantial performance drop. This indicates that the type of data and its distribution has a significant impact on the synthetisation success. An interesting observation is how unsupervised methods performed differently on annthyroid and credit card datasets when the datasets are synthesized with DataSynthesizer in the independent attribute mode. Whereas on the annthyroid dataset this mode shows the worst results among the other synthetic datasets, it achieves the best F2 score of all synthesisers on the credit card dataset. This might be caused by the nature of the features in this dataset, which are obtained via PCA, a dimensionality reduction technique.

5   Conclusion

Data privacy tries to balance non-disclosing management of sensitive data, while also preserving data utility. Synthetic data generation is one method that recently gained a lot of attention, and has shown to exhibit high utility for several tasks. In this paper, we used multiple metrics to assess the application fidelity of existing synthetic data creation strategies on datasets from the financial and health domain. We used state-of-the-art synthetic data generators for synthesizing the dataset; due to highly unbalanced datasets we applied several sampling approaches. Moreover, supervised, unsupervised and semi-supervised anomaly detection methods were used.

The results reveal that Synthpop overall outperforms the other synthetic data generators. As a guideline, it is thus a good overall choice to employ. In a few cases, especially on some of the supervised classifiers on the credit card data, the SDV outperformed Synthpop, though. Further, in the semi- and unsupervised settings on the same dataset, the DataSynthesizer yielded substantially better results. These might both correlate with the specific nature of most of the variables in this dataset. Thus, the best choice without any prior knowledge still remains Synthpop; however, if possible, a wider range of synthesizers should be tested before deciding on a specific approach, to account for particular attributes, distributions and correlations within a dataset.

Moreover, we have seen that XGB, Adaboost, and RandomForest outperform the other supervised approaches in our datasets, whereas Isolation Forest outperformed the other unsupervised techniques we used. For semi-supervised settings, the AutoEncoder was the best choice, and should be utilised first. While these two are thus recommended as guidelines for the semi- and unsupervised settings, for the supervised methods, we can generally recommend ensemble-based techniques as the ones mention above, of which in particular XGB was most successful in our evaluation.

Future work will focus on generalising our results beyond our current evaluation, in particular by addressing further datasets. Also other synthetisation methods, e.g. based on Generative Adversarial Networks (GANs), will be considered.

Acknowledgement. This work was partially funded by the Austrian Research Promotion Agency FFG under grants 877173 (GASTRIC) and 871267 (WellFort). SBA Research (SBA-K1) is a COMET center within the COMET - Competence Centers for Excellent Technologies program, funded by BMK, BMDW, and the federal state of Vienna. The COMET program is managed by FFG.

A   Appendix: Additional Results

A.1   Credit Card Dataset Results

Tables 8 and 9 show the precision respectively recall for semi-supervised and unsupervised techniques on original and synthetic datasets, whereas Tables 12 and 13 show the precision respectively recall for supervised methods.

A.2   Annthyroid Dataset Results

Tables 10 and 11 show the precision respectively recall for semi-supervised and unsupervised techniques on original and synthetic datasets, whereas Tables 14 and 15 show the precision respectively recall for supervised methods.

Table 8. Credit card: semi- & unsupervised results, precision

Dataset method	Real	SDV	DS-Corr.	DS-Ind.	SP	Avg
AutoEncoder	20.3	15.3	26.7	36.5	20.3	23.8
GMM	15.4	12.9	17.4	33.3	18.2	19.5
Isol.Forest	3.9	4.2	8.4	40.8	3.5	12.2
LOF	0.4	2.4	11.1	11.7	0.8	5.3
1-ClassSVM	0.3	0.4	0.9	1.2	0.3	0.6
Avg	8.1	7.0	12.9	24.7	8.6	

Table 9. Credit card: semi- & unsupervised results, recall

Dataset method	Real	SDV	DS-Corr.	DS-Ind.	SP	Avg
AutoEncoder	60.2	60.2	61.2	63.3	62.2	61.4
GMM	49.0	59.2	66.3	61.2	49.0	56.9
Isol.Forest	81.6	79.6	83.7	29.6	84.7	71.8
LOF	14.3	85.7	41.8	81.6	13.3	47.3
1-ClassSVM	96.9	95.9	93.9	91.8	96.9	95.1
Avg	60.4	76.1	69.4	65.5	61.2	

Table 10. Annthyroid: semi- & unsupervised results, precision

Dataset method	Real	SDV	DS-Corr.	DS-Ind.	SP	Avg
AutoEncoder	14.0	10.7	10.6	10.5	14.5	12.0
GMM	7.4	7.4	7.4	7.4	7.4	7.4
Isol.Forest	15.5	13.3	10.4	10.3	16.0	13.1
LOF	18.1	12.0	9.0	21.6	21.4	16.4
1-ClassSVM	9.1	8.8	8.9	9.2	9.0	9.0
Avg	12.8	10.4	9.3	11.8	13.7	

Table 11. Annthyroid: semi- & unsupervised results, recall

Dataset method	Real	SDV	DS-Corr.	DS-Ind.	SP	Avg
AutoEncoder	65.4	72.0	59.8	80.4	55.1	66.5
GMM	100	100	100	100	100	100
Isol.Forest	8.4	20.6	4.7	3.7	7.5	9.0
LOF	17.8	28.0	6.5	7.5	20.6	16.1
1-ClassSVM	58.9	28.0	57.0	32.7	59.8	47.3
Avg	50.1	49.7	45.6	44.9	48.6	

Table 12. Credit card: supervised results, precision (ROS: random oversampling, RUS: random undersampling, SM: SMOTE sampling)

Dataset	Real				SDV				DS-Corr.				SP				Avg
Sampl.	No	ROS	RUS	SM	No	ROS	RUS	SM	No	ROS	RUS	SM	No	ROS	RUS	SM	
AB	87.8	32.6	2.8	25.9	67.5	29.6	4.1	57.7	52.0	17.1	4.1	4.2	69.2	11.0	2.2	7.9	29.7
GNB	5.9	5.2	3.7	5.8	18.5	0.2	0.9	0.0	32.0	14.8	15.2	14.7	5.2	4.6	3.1	4.5	8.4
KNN	84.9	68.6	6.4	46.6	75.0	10.0	0.7	0.7	80.4	28.8	1.3	2.6	75.8	40.7	5.3	21.7	34.3
LSVC	84.0	7.0	1.6	8.5	74.8	33.9	22.3	35.6	75.0	7.3	8.8	8.1	81.7	3.7	3.3	3.7	28.7
LR	83.1	6.1	3.3	13.6	72.1	31.9	22.5	34.5	56.1	6.3	7.6	7.3	71.1	3.2	2.9	5.7	26.7
RF	94.3	95.1	4.5	82.7	71.7	0.0	1.4	0.9	66.7	100.0	9.6	75.0	80.9	81.1	2.5	60.6	51.7
XGB	94.0	89.0	3.3	85.0	74.8	78.6	2.0	76.5	63.0	66.3	4.0	61.4	84.1	77.0	2.7	57.9	57.5
Avg	76.3	43.4	3.7	38.3	64.9	26.3	7.7	29.4	60.7	34.4	7.2	24.8	66.9	31.6	3.1	23.1	

Table 13. Credit card: supervised results, recall (ROS: random oversampling, RUS: random undersampling, SM: SMOTE sampling)

Dataset	Real				SDV				DS-Corr.				SP				Avg
Sampl.	No	ROS	RUS	SM	No	ROS	RUS	SM	No	ROS	RUS	SM	No	ROS	RUS	SM	
AB	80.6	87.8	90.8	86.7	80.6	80.6	86.7	80.6	52.0	82.7	87.8	68.4	73.5	91.8	90.8	87.8	81.8
GNB	84.7	87.8	87.8	87.8	71.4	90.8	82.7	4.1	84.7	86.7	86.7	80.6	84.7	87.8	84.7	87.8	80.0
KNN	74.5	82.7	86.7	83.7	6.1	23.5	84.7	31.6	45.9	17.3	66.3	39.8	70.4	62.2	85.7	76.5	58.6
LSVC	80.6	91.8	94.9	90.8	81.6	81.6	81.6	81.6	24.5	89.8	87.8	89.8	59.2	90.8	92.9	91.8	82.0
LR	65.3	91.8	91.8	89.8	81.6	81.6	81.6	81.6	37.8	89.8	88.8	89.8	55.1	92.9	92.9	89.8	81.4
RF	83.7	78.6	90.8	82.7	33.7	0.0	86.7	1.0	49.0	3.1	89.8	24.5	73.5	74.5	91.8	81.6	59.1
XGB	80.6	82.7	91.8	86.7	81.6	78.6	85.7	79.6	34.7	56.1	88.8	63.3	75.5	78.6	92.9	82.7	77.5
Avg	78.6	86.2	90.7	86.9	62.4	62.4	84.3	51.5	46.9	60.8	85.1	65.2	70.3	82.7	90.2	85.4	

Table 14. Annthyroid: supervised results, precision (ROS: random oversampling, RUS: random undersampling, SM: SMOTE sampling)

Dataset	Real				SDV				DS-Corr.				SP				Avg
Sampl.	No	ROS	RUS	SM	No	ROS	RUS	SM	No	ROS	RUS	SM	No	ROS	RUS	SM	
AB	95.5	95.5	91.5	93.8	100.0	64.4	18.0	37.6	83.3	22.1	14.8	27.5	92.2	90.7	82.9	89.2	68.7
GNB	10.4	9.4	9.1	9.6	7.5	7.6	37.0	7.7	25.0	10.3	10.6	9.2	37.8	10.2	10.2	9.0	13.8
KNN	93.0	35.0	21.9	28.7	0.0	26.5	12.0	18.1	77.3	17.9	14.2	15.6	89.2	32.0	20.0	29.0	33.1
LSVC	90.2	45.7	38.4	31.9	0.0	16.4	24.1	21.8	90.9	37.4	27.9	35.7	83.3	73.0	61.0	47.7	45.3
LR	90.3	78.1	67.8	79.3	0.0	21.6	25.1	21.5	83.3	35.8	28.6	35.3	87.5	77.2	73.6	82.5	55.5
RF	97.2	96.4	87.7	96.4	0.0	0.0	27.6	53.7	85.7	81.0	18.0	48.6	95.5	95.5	82.3	95.5	66.3
XGB	97.2	96.4	87.0	96.4	0.0	83.3	23.2	38.3	76.9	73.3	15.3	41.3	94.5	94.6	85.6	95.5	68.7
Avg	82.0	65.2	57.6	62.3	15.4	31.4	23.8	28.4	74.6	39.7	18.5	30.4	82.9	67.6	59.4	64.1	

Table 15. Annthyroid: supervised results, recall (ROS: random oversampling, RUS: random undersampling, SM: SMOTE sampling)

Dataset	Real				SDV				DS-Corr.				SP				Avg
Sampl.	No	ROS	RUS	SM	No	ROS	RUS	SM	No	ROS	RUS	SM	No	ROS	RUS	SM	
AB	99.1	99.1	100.0	98.1	2.8	27.1	77.6	65.4	28.0	58.9	68.2	59.8	100.0	100.0	100.0	100.0	74.0
GNB	94.4	96.3	94.4	96.3	91.6	91.6	41.1	91.6	29.0	66.4	80.4	85.0	66.4	96.3	95.3	94.4	81.9
KNN	37.4	64.5	76.6	72.9	0.0	8.4	72.9	28.0	15.9	34.6	62.6	48.6	30.8	51.4	66.4	62.6	45.9
LSVC	51.4	99.1	78.5	95.3	0.0	66.4	79.4	62.6	9.3	63.6	67.3	61.7	56.1	93.5	87.9	86.9	66.2
LR	60.7	100.0	94.4	100.0	0.0	60.7	74.8	60.7	18.7	64.5	67.3	61.7	58.9	98.1	99.1	97.2	69.8
RF	99.1	99.1	100.0	99.1	0.0	0.0	90.7	20.6	28.0	31.8	70.1	47.7	99.1	99.1	100.0	100.0	67.8
XGB	98.1	99.1	100.0	99.1	0.0	4.7	85.0	21.5	28.0	30.8	68.2	48.6	97.2	98.1	100.0	100.0	67.4
Avg	77.2	93.9	92.0	94.4	13.5	37.0	74.5	50.1	22.4	50.1	69.2	59.0	72.6	90.9	92.7	91.6	

References

1. Abbad Ur Rehman, H., Lin, C.-Y., Mushtaq, Z., Su, S.-F.: Performance analysis of machine learning algorithms for thyroid disease. Arab. J. Sci. Eng. **46**(10), 9437–9449 (2021). https://doi.org/10.1007/s13369-020-05206-x
2. Acs, G., Melis, L., Castelluccia, C., De Cristofaro, E.: Differentially private mixture of generative neural networks. IEEE Trans. Knowl. Data Eng. **31**(6), 1109–1121 (2019). https://doi.org/10.1109/TKDE.2018.2855136

3. Brickell, J., Shmatikov, V.: The cost of privacy: destruction of data-mining utility in anonymized data publishing. In: ACM SIGKDD International Conference on Knowledge Discovery and Data Mining, KDD, Las Vegas, Nevada, USA. ACM Press (2008). https://doi.org/10.1145/1401890.1401904

4. Chandel, K., Kunwar, V., Sabitha, S., Choudhury, T., Mukherjee, S.: A comparative study on thyroid disease detection using K-nearest neighbor and Naive Bayes classification techniques. CSI Trans. ICT **4**, 313–319 (2017). https://doi.org/10.1007/s40012-016-0100-5

5. Chandola, V., Banerjee, A., Kumar, V.: Anomaly detection: a survey. ACM Comput. Surv. **41**(3), 1–58 (2009). https://doi.org/10.1145/1541880.1541882

6. Chawla, N.V., Bowyer, K.W., Hall, L.O., Kegelmeyer, W.P.: SMOTE: synthetic minority over-sampling technique. J. Artif. Intell. Res. **16**, 321–357 (2002). https://doi.org/10.1613/jair.953

7. Dankar, F.K., Ibrahim, M.K., Ismail, L.: A multi-dimensional evaluation of synthetic data generators. IEEE Access **10**, 11147–11158 (2022). https://doi.org/10.1109/ACCESS.2022.3144765

8. Dhankhad, S., Mohammed, E., Far, B.: Supervised machine learning algorithms for credit card fraudulent transaction detection: a comparative study. In: IEEE International Conference on Information Reuse and Integration, IRI, Salt Lake City, UT. IEEE, July 2018. https://doi.org/10.1109/IRI.2018.00025

9. Dornadula, V.N., Geetha, S.: Credit card fraud detection using machine learning algorithms. Procedia Comput. Sci. **165**, 631–641 (2019). https://doi.org/10.1016/j.procs.2020.01.057

10. Goix, N.: How to evaluate the quality of unsupervised anomaly detection algorithms? In: ICML Anomaly Detection Workshop, New York, NY, USA, July 2016

11. Goldstein, M., Uchida, S.: A comparative evaluation of unsupervised anomaly detection algorithms for multivariate data. PLoS One **11**(4) (2016). https://doi.org/10.1371/journal.pone.0152173

12. Hittmeir, M., Ekelhart, A., Mayer, R.: On the utility of synthetic data: an empirical evaluation on machine learning tasks. In: International Conference on Availability, Reliability and Security, ARES, Canterbury, CA, United Kingdom. ACM, August 2019. https://doi.org/10.1145/3339252.3339281

13. Hittmeir, M., Ekelhart, A., Mayer, R.: Utility and privacy assessments of synthetic data for regression tasks. In: 2019 IEEE International Conference on Big Data (Big Data), Los Angeles, CA, USA. IEEE, December 2019. https://doi.org/10.1109/BigData47090.2019.9005476

14. Hittmeir, M., Mayer, R., Ekelhart, A.: Utility and privacy assessment of synthetic microbiome data. In: Sural, S., Lu, H. (eds.) DBSec 2022. LNCS, vol. 13383, pp. 15–27. Springer, Cham (2022). https://doi.org/10.1007/978-3-031-10684-2_2

15. Hodge, V., Austin, J.: A survey of outlier detection methodologies. Artif. Intell. Rev. **22**(2), 85–126 (2004). https://doi.org/10.1023/B:AIRE.0000045502.10941.a9

16. Ibidunmoye, O., Hernández-Rodriguez, F., Elmroth, E.: Performance anomaly detection and bottleneck identification. ACM Comput. Surv. **48**(1), 1–35 (2015). https://doi.org/10.1145/2791120

17. Ioniţǎ, I., Ioniţǎ, L.: Prediction of thyroid disease using data mining techniques. BRAIN. Broad Res. Artif. Intell. Neurosci. **7**(3), 115–124 (2016)

18. Jansson, D., Medvedev, A., Axelson, H., Nyholm, D.: Stochastic anomaly detection in eye-tracking data for quantification of motor symptoms in Parkinson's disease. In: International Symposium on Computational Models for Life Sciences, Sydney, Australia (2013). https://doi.org/10.1063/1.4825001

19. Kim, G., Lee, S., Kim, S.: A novel hybrid intrusion detection method integrating anomaly detection with misuse detection. Expert Syst. Appl. **41**(4), 1690–1700 (2014). https://doi.org/10.1016/j.eswa.2013.08.066

20. Kong, J., Kowalczyk, W., Menzel, S., Bäck, T.: Improving imbalanced classification by anomaly detection. In: Bäck, T., et al. (eds.) PPSN 2020. LNCS, vol. 12269, pp. 512–523. Springer, Cham (2020). https://doi.org/10.1007/978-3-030-58112-1_35

21. Lazarevic, A., Kumar, V.: Feature bagging for outlier detection. In: ACM SIGKDD International Conference on Knowledge Discovery in Data Mining, KDD, Chicago, Illinois, USA. ACM Press (2005). https://doi.org/10.1145/1081870.1081891

22. Le Borgne, Y.A., Siblini, W., Lebichot, B., Bontempi, G.: Reproducible Machine Learning for Credit Card Fraud Detection - Practical Handbook. Université Libre de Bruxelles (2022). https://github.com/Fraud-Detection-Handbook/fraud-detection-handbook

23. Mayer, R., Hittmeir, M., Ekelhart, A.: Privacy-preserving anomaly detection using synthetic data. In: Singhal, A., Vaidya, J. (eds.) DBSec 2020. LNCS, vol. 12122, pp. 195–207. Springer, Cham (2020). https://doi.org/10.1007/978-3-030-49669-2_11

24. Maysanjaya, I.M.D., Nugroho, H.A., Setiawan, N.A.: A comparison of classification methods on diagnosis of thyroid diseases. In: International Seminar on Intelligent Technology and Its Applications, ISITIA, Surabaya. IEEE, May 2015. https://doi.org/10.1109/ISITIA.2015.7219959

25. Mittal, S., Tyagi, S.: Performance evaluation of machine learning algorithms for credit card fraud detection. In: International Conference on Cloud Computing, Data Science & Engineering. Confluence, Noida, India. IEEE, January 2019. https://doi.org/10.1109/CONFLUENCE.2019.8776925

26. Nowok, B., Raab, G.M., Dibben, C.: synthpop: bespoke creation of synthetic data in R. J. Stat. Softw. **74**(11), 1–26 (2016). https://doi.org/10.18637/jss.v074.i11

27. Patki, N., Wedge, R., Veeramachaneni, K.: The synthetic data vault. In: IEEE International Conference on Data Science and Advanced Analytics, DSAA, Montreal, QC, Canada. IEEE, October 2016. https://doi.org/10.1109/DSAA.2016.49

28. Ping, H., Stoyanovich, J., Howe, B.: DataSynthesizer: privacy-preserving synthetic datasets. In: International Conference on Scientific and Statistical Database Management, SSDBM, Chicago, IL, USA. ACM, June 2017. https://doi.org/10.1145/3085504.3091117

29. Purarjomandlangrudi, A., Ghapanchi, A.H., Esmalifalak, M.: A data mining approach for fault diagnosis: an application of anomaly detection algorithm. Measurement **55**, 343–352 (2014). https://doi.org/10.1016/j.measurement.2014.05.029

30. Raisinghani, S., Shamdasani, R., Motwani, M., Bahreja, A., Raghavan Nair Lalitha, P.: Thyroid prediction using machine learning techniques. In: Singh, M., Gupta, P.K., Tyagi, V., Flusser, J., Ören, T., Kashyap, R. (eds.) ICACDS 2019. CCIS, vol. 1045, pp. 140–150. Springer, Singapore (2019). https://doi.org/10.1007/978-981-13-9939-8_13

31. Rankin, D., Black, M., Bond, R., Wallace, J., Mulvenna, M., Epelde, G.: Reliability of supervised machine learning using synthetic data in health care: model to preserve privacy for data sharing. JMIR Med. Inform. **8**(7) (2020). https://doi.org/10.2196/18910

32. Rubin, D., Reiter, J., Rubin, D.: Statistical disclosure limitation. J. Off. Stat. **9**(2), 461–468 (1993)

33. Salman, K., Sonuç, E.: Thyroid disease classification using machine learning algorithms. J. Phys. Conf. Ser. **1963**(1) (2021). https://doi.org/10.1088/1742-6596/1963/1/012140

34. Samarati, P.: Protecting respondents identities in microdata release. IEEE Trans. Knowl. Data Eng. **13**(6), 1010–1027 (2001). https://doi.org/10.1109/69.971193
35. Sidiq, U., Mutahar Aaqib, S., Khan, R.A.: Diagnosis of various thyroid ailments using data mining classification techniques. Int. J. Sci. Res. Comput. Sci. Eng. Inf. Technol. **5**(1), 131–136 (2019). https://doi.org/10.32628/CSEIT195119
36. Sindhya, K.: Effective prediction of hypothyroid using various data mining techniques. Int. J. Res. Dev. **5**(2), 311–317 (2020)
37. Trivedi, N.K., Simaiya, S., Lilhore, U.K., Sharma, S.K.: An efficient credit card fraud detection model based on machine learning methods. Int. J. Adv. Sci. Technol. **29**(5), 3414–3424 (2020)
38. Tyagi, A., Mehra, R., Saxena, A.: Interactive thyroid disease prediction system using machine learning technique. In: International Conference on Parallel, Distributed and Grid Computing, PDGC, Solan, Himachal Pradesh, India. IEEE, December 2018. https://doi.org/10.1109/PDGC.2018.8745910
39. Zhang, W., He, X.: An anomaly detection method for medicare fraud detection. In: IEEE International Conference on Big Knowledge, ICBK, Hefei, China. IEEE, August 2017. https://doi.org/10.1109/ICBK.2017.47

SECI Model in Data-Based Procedure for the Assessment of the Frailty State in Diabetic Patients

František Babič[1(✉)], Viera Anderková[1], Zvonimir Bosnić[2], Mile Volarić[3], and Ljiljana Trtica Majnarić[2]

[1] Department of Cybernetics and Artificial Intelligence, Faculty of Electrical Engineering and Informatics, Letná 1/9, 04001 Košice-Sever, Slovakia
{frantisek.babic,viera.anderkova}@tuke.sk
[2] Department of Family Medicine, Faculty of Medicine, Josip Juraj Strossmayer University of Osijek, 31000 Osijek, Croatia
[3] Faculty of Medicine, Josip Juraj Strossmayer University of Osijek, 31000 Osijek, Croatia

Abstract. The manuscript provides a theoretical consideration of the process of knowledge creation in the area of the complex medical domain and the case of unanswered medical uncertainties. We used the SECI model to help explain the challenges that must be faced and the complex structure of the procedures that stay herein. Typically, the SECI model proposed by Nonaka in 1994 represents the best-known conceptual framework for understanding organization knowledge generation processes. In this model, the knowledge is continuously converted and created within user practices, collaboration, interaction, and learning. This paper describes an application of the SECI model to the data-based procedure for assessing the frailty state of diabetic patients. We focused on effectively supporting collaboration and knowledge transfer between participating data analysts and medical experts. We used Exploratory Data Analysis, cut-off values extraction, and regression to create new knowledge (combination) based on the expressed tacit ones (externalization). Also, we used internalization and socialization to design experiments and describe the results achieved in the discussion. Finally, we could conclude that effective knowledge transfer, conversion, and creation, are the basis of every data-based diagnostic procedure. In the case of the complex medical domain, the role of the medical expert is more important than usual, and this aspect of knowledge creation is mainly unconscious in the scientific literature.

Keywords: Knowledge creation · Expert knowledge · The complex medical problems · Frailty · Diabetic patients

1 Introduction

Research in chronic aging diseases is challenging because chronic diseases in one person usually appear as two or more coexisting diseases, that is called multimorbidity [1]. Some characteristics of multimorbidity, such as disease clustering (non-random associations),

A. Holzinger et al. (Eds.): CD-MAKE 2022, LNCS 13480, pp. 328–342, 2022.
https://doi.org/10.1007/978-3-031-14463-9_21

overlapping (between individuals), mutual interactions, and accumulation over time, contribute to the heterogeneity of phenotypes of older individuals in the population and to the complexity of the health problems of an individual patient with multimorbidity [2, 3]. Both prevalence and the level of multimorbidity (measured by the number of diagnoses) increase with age and it is associated with a decline in physical, cognitive and social functioning of older individuals [4, 5]. However, there are great variations among older population in the expression of chronic diseases and functional impairments, reflecting differences in rates of aging, so that real (biological) age may not be complementary with the chronological age [6]. The position of an individual on trajectories of aging is influenced by many factors, including genetic, behavioral, personal, and societal ones, as well as by the efficiency of the delivered care [7, 8].

Diabetes type 2 is a complex chronic disease that involves metabolic, inflammatory, and vascular disorders [9]. Giving its widespread and serious health consequences, this disease represents a major public health concern [10]. This is especially so because of the close association of diabetes type 2 with cardiovascular disease (CVD) - the leading cause of the population mortality [11]. Most patients with diabetes type 2 are of older age (60 years and more) [12]. The fact that diabetes type 2 is an age-related disease indicates heterogeneity of these patients and increased susceptibility for development of multiple comorbidities and geriatric conditions such as malnutrition, sarcopenia (muscle mass loss) and frailty. Taken together, that means that patients diagnosed with diabetes type 2 show different phenotypes that can be graded according to the risk for negative health outcomes.

Frailty is a geriatric condition that attracts attention of many researchers in the recent years because it has been confirmed that this condition increases vulnerability of older individuals for many negative outcomes, including hospitalization, falls, disabilities, low functioning, and mortality [13]. This is an insufficiently understood concept which recognition is based on different scoring systems [14]. One of the simplest and best validated scoring system for assessing frailty defines this state with symptoms of shrinking (decreased muscle mass and strength), slowness (slow walking) and weakness (low activity and subjective exhaustion) [14]. In the 5-item grading system, 3–5 disorders indicate frailty (mostly an irreversible state), and 1–2 disorders indicate pre-frailty (mostly a reversible state).

Frailty is considered as the state of decreased homeostatic reserves in multiple organs and physiologic systems and as the final common pathway during development of multimorbidity [15]. There is a close association between diabetes type 2 and frailty [16]. This association may be a result of the high level of multimorbidity that often accompanies diabetes type 2, especially including CVD and chronic kidney disease, which both conditions are known to be associated with frailty [17, 18]. Moreover, the overlap between frailty and other geriatric conditions that are associated with body's shrinking and low physical performance, such as malnutrition and sarcopenia, are considered those factors that in older individuals may hamper the benefits of preventive pharmacological treatment of chronic diseases [19–21].

Which are, exactly, the clinical characteristics of diabetic patients who are also frail, are not well-known. Planning research in this medical domain is challenging because the knowledge representation base is weak. A researcher (medical expert) has to integrate information and facts from different areas of knowledge, and in all phases of research, from establishing a research question, *via* data preparing and processing, to making conclusions. A medical expert critically considers new results, comparing them with his/her existing knowledge, through the processes of retrieving and forwarding and "breaking a path in the jungle", the whole time leading a constructive conversation with a data scientist, and also learning from him/her. This process of knowledge creation in the area of chronic complex diseases, and especially in the face of new, complex issues, such as frailty, is poorly known, but it requires the same prerequisites as the process of decision making for complex medical problems – a sufficiently large knowledge base, and the methodological framework that is semantically similar to human reasoning [22]. The medical expert draws ideas from his/her internalized knowledge content, organized in the form of heuristics, rules, and concept schemes, which is intrinsic to the human reasoning. The wealth of ideas, and the complexity of cognitive patterns and connections between them, which an individual medical researcher possesses, depends on his/her expertize, creativity, and some other, less known, personal characteristics.

In this study, we used only a limited number of variables and a selected view on the problem of frailty, and simple analytical procedures, to illustrate some aspects of reasoning in the knowledge creation process, in this complex, and insufficiently informed area of research. Concretely, we expected that the status frailty is associated with significant changes in clinical features of diabetic patients, that can explain the higher health-related risks of frail vs. non-frail patients. We used variables indicating laboratory cardiovascular risk factors, anthropometric measures, the main entities of CVD, and a measure of renal function decline, proposing that these variables, selected from the existing data set, could contribute the most to the description of the frailty status.

The paper is organized as follows: the first section introduces the theoretical principles, our motivation, and used methods. The second section describes obtained results in line with all knowledge creation processes. The discussion and conclusion summarized obtained knowledge and experience with effective knowledge transfer between data analysts and medical experts.

1.1 SECI Model

Both knowledge base and methodological approaches in the process of knowledge creation, in the complex and weakly defined medical domains, are poorly known. As the model which we considered to provide the most appropriate theoretical framework in this issue, is the SECI model (Fig. 1). The SECI model is a model of knowledge creation explaining how tacit and explicit knowledge are converted into organizational knowledge [23]:

- *Socialization* represents a process of knowledge sharing through practice, guidance, and observation.
- *Externalization* is the process of making tacit knowledge explicit with the aim to be shared by others, becoming the basis of new knowledge.
- *Combination* involves organizing and integrating knowledge, whereby different types of explicit knowledge are merged.

- *Internalization* represents the receiving and application of knowledge by an individual, enclosed by learning-by-doing.

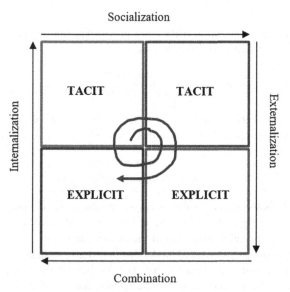

Fig. 1. SECI model.

1.2 Analytical Methods

John W. Tukey defined exploratory data analysis (EDA) as "detective work – or more precisely numerical detective work" [24]. It is mainly the philosophy of data analysis. EDA's primary goal is to examine the data for distribution, outliers, and anomalies to direct specific testing of the hypothesis. The EDA gives analysts unparalleled power for gaining insight into the data and their visualizations [25]. In solving complex medical problems, no one mathematical model can account for all steps of the cognitive reasoning process [22]. Based on the own experience, we state that only by using a combination of methods, it is possible to achieve a reasonable solution. Nevertheless, we chosen a simple research design, to rationalize the analytical steps in the knowledge creation procedure in the complex medical domain.

Date was used from two data sets, collected in the primary care setting, in the similar time frame, and with a similar study design. One data set contained only older (>50) diabetic patients (N = 170), and another data set contained older individuals (>60) of the general population. We used a part of this data set, that one which referred to diabetic patients (N = 52). The total number of subjects for this evaluation was 222 (M: 94, F: 128).

The sensitivity is a proportion of patients who test positive among all those who have the disease.

$$sensitivity = \frac{true\ positive}{true\ positive + false\ negative}$$

The specificity is a proportion of people who test negative among all those who do not have this disease.

$$specificity = \frac{true\ negative}{true\ negative + false\ positive}$$

The positive predictive value (PPV) is a probability that following a positive test result, that individual will truly have this disease.

$$PPV = \frac{true\ positive}{true\ positive + false\ positive}$$

The negative predictive value (NPV) is a probability that following a negative test result, that individual will truly not have this disease.

$$NPV = \frac{true\ negative}{true\ negative + false\ negative}$$

For diagnostic based on the continuous input variables, cut-off values represent dividing points on measuring scales where the test results are divided into target categories, typically positive or negative diagnosis of the investigated disease.

Logistic regression is a statistical model that in its basic form uses a logistic function to model a binary dependent variable [26]. The Akaike information criterion (AIC) is an estimator of prediction error and thereby relative quality of statistical models for a given set of data [27]. An odds ratio (OR) is a statistic that quantifies the strength of the association between two events. The analytical process has been guided by the clinical reasoning process of the medical expert. And *vice versa*, the medical expert has learnt on the results, which helped him/her make new cognitive associations within the bulk of internalized knowledge.

1.3   Related Work

In their recently published review paper, Al Saedi et al. have summarized the existing evidence on the commonly used biomarkers for diagnosing frailty and found a large set of biomarkers that they set up in several nosology groups, such as: musculoskeletal changes, serum markers, hormonal changes, stem-cell changes, metabolic markers, and inflammatory markers [28]. In our previous papers, we used a large data set and ML methods to identify clusters of a combination of two major age-related functional impairments – physical frailty and cognitive impairment, on older individuals from a primary care setting, and to associate comorbidity patterns with these clusters [29–31]. Knowledge is emerging on pathophysiology of frailty in diabetic patients, which might accelerate research on clinically relevant biomarkers of frailty and improve diagnostic criteria [32]. Diabetes type 2, although well-defined as a diagnostic entity, has been increasingly recognized as a heterogeneous and complex age-related disorder [9, 12]. However, it is insufficiently understood how particular factors influence differences in risks of these patients for the development of complications and bad health-related outcomes [33]. Many biochemical and molecular markers that are used in geriatric research show strong associations with sociodemographic and clinical features of the examined patient groups [34]. Hassler et al. conducted research to develop a prediction model of frailty from the large patient register and by testing different machine learning methods [35]. They concluded that data pre-processing is necessary for developing a model of good predictive performances. Although there is a tendency that data pre-paration pipeline becomes fully automatized, especially in the light of increasing availability of new, automated, deep-learning methods, the authors expressed their mistrust towards such possibility, considering the importance of the domain-knowledge for the transparency of the variables extracted in the model and model's practical utility. Almuayqil et al. investigated the effectiveness of knowledge management application in the healthcare sector in the Kingdom of Saudi Arabia [36]. They confirmed that the SECI model can encourage diabetes self-management and education. Centobelli et al. investigated the existing literature on the technological innovation of healthcare sector in terms of knowledge creation processes [37]. They confirmed that knowledge exchange between local services and hospitals, between patients and doctors, and health personnel is necessary, and it is a vital resource.

2 Results

The whole analytical process started with socialization based on the previous experience (data analysts) and daily routine and internalized knowledge (medical experts). Within the externalization, we provided the design of experiments, selection of dataset, and suitable analytical methods. The first results were the object of internalization for their evaluation and possible improvement. Further externalization was followed by a combination to provide a new set of results. And the individual step continued and repeated in the close collaboration between participating data analysts and medical experts. A very important prerequisite was setting up a common terminology to prevent misunderstandings and unnecessary experiments. Distinctly from the usual communication channels established between a medical expert and a data scientist, in the face of the need to solve a task from the complex and uncertain medical domain, the role played by the medical expert, and the quality and complexity of his/her internalized knowledge, gets a far bigger importance [38].

Almost half of patients diagnosed with diabetes type 2 were pre-frail (65) or frail (39). There was a clear gender bias in the frail patient group, with females (76.9%) dominating over males (23.1%). Frail patients were older than those pre-frail or non-frails; most of frail patients were older than 70 years and a part of them reached 80 and more years of age (Fig. 2).

Most diabetic patients had mildly reduced renal function (eGFR < 90–60); more female than male patients (Fig. 3).

In pre-frail (78.5%) and frail (71.8%) groups there were more patients with longer diabetes duration (≥ 5 years) than in the non-frail patient group (52.5%).

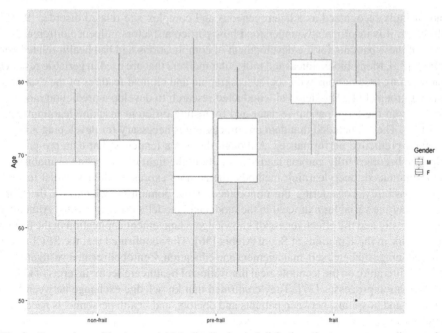

Fig. 2. Comparison of the Age variable's distribution in 3 diabetic patient groups according gender distribution.

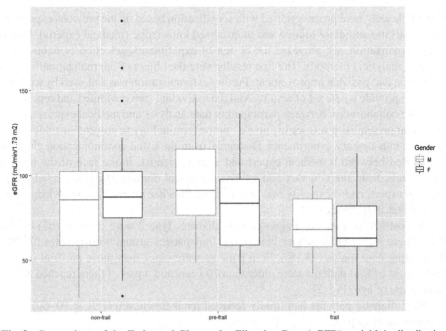

Fig. 3. Comparison of the Estimated Glomerular Filtration Rate (eGFR) variable's distribution in 3 diabetic patient groups according gender distribution.

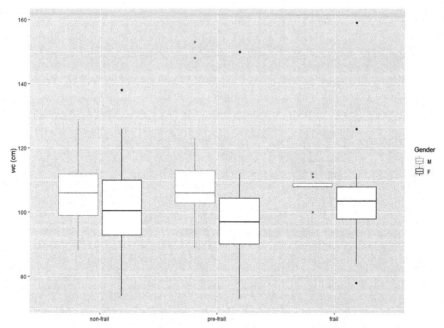

Fig. 4. Comparison of the Waist circumference (wc) variable's distribution in 3 diabetic patient groups according gender distribution.

In all diabetic patients, an average wc value indicated the presence of abdominal (visceral) obesity, i.e., >88 cm for women and >102 cm for men (Fig. 4). In pre-frail and frail patients, there is a tendency for the development of sarcopenia as indicated by mac values < 30 cm (Fig. 5).

The estimation of cut-off values of some selected variables has provided several rules (Table 1): 1) if diabetic patients are younger than 70 years, it can be said with high level of confidence that they are not frail - the same is visible also on Fig. 2) rarely, frail diabetic patients have waist circumference that is less than 110 cm - as visible also on Fig. 4; 3) diabetic patients usually have increased total serum cholesterol (>4.5) irrespective of the frailty status.

Finally, we investigated the possible relationships between the selected input variables (Age, eGFR, CAD, CHD, mac) within logistic regression (LR) models (frail + pre-frail vs. non-frail), separately for males and females. All input variables that are proposed as to be important in determining the frailty status, are closely associated with increased age – in both cases this variable was marked as significant based on the 0.05 level: male's model: AIC 116.31, Odds ratio 1.13 (1.07–1.20); female's model: AIC 167.14, Odds ratio 1.10 (1.05–1.15).

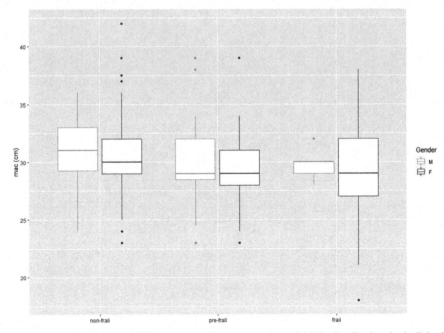

Fig. 5. Comparison of the Mid-arm circumference (mac) variable's distribution in 3 diabetic patient groups according gender distribution.

Table 1. Cut-off values of selected variables depending on the patient groups: non-frail, pre-frailty + frailty.

Variable	Value	Sensitivity	Specificity	PPV	NPV
Age	69	0.66	0.69	0.65	0.70
Estimated glomerular filtration rate	81	0.68	0.49	0.60	0.57
BMI	29.98	0.56	0.57	0.55	0.53
Waist circumference	110	0.32	0.78	0.62	0.50
Mid-arm circumference	30	0.69	0.54	0.63	0.61
Total cholesterol	4.5	0.75	0.31	0.55	0.52
LDL	3.0	0.61	0.47	0.57	0.52
HDL	1.5	0.39	0.64	0.49	0.55
Triglycerides	1.99	0.41	0.65	0.51	0.56

3 Discussion

The recent knowledge on pathophysiology factors that underlie physical frailty has revealed the existence of many loops that operate across different levels of the body's organization – from molecular and cellular to organ and system levels [19]. These patho-physiology mechanisms are associated with the aging process and can be potentiated with behavioral and social factors, such as sedentary behavior, poor nutrition, comorbidities, polypharmacy, and social isolation [14]. Therefore, an awareness is increasing among researchers in geriatric/gerontology medicine that multivariate methodologies should be an alternative approach to single-variable strategies when trying to get answers on research questions associated with aging and chronic age-related diseases [19].

In this study, we examined clinical characteristics of frail vs. non-frail older diabetic patients. Our intention was not searching for new characteristics of the frailty status or for the patterns of clinical variables that cluster with this condition, but rather, it was to show difficulties in the reasoning process that a researcher has to deal with when trying to reach understanding of the associations that are seen in multiple pairwise comparisons. Although we used individuals diagnosed with one well-defined disease label – diabetes type 2 – we have shown that by the simple selection of these patients with respect to the status frailty, it may reveal significant heterogeneity that exists among diabetic patients. Many variables were shown to influence the status frailty, including age, gender and the variables indicating comorbidities that share the common pathophysiology background with diabetes type 2, such as diagnoses of certain CVD entities, and different degrees of the renal function decline. We could conclude that frailty, in older diabetic patients, is associated with older age, female gender, the presence of CVD, lower renal function, and longer diabetes duration. These sociodemographic and clinical influencing factors are similar as stated in the literature [39]. However, the results gained by the data analytic procedure have not provided clear patterns, especially in the case of the regression models. In achieving the transparency and interpretability of the results, the role of the medical expert is of the utmost importance. To be able to gain a more comprehensive picture, as it is in the real-life situation, by viewing diabetic patients in their wholeness and plasticity, there was the need to integrate the results from different single-variable analyses and to explain them within the wider context of the existing knowledge. By striving to do so, we had to complement some single-variable comparisons with new ones, by means of "learning-by-doing", and *vice versa*, and by means of searching for more details in the context of single-variable associations. That has to be done so because when studying chronic age-related diseases and multimorbidity, single variables represent the elements that are drown from the networked reactions, where multi-collinearities and non-linear associations between paired variables are taking place [40]. Yet, by revealing some hidden layers of pairwise variable associations, and by partitioning patients into more narrowly defined subgroups, it is possible to achieve the sufficient clarification of the problem under investigation.

For example, the observed gender-related bias in frailty expression, with women showing higher predisposition for the development of frailty than men, can be explained by the fact that women live longer than men and the fact that CVD in women presents distinctly than in men [41]. In men, CVD usually presents as CAD and is caused by the coronary artery obstruction, which can lead to the premature death. On the contrary, women are more likely to get CHD (heart failure) without obstructive CAD, which develops to a greater extent later in the life-course, and is associated with sarcopenia and frailty [42, 43]. In addition, more intensive accumulation of CV risk factors in women in post-menopause, who present with recent onset diabetes type 2 (with less than 5 years of diabetes duration), compared to age-matched men, and a longer exposure of the kidneys to the effect of CV risk factors, may explain our results indicating higher susceptibility of women for lower renal function, which could also contribute to higher rates of frailty, seen in women [18, 44]. To reveal the hidden information that is associated with gender bias in the expression of frailty in older diabetic patients, we had to complement many pairwise comparisons with gender-related differences. We emphasize the importance of the medical expert's ability of making connections between the plain results and cognitive constructs of his/her own, which in the case of research in the complex medical domain requires knowledge combination from different areas of medical knowledge. When viewing this process in the light of the SECI model, there are intensive interactions and feed-back loops between the externalization, combination, and internalization phases of the knowledge creation process, and the need to integrate information from many steps in data analysis with the complex network of heuristics and patterns of the internalized (socialized) knowledge and experience of the medical expert.

Even when some variables seemed not to be influenced with the frailty status, like variables indicating serum lipids, these single variable-based comparisons should be used by a caution. For example, regarding serum total (and LDL) cholesterol, the values of these variables may be influenced by the hypolipidemic drugs, that are usually prescribed to diabetic patients but not to those aged 70 years and more, when, in turn, frailty starts to emerge at higher rates [45]. On the other hand, advancing in age and the presence of frailty, were both shown to be associated with lowering of the serum cholesterol levels [46]. This fact might have caused justification of the values of these variables across the three patient groups, as seen in our results.

There is a growing consensus among researchers in geriatric/gerontology research area that multiple data clustering may provide a more holistic view of diabetes type 2 and a research framework for delivering personalized care to these patients than single clinical features [47, 48]. Also our research group has recently published a paper in which we combined data of different types to identify clusters of older diabetic patients, which might have a potential for the prediction of outcomes and personalized treatments [49].

Based on the long-term experience in preparing multivariate research associated with chronic diseases and multimorbidity, our research group state that the sufficient understanding of the research problem is possible to get only by using multiple methods for data analysis, by repeating the analyses and retrieving from the knowledge base, in terms of the explainable AI, by a step-by-step supplementation of the results, and by achieving an efficient communication between the researcher and the data analyst [50]. The supra-level of the procedure interactions is needed when the research question is used from the complex and largely unknown medical domain. The role of the medical expert, in providing internalized knowledge, and in participating in the analytical process, is of the utmost importance. That is, what we wanted to emphasize, and what is still unconscious in the scientific literature.

We elaborated on the problem of distinguishing between what has been found by statistical analysis and the scientific truth [51]. This problem comes to mind when analyzing the research question associated with complex chronic diseases, where the existence of the multiple (albeit previously undefined) subject subgroups, and multiple possible (largely unknown) confounding factors, may affect the results of pairwise comparisons. The average values provided by statistical analyses do not fit all subjects, taken alone or as subgroups, as different subject groups are likely to be lumped together. In a situation when also the knowledge base is weak to allow research hypotheses to be more clearly defined in advance (and the possible input space is huge and weekly defined), a researcher (the medical expert) must jump here and there, comparing different (existent) knowledge bases (and his/her existed heuristics), with data-driven results, to fuel the research protocol further on.

4   Conclusion

To move the progress in research in chronic diseases and multimorbidity further on, it is necessary to establish a better understanding between researchers and data scientists, and to strive, in cooperation, to develop research protocols and new multivariate algorithms. The role of the medical expert's expertise, and his/her personal ability in making connections between different knowledge bases and in guiding the flow of the experiments that are being performed, is still unconscious part of the knowledge creation procedure in the complex medical domains.

This conclusion follows from the fact that the SECI model can be deployed effectively in the domain of medical diagnostics support. The primary goal of this model is new knowledge creation, which also corresponds to the motivation behind the medical diagnostic procedures. Based on the obtained experience, all four knowledge processes are important and complement to each other. For example, the combination (data analyst design and performs the experiments; expert provides the reliable interpretations) would not be possible without externalization (expert specifies the tasks and conditions based on his/her experience and existing tacit knowledge; data analyst interpret the results from an analytical point of view), internalization (expert understand the results provided in the simple understandable form; data analyst get new knowledge needed for more tailored experiments), and socialization (expert evaluates the results according to his experience or existing tacit knowledge; data analyst think about the improved design of the experiments based on the expert's feedback).

Our future work will be focused on the interpretability of the decision support models generated by suitable machine learning methods in chronic complex diseases and unanswered medical uncertainties.

Acknowledgements. The work was partially supported by The Slovak Research and Development Agency under grants no. APVV-20-0232; The Scientific Grant Agency of the Ministry of Education, Science, Research and Sport of the Slovak Republic under grant no. VEGA 1/0685/2; and the University of Osijek through the project IP2 - 2021 "Low Resilience to Chronic Stress and Chronic Aging Diseases".

References

1. Ogura, S., Jakovljevic, M.M.: Editorial: global population aging - health care, social and economic consequences. Front. Public Heal. **6**, 335 (2018). https://doi.org/10.3389/fpubh. 2018.00335
2. Whitty, C.J.M., et al.: Rising to the challenge of multimorbidity. BMJ (Clin. Res. Edn.) **368**, l6964 (2020). https://doi.org/10.1136/bmj.l6964
3. Nardi, R., Scanelli, G., Corrao, S., Iori, I., Mathieu, G., Cataldi Amatrian, R.: Co-morbidity does not reflect complexity in internal medicine patients. Eur. J. Intern. Med. **18**(5), 359–368 (2007). https://doi.org/10.1016/j.ejim.2007.05.002
4. Barnett, K., Mercer, S.W., Norbury, M., Watt, G., Wyke, S., Guthrie, B.: Epidemiology of multimorbidity and implications for health care, research, and medical education: a cross-sectional study. Lancet **380**(9836), 37–43 (2012). https://doi.org/10.1016/S0140-6736(12)60240-2
5. Hanlon, P., Nicholl, B.I., Jani, B.D., Lee, D., McQueenie, R., Mair, F.S.: Frailty and pre-frailty in middle-aged and older adults and its association with multimorbidity and mortality: a prospective analysis of 493 737 UK Biobank participants. Lancet. Public Heal. **3**(7), e323–e332 (2018). https://doi.org/10.1016/S2468-2667(18)30091-4
6. Franceschi, C., et al.: The continuum of aging and age-related diseases: common mechanisms but different rates. Front. Med. **5**, 61 (2018). https://doi.org/10.3389/fmed.2018.00061
7. CalderónLarrañaga, A., et al.: Multimorbidity and functional impairment-bidirectional interplay, synergistic effects and common pathways. J. Intern. Med. **285**(3), 255–271 (2019). https://doi.org/10.1111/joim.12843
8. Majnarić, L.T., et al.: Low psychological resilience in older individuals: an association with increased inflammation, oxidative stress and the presence of chronic medical conditions. Int. J. Mol. Sci. **22**(16), 8970 (2021). https://doi.org/10.3390/ijms22168970
9. Roden, M., Shulman, G.I.: The integrative biology of type 2 diabetes. Nature **576**(7785), 51–60 (2019). https://doi.org/10.1038/s41586-019-1797-8
10. Cho, N.H., et al.: IDF diabetes atlas: global estimates of diabetes prevalence for 2017 and projections for 2045. Diabetes Res. Clin. Pract. **138**, 271–281 (2018). https://doi.org/10.1016/j.diabres.2018.02.023
11. Mitchell, S., et al.: A roadmap on the prevention of cardiovascular disease among people living with diabetes. Glob. Heart **14**(3), 215–240 (2019). https://doi.org/10.1016/j.gheart. 2019.07.009
12. Bellary, S., Kyrou, I., Brown, J.E., Bailey, C.J.: Type 2 diabetes mellitus in older adults: clinical considerations and management. Nat. Rev. Endocrinol. **17**(9), 534–548 (2021). https://doi.org/10.1038/s41574-021-00512-2
13. Vermeiren, S., et al.: Frailty and the prediction of negative health outcomes: a meta-analysis. J. Am. Med. Dir. Assoc. **17**(12), 1163.e1–1163.e17 (2016). https://doi.org/10.1016/j.jamda. 2016.09.010
14. Dent, E., et al.: Physical frailty: ICFSR international clinical practice guidelines for identification and management. J. Nutr. Health Aging **23**(9), 771–787 (2019). https://doi.org/10. 1007/s12603-019-1273-z
15. Fried, L.P., et al.: Nonlinear multisystem physiological dysregulation associated with frailty in older women: implications for etiology and treatment. J. Gerontol. A. Biol. Sci. Med. Sci. **64**(10), 1049–1057 (2009). https://doi.org/10.1093/gerona/glp076
16. Sinclair, A.J., Rodriguez-Mañas, L.: Diabetes and frailty: two converging conditions? Can. J. Diabetes **40**(1), 77–83 (2016). https://doi.org/10.1016/j.jcjd.2015.09.004
17. Kleipool, E.E., et al.: Frailty in older adults with cardiovascular disease: cause, effect or both? Aging Dis. **9**(3), 489–497 (2018). https://doi.org/10.14336/AD.2017.1125

18. Walker, S.R., Wagner, M., Tangri, N.: Chronic kidney disease, frailty, and unsuccessful aging: a review. J. Ren. Nutr. Off. J. Counc. Ren. Nutr. Natl. Kidney Found. **24**(6), 364–370 (2014). https://doi.org/10.1053/j.jrn.2014.09.001

19. Calvani, R., et al.: The 'BIOmarkers associated with Sarcopenia and PHysical frailty in EldeRly pErsons' (BIOSPHERE) study: rationale, design and methods. Eur. J. Int. Med. **56**, 19–25 (2018). https://doi.org/10.1016/j.ejim.2018.05.001

20. Kurkcu, M., Meijer, R.I., Lonterman, S., Muller, M., de van der Schueren, M.A.E.: The association between nutritional status and frailty characteristics among geriatric outpatients. Clin. Nutr. ESPEN. **23**, 112–116 (2018). https://doi.org/10.1016/j.clnesp.2017.11.006

21. Onder, G., Vetrano, D.L., Marengoni, A., Bell, J.S., Johnell, K., Palmer, K.: Accounting for frailty when treating chronic diseases. Eur. J. Intern. Med. **56**, 49–52 (2018). https://doi.org/10.1016/j.ejim.2018.02.021

22. Bocklisch, F., Hausmann, D.: Multidimensional fuzzy pattern classifier sequences for medical diagnostic reasoning. Appl. Soft Comput. **66**, 297–310 (2018). https://doi.org/10.1016/j.asoc.2018.02.041

23. Nonaka, I.: A dynamic theory of organizational knowledge creation. Organ. Sci. **5**(1), 14–37 (1994). http://www.jstor.org/stable/2635068

24. Cox, V.: Translating Statistics to Make Decisions. Apress, New York (2017). https://doi.org/10.1007/978-1-4842-2256-0

25. Komorowski, M., Marshall, D.C., Salciccioli, J.D., Crutain, Y.: Secondary Analysis of Electronic Health Records, pp. 1–427 (2016). https://doi.org/10.1007/978-3-319-43742-2

26. Tolles, J., Meurer, W.J.: Logistic regression: relating patient characteristics to outcomes. JAMA **316**(5), 533–534 (2016). https://doi.org/10.1001/jama.2016.7653

27. Akaike, H.: A new look at the statistical model identification. IEEE Trans. Automat. Contr. **19**(6), 716–723 (1974). https://doi.org/10.1109/TAC.1974.1100705

28. Al Saedi, A., Feehan, J., Phu, S., Duque, G.: Current and emerging biomarkers of frailty in the elderly. Clin. Interv. Aging. **14**, 389–398 (2019). https://doi.org/10.2147/CIA.S168687

29. Majnarić, L.T., Bekić, S., Babič, F., Pusztová, Ľ, Paralič, J.: Cluster analysis of the associations among physical frailty, cognitive impairment and mental disorders. Med. Sci. Monit. Int. Med. J. Exp. Clin. Res. **26**, e924281 (2020). https://doi.org/10.12659/MSM.924281

30. Babić, F., Trtica Majnarić, L., Bekić, S., Holzinger, A.: Machine learning for family doctors: a case of cluster analysis for studying aging associated comorbidities and frailty. In: Holzinger, A., Kieseberg, P., Tjoa, A.M., Weippl, E. (eds.) Machine Learning and Knowledge Extraction. LNCS, vol. 11713, pp. 178–194. Springer, Cham (2019). https://doi.org/10.1007/978-3-030-29726-8_12

31. Bekić, S., Babič, F., Pavlišková, V., Paralič, J., Wittlinger, T., Majnarić, L.T.: Clusters of physical frailty and cognitive impairment and their associated comorbidities in older primary care patients. Healthcare. **9**(7), 891 (2021). https://doi.org/10.3390/healthcare9070891

32. Sinclair, A.J., Abdelhafiz, A.H., Rodriguez-Manas, L.: Frailty and sarcopenia-newly emerging and high impact complications of diabetes. J. Diabetes Compl. **31**(9), 1465–1473 (2017)

33. Howard, R., Scheiner, A., Kanetsky, P.A., Egan, K.M.: Sociodemographic and lifestyle factors associated with the neutrophil-to-lymphocyte ratio. Ann. Epidemiol. **38**, 11-21.e6 (2019). https://doi.org/10.1016/j.annepidem.2019.07.015

34. Zoungas, S., et al.: Impact of age, age at diagnosis and duration of diabetes on the risk of macrovascular and microvascular complications and death in type 2 diabetes. Diabetologia **57**(12), 2465–2474 (2014). https://doi.org/10.1007/s00125-014-3369-7

35. Hassler, A.P., Menasalvas, E., García-García, F.J., Rodríguez-Mañas, L., Holzinger, A.: Importance of medical data preprocessing in predictive modeling and risk factor discovery for the frailty syndrome. BMC Med. Inform. Decis. Mak. **19**(1), 33 (2019). https://doi.org/10.1186/s12911-019-0747-6

36. Almuayqil, S., Atkins, A.S., Sharp, B.: Application of the SECI model using web tools to support diabetes self-management and education in the kingdom of Saudi Arabia. Intell. Inf. Manag. **09**(05), 156–176 (2017). https://doi.org/10.4236/iim.2017.95008
37. Centobelli, P., Cerchione, R., Esposito, E., Riccio, E.: Enabling technological innovation in healthcare: a knowledge creation model perspective (2021)
38. Rokošná, J., Babič, F., Majnarić, L.T., Pusztová, L.: Cooperation between data analysts and medical experts: a case study. In: Holzinger, A., Kieseberg, P., Tjoa, A.M., Weippl, E. (eds.) Machine Learning and Knowledge Extraction. LNCS, vol. 12279, pp. 173–190. Springer, Cham (2020). https://doi.org/10.1007/978-3-030-57321-8_10
39. Chen, X., Mao, G., Leng, S.X.: Frailty syndrome: an overview. Clin. Interv. Aging **9**, 433–441 (2014). https://doi.org/10.2147/CIA.S45300
40. Majnarić, L.T., Babič, F., O'Sullivan, S., Holzinger, A.: AI and big data in healthcare: towards a more comprehensive research framework for multimorbidity. J. Clin. Med. **10**(4), 766 (2021). https://doi.org/10.3390/jcm10040766
41. Reynolds, H.R., et al.: Mechanisms of myocardial infarction in women without angiographically obstructive coronary artery disease. Circulation **124**(13), 1414–1425 (2011). https://doi.org/10.1161/CIRCULATIONAHA.111.026542
42. AlBadri, A., et al.: Inflammatory biomarkers as predictors of heart failure in women without obstructive coronary artery disease: a report from the NHLBI-sponsored Women's Ischemia Syndrome Evaluation (WISE). PLoS One **12**(5), e0177684 (2017). https://doi.org/10.1371/journal.pone.0177684
43. Beltrami, M., Fumagalli, C., Milli, M.: Frailty, sarcopenia and cachexia in heart failure patients: different clinical entities of the same painting. World J. Cardiol. **13**(1), 1 (2021)
44. Šabanović, Š., et al.: Metabolic syndrome in hypertensive women in the age of menopause: a case study on data from general practice electronic health records. BMC Med. Inform. Decis. Mak. **18**(1), 24 (2018). https://doi.org/10.1186/s12911-018-0601-2
45. Trtica Majnarić, L., Bosnić, Z., Kurevija, T., Wittlinger, T.: Cardiovascular risk and aging: the need for a more comprehensive understanding. J. Geriatr. Cardiol. **18**(6), 462–478 (2021). https://doi.org/10.11909/j.issn.1671-5411.2021.06.004
46. Strain, W.D., Down, S., Brown, P., Puttanna, A., Sinclair, A.: Diabetes and frailty: an expert consensus statement on the management of older adults with type 2 diabetes. Diabetes Therapy **12**(5), 1227–1247 (2021). https://doi.org/10.1007/s13300-021-01035-9
47. Ahlqvist, E., et al.: Novel subgroups of adult-onset diabetes and their association with outcomes: a data-driven cluster analysis of six variables. Lancet. Diabetes Endocrinol. **6**(5), 361–369 (2018). https://doi.org/10.1016/S2213-8587(18)30051-2
48. Dennis, J.M., Shields, B.M., Henley, W.E., Jones, A.G., Hattersley, A.T.: Disease progression and treatment response in data-driven subgroups of type 2 diabetes compared with models based on simple clinical features: an analysis using clinical trial data. Lancet. Diabetes Endocrinol. **7**(6), 442–451 (2019). https://doi.org/10.1016/S2213-8587(19)30087-7
49. Bosnic, Z., et al.: Clustering inflammatory markers with sociodemographic and clinical characteristics of patients with diabetes type 2 can support family physicians' clinical reasoning by reducing patients' complexity. Healthcare **9**(12), 1687 (2021). https://doi.org/10.3390/healthcare9121687
50. Longo, L., Goebel, R., Lecue, F., Kieseberg, P., Holzinger, A.: Explainable artificial intelligence: concepts, applications, research challenges and visions. In: Holzinger, A., Kieseberg, P., Tjoa, A., Weippl, E. (eds.) Machine Learning and Knowledge Extraction. CD-MAKE 2020. LNCS, vol. 12279. Springer, Cham (2020). https://doi.org/10.1007/978-3-030-57321-8_1
51. Gelman, A., Loken, E.: The garden of forking paths: why multiple comparisons can be a problem, even when there is no 'fishing expedition' or 'p-hacking' and the research hypothesis was posited ahead of time ∗ (2019)

Comparing Machine Learning Correlations to Domain Experts' Causal Knowledge: Employee Turnover Use Case

Eya Meddeb[1,2]($\boxtimes$) (iD), Christopher Bowers[1,2] (iD), and Lynn Nichol[2]

[1] Department of Computing, University of Worcester, Worcester WR1 3AS, UK
[2] Worcester Business School, University of Worcester, Worcester WR1 3AS, UK
{e.meddeb,c.bowers,l.nichol}@worc.ac.uk

Abstract. This paper addresses two major phenomena, machine learning and causal knowledge discovery in the context of human resources management. First, we examine previous work analysing employee turnover predictions and the most important factors affecting these predictions using regular machine learning (ML) algorithms, we then interpret the results concluded from developing and testing different classification models using the IBM Human Resources (HR) data. Second, we explore an alternative process of extracting causal knowledge from semi-structured interviews with HR experts to form expert-derived causal graph (map). Through a comparison between the results concluded from using machine learning approaches and from interpreting findings of the interviews, we explore the benefits of adding domain experts' causal knowledge to data knowledge. Recommendations are provided on the best methods and techniques to consider for causal graph learning to improve decision making.

Keywords: Machine learning · Causal knowledge · Decision-making · Human resource management

1 Introduction

Causal knowledge involves the awareness and understanding of real-world cause and effect relationships to predict the outcome of an action and the mechanism of transition from one state to another through different interventions [28,45]. Predictions based on causal knowledge are ubiquitous across many fields, yet it is still not extensively investigated in management areas.

Causal predictive models can answer the "What if" questions which typically arise in the context of strategic business problems. For example, human resources managers can test if increased teleworking would exert a positive influence on employee productivity and wellbeing before taking any actions [60], hence, causal understanding plays an important role in testing the outcomes of different scenarios to improve strategic business decisions [28].

© IFIP International Federation for Information Processing 2022
Published by Springer Nature Switzerland AG 2022
A. Holzinger et al. (Eds.): CD-MAKE 2022, LNCS 13480, pp. 343–361, 2022.
https://doi.org/10.1007/978-3-031-14463-9_22

"The goal of many sciences is to understand the mechanisms by which variables came to take on the values they have (that is, to find a generative model), and to predict what the values of those variables would be if the naturally occurring mechanisms were subject to outside manipulations" [57].

This paper focuses on a specific use case, employee turnover prediction. The term 'turnover' means that the employee leaves the company permanently and ends the relationship with the organisation. Scholars in this field correctly defined it as the rotation of employees around the market; between firms, jobs, and occupations; and between the states of employment and unemployment [30]. The main target for companies is to maintain a stable turnover rate since a high or low rate can be considered as an indicator of an issue within the organisation [55]. Although it is a topic relying on decision-making, it is rarely investigated from a causal knowledge perspective. In this paper, the authors highlight the differences between the results that can be concluded from regular machine learning models using a specific dataset and the actual causal knowledge concluded from semi-structured interviews with HR experts.

2   Predicting Employee Turnover Using Machine Leaning

The number of research contributions that aim at supporting the practical adoption of human resource management (HRM) using ML has been rapidly growing in the last couple of years [58,62]. These contributions refer to various HRM activities and processes, such as predicting and evaluating employee performance, predicting employee turnover, etc. [36]. Intelligent algorithms based on ML, can help in resolving some of the mentioned challenges as well as in increasing efficiency and effectiveness of HRM. However, managers should be mindful that the purpose of integrating these technological capabilities is not to replace humans, but rather to improve the decision making around people [24].

2.1   Related Work

The literature review is split into two sections, the first section is focusing on the combined techniques and ML algorithms used to predict employee turnover and the second one is highlighting features importance analysis.

Generally, common classification algorithms are used for turnover prediction including decision tree [3,15] and ensemble learning such as random forest and boosted tree methods [19,29,39,62]. Recent research in employee turnover prediction focus more on introducing new methods and combined techniques to improve models performance and results analysis. In [63], the authors suggest a novel time dependent method to predict the turnover called CoxRF by combining ensemble learning and survival analysis. Their results prove that this combination improved the accuracy of the model compared with the other regular classification algorithms. A new graph embedding method is presented in

[26] including new properties and clues from both internal and external views, the authors in this paper focus on the external-market view by developing a heterogeneous graph that connects the employees with the external job markets through shared skills to get the External-Market representation of each employee. Their results show that the model can achieve strong results in variety of situations compared with previous methods. In [12], the researchers propose a graph embedding method with temporal information called dynamic bipartite graph embedding (DBGE) to learn low-dimensional vector representation of employees and companies. After applying Learned features and basic features to machine leaning using multiple datasets, the results confirm that the proposed method is more effective in both employee turnover prediction and link prediction. Another recent study by [33] investigates the employee prediction from a causal point of view using European survey datasets. First they test classical ML techniques, then, they use structural causal modeling to validate the ranking of features importance concluded from ML algorithms. Light Gradient Boosting and logistic regression show the best performance.

For Features importance ranking, it is mostly determined after testing different models and choosing the best performance algorithm. Features are commonly derived from internal HR metrics (data captured and archived by HR departments). This step helps to understand the strongest correlations between the features implemented within the dataset and the employee turnover. For example, in [63], the results indicate that gender is one of the most relevant features to consider in predicting employee turnover followed by economic indicator Gross Domestic Product (GDP). The results in [62] confirm that last pay raise and job tenure are the most important features affecting the turnover prediction. However, the paper by [55] shows that features such as social interaction ability, age and marital status are insignificant to predict the turnover and that the number of previous job changes and knowledge about the working conditions are highly correlated to the possibility of turnover, similarly, the authors in [12] highlight the importance of including users' historical job records. In [26], the authors show the importance of considering job market condition beside internal factors in improving employee turnover prediction. Another layer is added in [33] to rank the most important features using structural causal modeling analysis to validate ML correlations. The results show that sustainable employability, employership, and attendance stability are among the top determinants of employee turnover. This demonstrates that features importance within an HR dataset is unique and so may differ between organisations. Therefore, for each organisation with a different dataset, the entire logic must be adapted based on the strongest correlations found within that specific data.

2.2   Testing Classification Algorithms on IBM HR Data

In this section, we go through the regular process of using supervised machine learning algorithms to predict employee turnover using the IBM HR data [18], which is commonly used in previous research [16,33]. A general overview of the analysis process and the obtained results is provided in the following section.

Data Pre-processing. The HR dataset provided by IBM includes 31 features, we used an improved version with 32 features. It is a clean dataset and has no missing values, though some common data pre-processing steps are applied to improve the quality of the provided data such as: removing outliers, using the Box-cox method [51] to normalize the data distribution and using Label encoding to transform non-numerical categories to numerical values [46]. An issue that should be noted with this dataset, it is very imbalanced with 1233 no turnover versus 237 turnover data points, therefore, an over sampling approach can be considered to avoid any unintended sample bias. The approach introduced by [13] called Synthetic Minority Over-sampling (SMOTE) is used for this case. Their results show that the SMOTE approach can improve the accuracy of classifiers for a minority class [21].

Classifiers Performance Analysis. The following supervised machine learning algorithms are tested:

- Decision tree (DT) is a supervised machine learning algorithm which can be used for classification or regression models in a tree-like structure. It is an intuitive graphical method that uses statistical probability analysis and represents a mapping between object attributes and object values. Each node in the tree represents the judgment condition of the object attribute, and its branch represents the object that meets the node condition. The leaf node of the tree represents the prediction result to which the object belongs [62, 63].
- Random Forests (RF) is an ensemble approach that provides an improvement over the basic decision tree structure by combining a group of weak learners to form a stronger learner. Ensemble methods utilise a divide-and-conquer approach to improve the algorithm performance [62].
- Support vector classifier (SVC) is commonly used as a discriminative classifier to assign new data samples to one of two possible categories. The basic idea of SVC is to define a hyperplane which separates the n-dimensional data into two classes, wherein the hyperplane maximizes the geometric margin to the nearest data points, so-called support vectors [62].
- Logistic Regression (LR) is a traditional classification algorithm involving linear discriminants. The primary output is a probability that the given input point belongs to a certain class. Based on the value of the probability, the model creates a linear boundary separating the input space into two regions [62].
- Naïve Bayes (GaussianNB) is a probabilistic approach that uses the Bayes Theorem. This Theorem describes the occurrence probability of an event based on the prior knowledge of related features. Naïve Bayes classifiers first learn joint probability distribution of their inputs by utilising the conditional independence assumption. Then, for a given input, the methods produce an output by computing the maximum posterior probability with Bayes Theorem [62].

- K-Nearest Neighbours (KNN), the idea for classification is to identify the K data points in the training data that are closest to the new instance and classify this new instance by a majority vote of its K neighbours [62].
- Boosting Methods: four boosting methods are tested for this use case; Gradient Boosting Trees (GBT), Extreme Gradient Boosting (XGB), Adaptive boosting (AdaBoost) and Light Gradient Boosting (LGBM). GBT is an ensemble machine learning method proposed in 2001 by Friedman [22] for regression and classification purposes. The difference between RF and GBT is that GBT models learn sequentially. In GBT, a series of trees are built, and each tree attempts to correct the mistakes of the previous one in the series [62]. XGBoost is a tree-based method as well that was introduced in 2014 by Chen [14]. It is a scalable and accurate implementation of gradient boosted trees, explicitly designed for optimizing the computational speed and model performance [62]. AdaBoost is a widely used boosting method based on decision stump, an assigned threshold, and the prediction is made according to this threshold. It includes multiple algorithms built sequentially to update their weights and take a separate role in making the most accurate estimate [31]. LightBoost was initially released by Microsoft in 2017, the key difference from other methods is that it splits the tree based on leaves, therefore, it can detect and stop the units needed by point shooting. It is a very effective method in terms of reducing the error and hence increasing the accuracy and speed [62].

In this research, the positive class (P) is when employees leave, and the negative class (N) is when they stay (see Table 1). To compare the performance of classifiers six evaluation metrics are introduced: (1) Precision is defined as the number of true positives divided by the sum of true positives and false positives; (2) Recall is defined as the number of true positives divided by the sum of true positives and false negatives; (3) F1 score is defined as the harmonic mean of precision and recall; (4) Accuracy is defined as the percentage of the correctly classified data by the model; (5) Area Under the Receiver Operating Characteristics curve (AUCROC), yields further insights in classifier performance at different classification thresholds [62,63]. (6) Area Under the Precision-Recall curve (AUCPR) is the average of the precision weighted by the probability of a given threshold [10]. 5-fold cross validation is applied and the mean values for all the metrics is calculated (see Table 2).

Table 1. Binary classifications - confusion matrix

Actual class	Predicted class	
	Turnover	No turnover
Turnover	True positive (TP)	False negative (FN)
No turnover	False positive (FP)	True negative (TN)
Total	T_1	T_2

$$Precision = \frac{TP}{TP + FP} \tag{1}$$

$$Recall = \frac{TP}{TP + FN} \tag{2}$$

$$F1 = \frac{2Precision \times Recall}{Precision + Recall} = \frac{2TP}{2TP + FP + FN} \tag{3}$$

$$Accuracy = \frac{TP + TN}{TP + FP + TN + FN} \tag{4}$$

$$AUCROC = \frac{\sum\limits_{ins_i \in T_1} rank_{ins_i} - \frac{T_1 \times (T_1 + 1)}{2}}{T_1 \times T_2} \tag{5}$$

where $rank_{ins_i}$ represents the order index of the sample i.

$$AUCPR = \int_{-\infty}^{+\infty} Precision(c)\, dP(Y = \leq c), -\infty < c < +\infty \tag{6}$$

where Y is the output values for the positive examples and P is the probability of the given threshold.

Table 2. Performance of classifiers

Classification model	Precision	Recall	F1 score	Accuracy	AUCROC	AUCPR
GaussianNB	0.613	0.691	0.609	0.699	0.691	0.519
LR	0.66	**0.751**	0.674	0.766	**0.722**	**0.549**
KNN	0.598	0.676	0.577	0.654	0.707	0.538
SVC	0.711	0.698	0.702	0.844	0.705	0.539
DT	0.627	0.664	0.635	0.77	0.696	0.526
RF	0.761	0.624	0.654	0.859	0.685	0.527
XGBoost	**0.797**	0.673	0.708	**0.872**	0.684	0.536
LGBM	0.788	0.657	0.691	0.868	0.68	0.541
AdaBoost	0.722	0.717	0.718	0.848	0.684	0.544
GBT	0.758	0.712	**0.729**	0.865	0.687	0.548

Previous research recommends using ROC curves for evaluation rather than accuracy in case of an imbalanced data which can provide another view of the classifier's performance [27,35,49,62]. However, some studies [10,11] confirm that it is recommended to use ROC curves only when there are roughly equal numbers of observations for each class and Precision-Recall curves when there is a moderate to large class imbalance. For this use case, both AUCPR and AUCROC values are considered for evaluation. Hence, LR shows the best AUCPR and AUCROC scores followed by GBT and Adaboost considering the AUCPR score (see Table 2, Fig. 1 and Fig. 2).

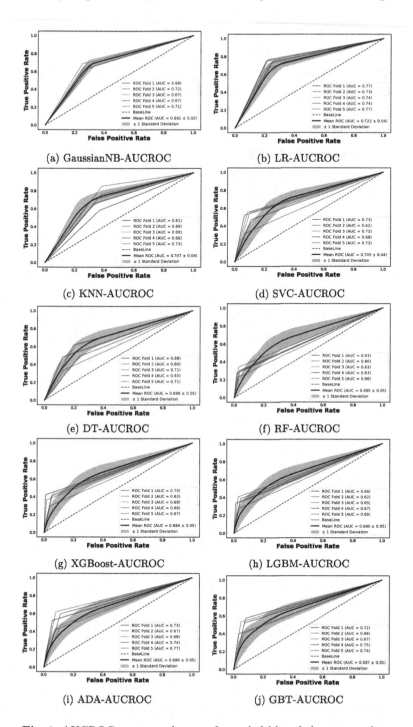

Fig. 1. AUCROC curves and scores for each fold and the mean values

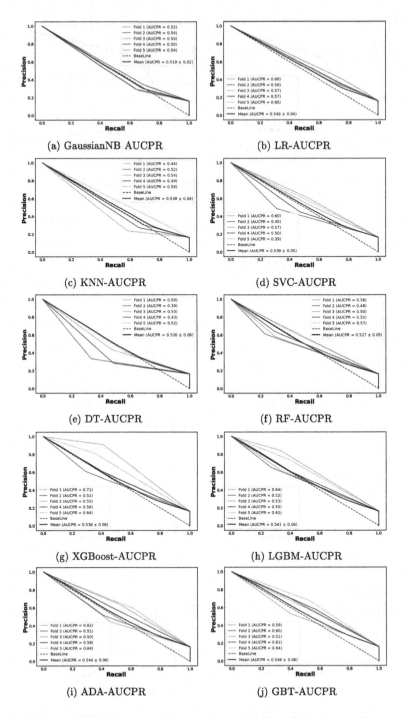

Fig. 2. AUCPR curves and scores for each fold and the mean values

Feature Importance Interpretation. L2 regularized LR is used to assign coefficients based on the importance of features (LR performed better with L2 penalty). Generally, it is used to determine the influence of specific features on the prediction performance. Consider data-points (x_i, y_i), i = 1,..., N, where N is the number of observations in data and $x_i \in \mathbb{R}^d$, d is the number of features in data (31 in this case), and $y_i \in 0, 1$ is a binary class label (no turnover, turnover). For classification, the probability of an observation × belonging to class y is given by the following Eq. (7) [50].

$$P(y|x) = \frac{1}{1 + e^{-(\beta_0 + \beta^T x)}} \tag{7}$$

where β is a vector containing d feature coefficients and β_0 is the intercept term.

The cost function to be minimized can be formulated as the negative of the regularized log-likelihood function (8) [50]:

$$L(\beta_0, \beta) = C[-\sum_{i=1}^{N}[y_i log(P(y|x)) + (1 - y_i)log(1 - P(y|x))]] + \frac{1}{2}\sum_{j=1}^{d}\beta_j^2 \tag{8}$$

The last term in the equation is a regularization parameter that is simply the sum of L2 norms of the features coefficients and $C = \frac{1}{\lambda}$ is the inverse of the regularization strength (in this example C = 0.1). The coefficients β of the logistic function can be learned by minimizing the log-likelihood function Loss (cost function) through gradient descent [48]. These coefficients can be used to interpret features importance rank [50]. The features are ranked as shown in Fig. 3, using the test and train data from the last fold, "Over time" followed by "years since last promotion" and "department" are the most important features affecting employee turnover.

3   Causal Knowledge Relevance in Improving Employee Turnover Prediction

Machine learning applied to real-world problems are most commonly in the form of black-box data-driven predictive models. These methods have been highly successful in practice but create challenges when applied to decision making and policy evaluation [4]. Some open research questions in the machine learning community may be better understood and tackled within a causal framework, domain generalisation, and adversarial robustness [52]. Therefore, having testing tools/models which depend on causal knowledge is needed more than ever to evaluate the outcome of different decisions before taking any actions [28,56].

In the area of employee turnover prediction, having a model make predictions based on causal knowledge will enable HR managers to test the outcome of their decision before taking any actions, to identify the real reasons behind turnover instability and to have more flexibility in setting multiple solutions that fit the organisations long term interest.

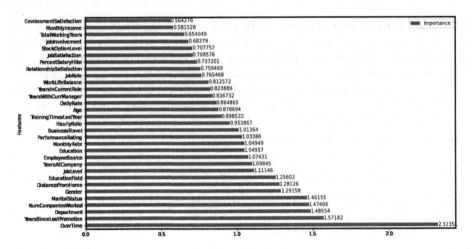

Fig. 3. Features importance using LR

Merging causal knowledge and machine learning can be tackled from different perspectives, either from the perspective of estimating causal effects of a specific decision using different machine learning techniques, the paper by [4] mentions multiple examples in this research area which can be called causal inference, or from the perspective of causal graph representation (causal learning), especially if the researchers are aiming for a generalised model [40,52]. Causal models in business are usually based on the accumulated knowledge and the shared beliefs of the organisation, depending on the selected mode of value creation and the business environment [28]. However, in this study the researchers are taking a different perspective by going outside the shared beliefs of the organisation to the shared beliefs of the profession, aiming for a model generalisation.

3.1 Causal Knowledge Discovery

Previous research about causal learning or causal maps modelling specifies what kind of knowledge and data need to be collected and which applied experiments must be performed to inform theory. Moreover, different studies offer remedies if data is imperfect due to limited perception and selective observation [8,17] and, propose tools to transport insights between various contexts [9,34,43,44] in order to have an effective theory [28]. The need for theoretical causal modelling to establish information transfer across layers of the causal hierarchy underscores the importance of domain experts in integrating causal inference into data science and constitutes a substantial opportunity for human-machine cooperation [28]. Since the target of this research is to improve HR practices in general, as a first step, the authors choose to capture causal maps from semi-structured interviews with different HR practitioners based on their experience. Their role as domain experts is critical for leveraging existing decision-making algorithms [54]. The question arises of how human knowledge and machine learning can work together to improve the generation of insights from business analytics [53]. In [5], the authors mention that automated prediction algorithms cannot leave domain experts out of the loop, therefore, they are considered the starting point for this process of causal knowledge discovery. In the next sections, a qualitative approach is outlined that utilised expert domain knowledge through the capture of multiple causal maps.

Participants. Participants were selected using purposeful sampling, a technique widely used in qualitative research for the identification and selection of information-rich cases for the most effective use of limited resources. This involves identifying and selecting individuals or groups of individuals that are especially knowledgeable about or experienced with a phenomenon of interest, which is human resources practices in this case [41]. It is highly subjective and requires careful consideration of qualifying criteria each participant must meet to be considered. The main criteria in this case was at least two years of experience in their roles as HR managers and to be from anywhere within the UK or Europe (following the General Data Protection Regulation (GDPR)). When data saturation is reached (no new information concluded or added from the interviews), the interviews stopped. It is important to get the perspectives of more than just a few people to have variety in the gathered data [1,25].

Materials. To conduct the interviews, an interview guide was developed, this guide was subject to change throughout the interviews. The interviews were refined based on the new insights and the interview guide included the possibility to update at three stages after the first interview, after the first round of interviews, and periodically thereafter [1,23].

Procedure. The interviews were run during the Covid-19 pandemic; therefore, they were all conducted online using Microsoft Teams[1] and Lucid[2] to facilitate the process of building a causal graph, each participant is required to fill a causal map. The sessions were recorded, then transcribed in a non-verbatim way.

Thematic analysis is used during the entire interview process, it is a qualitative method used to examine the ways that people make meaning out of their experiences, as well as how they construct their social worlds through meaning-making, but also it retains the focus on the ways in which these experiences are informed by their material experiences and contexts [20]. This process involves the critical review of responses to determine appropriate coding and the formation of themes from these codes. This technique is commonly used for qualitative data analysis since it provides structure and integrates reflexivity to the research [37].

---

[1] https://www.microsoft.com/en-us/microsoft-teams/group-chat-software.
[2] https://lucid.co/.

3.2   Causal Links and Factors Importance Interpretation

In these interviews the researchers tried to capture the most important internal factors (HR metrics) and external factors affecting employee turnover, as well as the most important cause-effect links among these factors based on domain-experts knowledge. Until now seven different interviews were conducted, each factor and each causal link is given a Weight Factor (WF) based on how often it was mentioned in the causal maps divided by the total number of maps. 71% means they appear in 5 maps, 57% means they appear in 4 maps, 42% means they appear in 3 maps, 28% means they appear in 2 maps and 14% means they appear in one map only. 23 external factors and 48 internal factors (HR metrics) are mentioned so far. For the internal ones, employee performance has the highest WF, 71%, followed by sickness leave and holidays (short or long), job conflicts, salaries and benefits, engaging with the company with a WF of 57%. For the external factors, health and wellbeing as a feature has the highest WF, 71%, followed by job market opportunities and pandemic consequences which have 57% as a WF.

Same logic is followed for causal links, 171 are mentioned up to this point. The causal link Performance → Turnover prediction has the highest WF, 71%, followed by the causal links job market opportunities → Turnover prediction, health and wellbeing → Turnover prediction, Sickness leave → Turnover prediction and holidays (short or long) → Turnover prediction with a WF 57%.

4   Results Comparison

Correlations and causation are two different types of relationships, while they can exist at the same time, correlation does not imply causation. The word correlation means that two quantities vary together, action A relates to action B but one event does not cause the other. However, causation is transitive, it explicitly applies to cases where action A causes outcome B [38].

Based on domain experts' knowledge shown in Fig. 4a, the most important causal links affecting employee turnover are performance, health and wellbeing, Job market opportunities, holiday and sickness leave but working hours (including over time) is mentioned twice and job tenure is mentioned only once, therefore, they have low weight factors compared with the other links. For correlations using IBM HR data, Fig. 4b shows that over time and total working years (tenure) have the highest correlations coefficients with the turnover as a feature. These correlations coefficients are very dependent on datasets, hence, if a different dataset is used, the rank might change completely. Even though this comparison might be biased since some of the features mentioned by experts are not included in the IBM HR dataset, which includes common features captured by HR departments, HR practitioners need to think about including new metrics to improve decision making such as how to measure identification with company. Moreover, HR experts taking part in this research confirm that despite working in different fields, some beliefs regarding maintaining a stable turnover are the same, thus, factors can be divided into core factors which should be included in every model predicting the turnover and organisation/field dependent factors.

They also confirm that they are more interested in capturing the core factors and in having an effect estimation (causal effect estimation) whenever one of these factors changes to take a suitable solution in time or to test the outcome of different decisions before taking any action.

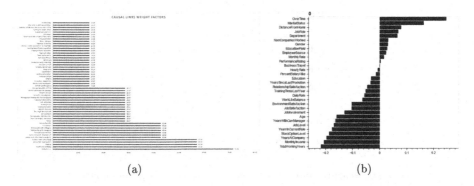

(a)   (b)

Fig. 4. Figure 4a shows results concluded from semi-structured interviews and Fig. 4b shows features correlation with turnover using IBM HR data

Tables 3 and 4 show the most important 10 factors using LR features importance and WFs concluded from HR experts' knowledge respectively. The IBM HR data includes mostly internal data, but HR experts confirmed during the interviews that external factors are as important as the internal ones in predicting employee turnover, however, they are more expensive to capture.

Several factors mentioned by HR experts are not usually included in predicting employee turnover, especially external factors such as transport services, city size and location, as well as health and wellbeing which has one of the highest weight factors as a feature (more details can be seen in Table 4), these factors can be considered core factors. Although, there are some studies that did mention the importance of including external factors such as job market and GDP indicator [26,63] but the majority is focusing only on internal data. Hence, external factors should be given the same importance as the internal ones when predicting the employee turnover, even if the data might seem harder to get.

5   Discussion and Recommendation for Future Research

Combining the theory of causality with machine learning techniques is an integral part of scientific inquiry, spanning a wide range of questions such as understanding behaviour in online systems, assessing the effect of social policies, or investigating the risk factors for diseases. Causal Machine Learning algorithms have the potential to highly impact society and businesses by answering what-if questions, allowing model generalisation, and enabling policy-evaluation in real-world contexts [2,7]. The comparison of results shows the difference between using regular machine learning correlations and domain experts' causal knowledge to tackle the problem of employee turnover prediction. Even though, some

Table 3. Features importance based on LR

Features	Coefficients
OverTime	2.32
YearsSinceLastPromotion	1.57
Department	1.48
NumCompaniesWorked	1.47
MaritalStatus	1.46
Gender	1.29
DistanceFromHome	1.28
EducationField	1.25
JobLevel	1.11
YearsAtCompany	1.09

Table 4. Features importance based on domain experts knowledge

Features	WF (%)
Health and wellbeing	71.43
Performance	71.43
Sickness leave (short/long)	57.14
Holidays (short/long)	57.14
Job conflicts	57.14
Salaries/benefits	57.14
Pandemic consequences	57.14
Engaging/Identification with the company	57.14
Job market opportunities	57.14
Transport services	42.86

of the factors mentioned by HR experts do not have specific metrics to measure them, the causal map can help to spot the metrics that are causally linked to these factors to quantify them, such as health and wellbeing of employees which can be detected through the frequency of short sickness leave, and satisfaction with leadership which can be captured through performance. A second step to take is how to build a predictive model based on a causal graph. Multiple methods can be considered for this step such as causal transfer random forest (CTRF) by [61] which focus on robustness of the prediction models to distributional shifts between training and testing data, causal Bayesian modelling using ML algorithms and domain knowledge assumptions [59]. Although correlation does not mean causality but it can be used to build and validate causal hypotheses [32,33]. Different techniques can be used based on the type of the study randomized or observational and the kind of data provided either from domain experts, experiments or both.

Some limitation and potential future area of research regarding, causal learning are highlighted in [52], such as Non-Linear Causal Relations at Scale and how future work should consider understanding the conditions under which non-linear causal relations can be learned, which training frameworks allow to best exploit the scalability of machine learning approaches and providing compelling evidence on the advantages over (non-causal) statistical representations in terms of generalisation, repurposing, and transfer of causal modules on real-world tasks, Learning Causal Variables, and understanding the Biases of Existing Deep Learning Approaches, precisely to understand how design choices in pre-training (e.g., which datasets/tasks) positively impact both transfer and robustness downstream in a causal sense.

6   Conclusion

In this paper, we conduct a comparison between machine learning correlations and domain experts' causal knowledge within the use case of employee turnover prediction. Multiple classifications algorithms are tested using IBM HR data, LR performs the best, thus, it is used to conclude the most important correlations. A qualitative approach is highlighted as well to capture causality from domain experts' knowledge. The results show that there is a huge difference between relying on machine learning correlations and experts' causal knowledge, which in a sense can explain why HR practitioners avoid using employee turnover prediction models despite their good performances. Finally, some recommendation on which hybrid techniques to consider and the limitations that exist in this research area are mentioned.

Focusing only on machine learning tools today refrain from making explicit assumptions, therefore, they are unsuitable for the task of causal inference [42]. Their main objective is to maximize out-of-sample fit in a hold-out sample, which provides a standard objective of evaluation. However, causal inference methods, with their requirement to incorporate expert domain knowledge not inferred from data directly, are perceived to be more elusive. Hence, merging causal assumptions with machine learning might lead to substantially different conclusions, which adds a layer of subjectivity to the analysis [28], and a potential for improved generalisation of predictive models. Yet, computational causality methods are still in their infancy, particularly learning causal structures from data, which is only feasible in limited situations, thus, it can be challenging to adapt causal knowledge to different areas of research [47]. Scholars and practitioners dealing with causal inference research, still struggle to identify causal effects, to measure the confounders in a particular setting, to select the right outcome variables, and to derive the accurate strategies from causal relationships [28]. Moreover, the robustness validation of such a model is not well developed as it is supposed to be, researchers have to go through complex multiple methods to confirm the validity of their results, consequently, further investigation is still needed in this research area [6]. Another area for discussion is how to include feedback loops, some of the HR participants in this research confirmed that when employee turnover is high, it can worsen the factors that caused it in first place, hence, considering feedback loops can give a new perspective to improve employee turnover prediction.

References

1. Adams, W.C., et al.: Conducting semi-structured interviews. In: Wholey, J., Hatry, H., Newcomer, K. (eds.) Handbook of Practical Program Evaluation, vol. 4, pp. 492–505. John Wiley & Sons, Inc., Hoboken (2015)
2. Aglietti, V., Damoulas, T., Álvarez, M., González, J.: Multi-task causal learning with Gaussian processes. arXiv preprint arXiv:2009.12821 (2020)
3. Al-Radaideh, Q.A., Al Nagi, E.: Using data mining techniques to build a classification model for predicting employees performance. Int. J. Adv. Comput. Sci. Appl. **3**(2) (2012). https://doi.org/10.14569/IJACSA.2012.030225, http://dx.doi.org/10.14569/IJACSA.2012.030225
4. Athey, S.: Machine learning and causal inference for policy evaluation. In: Proceedings of the 21th ACM SIGKDD International Conference on Knowledge Discovery and Data Mining, pp. 5–6. Association for Computing Machinery, New York, NY, USA (2015). https://doi.org/10.1145/2783258.2785466, https://doi.org/10.1145/2783258.2785466
5. Athey, S.: 21.The Impact of Machine Learning on Economics. In: The Economics of Artificial Intelligence, pp. 507–552. University of Chicago Press, Chicago (2019). https://doi.org/10.7208/chicago/9780226613475.001.0001, https://www.nber.org/books-and-chapters/economics-artificial-intelligence-agenda
6. Athey, S., Imbens, G.: A measure of robustness to misspecification. Am. Econ. Rev. **105**(5), 476–480 (2015). https://doi.org/10.1257/aer.p20151020, https://www.aeaweb.org/articles?id=10.1257/aer.p20151020
7. Barbiero, P., Squillero, G., Tonda, A.: Modeling generalization in machine learning: a methodological and computational study. arXiv preprint arXiv:2006.15680 (2020)
8. Bareinboim, E., Pearl, J.: Controlling selection bias in causal inference. In: Artificial Intelligence and Statistics, pp. 100–108. PMLR (2012). https://proceedings.mlr.press/v22/bareinboim12.html
9. Bareinboim, E., Pearl, J.: Transportability of causal effects: completeness results. In: Proceedings of the AAAI Conference on Artificial Intelligence, AAAI 2012, vol. 26, pp. 698–704 (2012)
10. Boyd, K., Eng, K.H., Page, C.D.: Area under the precision-recall curve: point estimates and confidence intervals. In: Blockeel, H., Kersting, K., Nijssen, S., Železný, F. (eds.) ECML PKDD 2013. LNCS (LNAI), vol. 8190, pp. 451–466. Springer, Heidelberg (2013). https://doi.org/10.1007/978-3-642-40994-3_29
11. Brownlee, J.: How to use ROC curves and precision-recall curves for classification in python. https://machinelearningmastery.com/roc-curves-and-precision-recall-curves-for-classification-in-python/ (2018). Accessed 10 Oct-2021
12. Cai, X., Shang, J., Jin, Z., Liu, F., Qiang, B., Xie, W., Zhao, L.: DBGE: employee turnover prediction based on dynamic bipartite graph embedding. IEEE Access **8**, 10390–10402 (2020)
13. Chawla, N.V., Bowyer, K.W., Hall, L.O., Kegelmeyer, W.P.: Smote: synthetic minority over-sampling technique. J. Artif. Intell. Res. **16**, 321–357 (2002)
14. Chen, T., Guestrin, C.: XGBoost: a scalable tree boosting system. In: Proceedings of the 22nd ACM SIGKDD International Conference on Knowledge Discovery and Data Mining, pp. 785–794 (2016)

15. Chien, C.F., Chen, L.F.: Data mining to improve personnel selection and enhance human capital: a case study in high-technology industry. Exp. Syst. Appl. **34**(1), 280–290 (2008). https://doi.org/10.1016/j.eswa.2006.09.003, https://www.sciencedirect.com/science/article/pii/S0957417406002776

16. Chowdhury, S., Joel-Edgar, S., Dey, P.K., Bhattacharya, S., Kharlamov, A.: Embedding transparency in artificial intelligence machine learning models: managerial implications on predicting and explaining employee turnover. Int. J. Hum. Resour. Manag. 1–32 (2022)

17. Correa, J.D., Tian, J., Bareinboim, E.: Identification of causal effects in the presence of selection bias. In: Proceedings of the AAAI Conference on Artificial Intelligence, vol. 33, pp. 2744–2751 (2019)

18. DGOKE1: IBM HR Dataset: exploratory data analysis. https://www.kaggle.com/code/dgokeeffe/ibm-hr-dataset-exploratory-data-analysis/data (2017). Accessed 17 June 2022

19. Duan, Y.: Statistical analysis and prediction of employee turnover propensity based on data mining. In: 2022 International Conference on Big Data, Information and Computer Network (BDICN), pp. 235–238 (2022). https://doi.org/10.1109/BDICN55575.2022.00052

20. Evans, C., Lewis, J.: Analysing Semi-Structured Interviews Using Thematic Analysis: Exploring Voluntary Civic Participation Among Adults. SAGE Publications Limited, London (2018)

21. Farzaneh, F.: Attrition-binary classification of imbalanced data. https://www.kaggle.com/code/oceands/attrition-binary-classification-of-imbalanced-data/notebook (2021). Accessed 09 Oct 2021

22. Friedman, J.H.: Greedy function approximation: a gradient boosting machine. Ann. Stat. **29**, 1189–1232 (2001)

23. Galletta, A.: Mastering the Semi-Structured Interview and Beyond. New York University Press, New York (2013)

24. Garg, S., Sinha, S., Kar, A.K., Mani, M.: A review of machine learning applications in human resource management. Int. J. Prod. Perform. Manag. **23** (2021)

25. Guest, G., Bunce, A., Johnson, L.: How many interviews are enough? an experiment with data saturation and variability. Field Methods **18**(1), 59–82 (2006)

26. Hang, J., Dong, Z., Zhao, H., Song, X., Wang, P., Zhu, H.: Outside. In: Market-aware heterogeneous graph neural network for employee turnover prediction. In: Proceedings of the Fifteenth ACM International Conference on Web Search and Data Mining, pp. 353–362 (2022)

27. Huang, J., Ling, C.X.: Using AUC and accuracy in evaluating learning algorithms. IEEE Trans. Knowl. Data Eng. **17**(3), 299–310 (2005)

28. Hünermund, P., Kaminski, J., Schmitt, C.: Causal Machine Learning And Business-Decision Making (2021)

29. Jain, R., Nayyar, A.: Predicting employee attrition using XGBoost machine learning approach. In: 2018 International Conference on System Modeling & Advancement in Research Trends (SMART), pp. 113–120. IEEE (2018)

30. Joarder, M.H.: The role of HRM practices in predicting faculty turnover intention: empirical evidence from private universities in Bangladesh. South East Asian J. Manag. **5** (2012)

31. Kovan, I.: An overview of boosting methods: CatBoost, XGBoost, AdaBoost, LightBoost, Histogram-based gradient boost. https://towardsdatascience.com/an-overview-of-boosting-methods-catboost-xgboost-adaboost-lightboost-histogram-based-gradient-407447633ac1 (2021). Accessed 3 Mar 2022

32. Kumova, B.I., Saller, D.: Mining causal hypotheses in categorical time series by iterating on binary correlations. In: Holzinger, A., Kieseberg, P., Tjoa, A.M., Weippl, E. (eds.) CD-MAKE 2021. LNCS, vol. 12844, pp. 99–114. Springer, Cham (2021). https://doi.org/10.1007/978-3-030-84060-0_7

33. Lazzari, M., Alvarez, J.M., Ruggieri, S.: Predicting and explaining employee turnover intention. Int. J. Data Sci. Anal. **33**(9), 911–923 (2022)

34. Lee, S., Correa, J., Bareinboim, E.: General transportability-synthesizing observations and experiments from heterogeneous domains. In: Proceedings of the AAAI Conference on Artificial Intelligence, vol. 34, pp. 10210–10217 (2020)

35. Ling, C.X., Huang, J., Zhang, H., et al.: AUC: a statistically consistent and more discriminating measure than accuracy. In: IJCAI, vol. 3, pp. 519–524 (2003)

36. Ma, X., Zhang, Y., Song, Y., Wang, E., Yao, F., Zhang, Z.: Application of data mining in the field of human resource management: a review. In: 1st International Symposium on Economic Development and Management Innovation (EDMI 2019), pp. 222–227. Atlantis Press (2019)

37. Mackieson, P., Shlonsky, A., Connolly, M.: Increasing rigor and reducing bias in qualitative research: A document analysis of parliamentary debates using applied thematic analysis. Qual. Soc. Work. **18**(6), 965–980 (2019)

38. Madhavan, A.: Correlation vs causation: understand the difference for your product. https://amplitude.com/blog/causation-correlation (2019). Accessed 6 Mar 2022

39. Maria-Carmen, L.: Classical machine-learning classifiers to predict employee turnover. In: Education, Research and Business Technologies, pp. 295–306. Springer, Singapore (2022). https://doi.org/10.1007/978-981-16-8866-9_25

40. Moraffah, R., Karami, M., Guo, R., Raglin, A., Liu, H.: Causal interpretability for machine learning-problems, methods and evaluation. ACM SIGKDD Explor. News **22**(1), 18–33 (2020)

41. Palinkas, L.A., Horwitz, S.M., Green, C.A., Wisdom, J.P., Duan, N., Hoagwood, K.: Purposeful sampling for qualitative data collection and analysis in mixed method implementation research. Adm. Policy Mental Health Serv. Res. **42**(5), 533–544 (2015)

42. Pearl, J.: The seven tools of causal inference, with reflections on machine learning. Commun. ACM **62**(3), 54–60 (2019)

43. Pearl, J., Bareinboim, E.: Transportability of causal and statistical relations: a formal approach. In: Twenty-Fifth AAAI Conference on Artificial Intelligence (2011)

44. Pearl, J., Bareinboim, E.: External validity: from do-calculus to transportability across populations. Stat. Sci. **29**(4), 579–595 (2014)

45. Pearl, J., Mackenzie, D.: The Book of Why: The New Science of Cause and Effect, 1st edn., Basic Books, New York (2018)

46. Pedregosa, F., et al.: Scikit-learn: machine learning in python. J. Mach. Learn. Res. **12**, 2825–2830 (2011)

47. Peters, J., Janzing, D., Schölkopf, B.: Elements of Causal Inference: Foundations and Learning Algorithms. The MIT Press, Cambridge (2017)

48. Pickus, S.: Logistic-regression-classifier-with-l2-regularization, April 2014. https://github.com/pickus91/Logistic-Regression-Classifier-with-L2-Regularization

49. Raschka, S.: Python Machine Learning. Packt Publishing Ltd., Birmingham (2015)

50. Saarela, M., Jauhiainen, S.: Comparison of feature importance measures as explanations for classification models. SN Appl. Sci. **3**(2), 1–12 (2021). https://doi.org/10.1007/s42452-021-04148-9

51. Sakia, R.M.: The box-cox transformation technique: a review. J. R. Stat, Soc. Ser. D **41**(2), 169–178 (1992)

52. Schölkopf, B., et al.: Towards causal representation learning. arXiv preprint arXiv:2102.11107 (2021)

53. Sharma, R., Mithas, S., Kankanhalli, A.: Transforming decision-making processes: a research agenda for understanding the impact of business analytics on organizations. Eur. J. Inf. Syst. **23**(4), 433–441 (2014)

54. Shrestha, Y.R., Ben-Menahem, S.M., Von Krogh, G.: Organizational decision-making structures in the age of artificial intelligence. Calif. Manage. Rev. **61**(4), 66–83 (2019)

55. Sikaroudi, E., Mohammad, A., Ghousi, R., Sikaroudi, A.: A data mining approach to employee turnover prediction (case study: Arak automotive parts manufacturing). J. Ind. Syst. Eng. **8**(4), 106–121 (2015)

56. Simon, H.A.: On the concept of organizational goal. Admin. Sci. Q. **9**,1–22 (1964)

57. Spirtes, P.: Introduction to causal inference. J. Mach. Learn. Res. **11**(5) (2010)

58. Strohmeier, S., Piazza, F.: Domain driven data mining in human resource management: A review of current research. Expert Syst. Appl. **40**(7), 2410–2420 (2013)

59. Tang, X., Chen, A., He, J.: A modelling approach based on Bayesian networks for dam risk analysis: integration of machine learning algorithm and domain knowledge. Int. J. Dis. Risk Reduct. **71**, 102818 (2022)

60. Vega, R.P., Anderson, A.J., Kaplan, S.A.: A within-person examination of the effects of telework. J. Bus. Psychol. **30**(2), 313–323 (2015)

61. Zeng, S., Bayir, M.A., Pfeiffer III, J.J., Charles, D., Kiciman, E.: Causal transfer random forest: combining logged data and randomized experiments for robust prediction. In: Proceedings of the 14th ACM International Conference on Web Search and Data Mining, pp. 211–219 (2021)

62. Zhao, Y., Hryniewicki, M.K., Cheng, F., Fu, B., Zhu, X.: Employee turnover prediction with machine learning: a reliable approach. In: Arai, K., Kapoor, S., Bhatia, R. (eds.) IntelliSys 2018. AISC, vol. 869, pp. 737–758. Springer, Cham (2019). https://doi.org/10.1007/978-3-030-01057-7_56

63. Zhu, Q., Shang, J., Cai, X., Jiang, L., Liu, F., Qiang, B.: CoxRF: employee turnover prediction based on survival analysis. In: 2019 IEEE SmartWorld, Ubiquitous Intelligence & Computing, Advanced & Trusted Computing, Scalable Computing & Communications, Cloud & Big Data Computing, Internet of People and Smart City Innovation (SmartWorld/SCALCOM/UIC/ATC/CBDCom/IOP/SCI), pp. 1123–1130. IEEE (2019)

Machine Learning and Knowledge Extraction to Support Work Safety for Smart Forest Operations

Ferdinand Hoenigsberger[1]([⊠])[iD], Anna Saranti[1][iD], Alessa Angerschmid[1][iD],
Carl Orge Retzlaff[1][iD], Christoph Gollob[1][iD], Sarah Witzmann[1][iD],
Arne Nothdurft[1][iD], Peter Kieseberg[2][iD], Andreas Holzinger[1][iD],
and Karl Stampfer[1][iD]

[1] University of Natural Resources and Life Sciences, Vienna, Austria
ferdinand.hoenigsberger@boku.ac.at
[2] University of Applied Sciences St.Poelten, St.Poelten, Austria

Abstract. Forestry work is one of the most difficult and dangerous professions in all production areas worldwide - therefore, any kind of occupational safety and any contribution to increasing occupational safety plays a major role, in line with addressing sustainability goal SDG 3 (good health and well-being). Detailed records of occupational accidents and the analysis of these data play an important role in understanding the interacting factors that lead to occupational accidents and, if possible, adjusting them for the future. However, the application of machine learning and knowledge extraction in this domain is still in its infancy, so this contribution is also intended to serve as a starting point and test bed for the future application of artificial intelligence in occupational safety and health, particularly in forestry. In this context, this study evaluates the accident data of Österreichische Bundesforste AG (ÖBf), Austria's largest forestry company, for the years 2005–2021. Overall, there are 2481 registered accidents, 9 of which were fatal. For the task of forecasting the absence hours due to an accident as well as the classification of fatal or non-fatal cases, decision trees, random forests and fully-connected neuronal networks were used.

Keywords: Occupational accident · Artificial Intelligence · Explainable AI · Forestry · Machine learning · Explainability · Human-in-the-Loop

1 Introduction

Even though mechanization significantly reduced the number of accidents and fatalities in forestry [26], they remain on a high level. The 2021 accident statistics for forestry work in Austria [2] recorded 1189 accidents, 21 of which were fatal. Forestry work is thereby still one of the most dangerous and difficult occupations in all fields of production and also appears in the list of heavy physical labor (§ 1 Abs 1 Z 4 SchwerarbeiterVO, BGBl. II 104/2006 as amended by BGBl. II 413/2019). New approaches are needed to further reduce the number of accidents.

© IFIP International Federation for Information Processing 2022
Published by Springer Nature Switzerland AG 2022
A. Holzinger et al. (Eds.): CD-MAKE 2022, LNCS 13480, pp. 362–375, 2022.
https://doi.org/10.1007/978-3-031-14463-9_23

Extraordinarily high hopes are therefore being placed in digital transformation to support the sustainable development goals [13], and its applications to make a major step toward improving ergonomics and occupational safety addressing the sustainability goal SDG 3 [37]. Detailed records of work accidents and the analysis of these data play an important role (and are therefore also legally required) in order to understand the interacting factors that lead to occupational accidents and, if possible, adjust them for the future. Various studies have looked at and analyzed different aspects of such accident records. Tsioras et al. (2014) [34] for example evaluated accidents during timber harvesting in Austria. Such studies were also completed for other countries (e.g. Poland [10], Slovakia [3, 15], Italy [17], or Brazil [18]) as well as comparisons between different countries [1]. One focus of research centers on accidents during the work with chainsaws respectively during motor-manual tree felling [5, 19, 24]. All these approaches and contributions reach their limits where a multitude of factors and influences interact. Therefore, this contribution compares ML methods with "conventional" statistical methods. Today, machine learning and knowledge extraction permeate practically all application domains and especially in the domain of forestry it has great future potential [12]. The use of ML owes it especially to the application in the field of accident prediction to two factors: on the one hand due to its explanatory capacity and with that the ability to gain insights into major causes of accidents [21], and secondly with its predictive capacity [22], enabling better accident prevention. Other professions, mainly those with high accident risk or high accident severity, also start to exploit the capabilities of machine learning for the analysis of occupational accidents. For example, the analysis of serious occupational accidents in the chemical industry [33], or in the steel industry [28]. For electrical engineering, Oyedele et al. (2021) [25] used deep learning and boosted trees algorithms to test the possibilities and effectiveness for the use in the area of safety risk management. Refer to [8, 23, 29, 35] for other examples of the application of ML in the field of accident prediction.

In general, however, the application of machine learning and knowledge extraction in this domain is still in its infancy. Therefore, this contribution is intended to serve as an important starting point and test bed not only for the application of machine learning in occupational safety and health, especially in forestry, but also for new developments, especially in emerging subfields of artificial intelligence, including explainable AI and counterfactual explanations ("what if ... questions") - especially with respect to a future trustworthy AI [11]. In this context, this study evaluates accident data from Österreichische Bundesforste AG (ÖBf), Austria's largest forestry company, for the years 2005–2021 using "conventional" statistical methods as well as machine learning algorithms to analyze and contrast the context of occupational accidents and lay the foundation for further work.

2   Dataset

2.1   Description

The dataset consists of tabular data containing information about occupational accidents. In principle, an accident can be fatal or non-fatal; in the latter case, the injured person will not be able to work for a particular number of hours or days. The information that is most descriptive from domain knowledge and previous models [33,34] is mainly contained in the columns that specify the time on the day on which the accident happened (given in minutes), the age of the person, and the day in the week (starting with 1 = Monday). Furthermore, there is a differentiation between workers and employees in that manner, since severe accidents rather happen to workers. The cause of the accident, working sector, and body part are also provided with an appropriate encoding. It is apparent that there are associations between the columns and potential causal relationships, but for the scope of this research work, only their linear and non-linear correlations were explored.

The dataset was explored and analyzed with the use of Python libraries `pandas`: https://pandas.pydata.org/, preprocessed mainly with https://scikit-learn.org and visualized with `matplotlib`: https://matplotlib.org/ and `Bokeh`: https://bokeh.org/:.

2.2 Preprocessing

Overall, there are 2481 registered accidents from 01 October 2005 until 21 December 2021. Of those, only 9 were fatal, which does not provide enough information on the influencing and decisive factors of those accidents. Nevertheless, a classification between non-fatal and fatal accidents is of great importance and can be supported with the gathering of more relevant data. A dedicated pre-processing stage consisted of the removal of non-filled entries and invalid content rows. In total, there were 7 input feature columns that were used for prediction: "worker or employee", "age at accident", "day of the week", "time in the day", "working sector", "accident cause" and "injured body part". After the removal of invalid entries (empty string, invalid value as f.e. working sector 51 where the valid ones are only between 1 and 35) there were 1965 rows with valid data left.

Before using the data as input for the predictive models, histograms that visualized their contents were computed. It is important to note that the cardinality of the domain of the input feature is playing a substantial role in its importance. Continuous features and categorical features with many categories have a bias towards being more descriptive than ones with less categories [14]. In the case of the input features "injured body part", "working sector" and "accident cause", grouping of values was used. For example, all accidents that affected the head and neck area were grouped together.

Both linear and non-linear correlations between all pairs of features were computed before they were used as input to the predictive models. The linear correlations were measured with the Pearson coefficient [9] and the non-linear with Mutual Information (MI) [20]. The highest linear correlation was found between working sector and cause of the accident (0.26), the second-highest between working sector and worker or employee (−0.26) and the third between cause and body part (−0.13). All other pairs had values < 0.1. The MI is correspondingly 0.25, 0.19 and 0.06 for the aforementioned pairs, and anything else is < 0.1. Those values are supported by domain knowledge and by previous research work. No substantial correlations were found in any of the pairs, so it was not necessary to remove redundant input features to trim the model, or even apply some dimensionality reduction before the training [9,36].

3   Predictive Models

3.1   Regression Task

Decision Tree. The main predictive task is - in case of a non-fatal accident - to predict the number of hours the injured person will not be able to work, because of the need to recover. There is a differentiation between workers and employees in that manner; the hours that employees need for recovery are quite different from the ones of the workers. Figure 1 shows the difference between the two distributions of values; this was to be expected, since manual work tends to cause more and more serious accidents. We created one model that picks up commonalities and differences between those two groups, but we are well aware that in the future, if necessary, we could create two different detailed models.

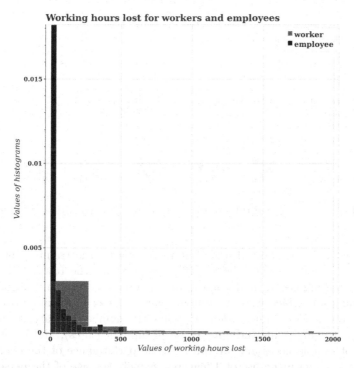

Fig. 1. Difference in the distributions of recovery working hours after an accident in workers and employees. The figure is cropped on the right side, to zoom on the vast majority of the values that are < 2000.

This is a regression problem [6] and the use of a Decision Tree (DT) [9] is a straightforward, explainable-per-design method for such a task. It is not necessary to scale the numeric or continuous features; nevertheless, as far as non-ordinal input features are concerned - such as the working sector, the cause of the accident and the injured body part - there is already a mapping to integer numbers. For example, in the body part feature, the ears are encoded with the number 102, the eyes with 113, 123, 133 to denote right, left or both eyes, the face with 104, the teeth with 105. As one can observe, those body parts are not in an ordinal relationship with each other, that means a predictive model should be prevented to assume so or base its decision and generalization on this. Furthermore, as mentioned in the previous Sect. 2.2, on those features, grouping of conceptually similar values was applied and composed one category. The one-hot encoding was used [36], where each category was encoded by a vector containing only one 1 while the rest of the entries are 0 and being orthonormal to the rest. This still contains issues in this case, since there is a column for each category, the feature values matrix is quite sparse and large.

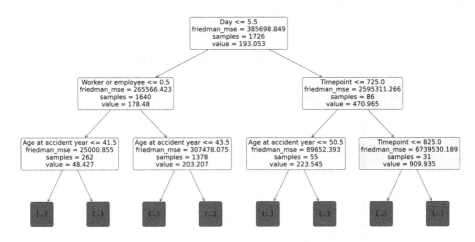

Fig. 2. The learned rules of the decision tree trained with 4 features.

The dataset was shuffled and split in two parts, the training set contains 80% of the original data, whereas 20% are used for the test set. There was a grid search applied to find out the best parameters for the decision tree on the training set, which were maximum depth of tree: 5, maximum number of features to consider: "auto" (denoting that the number of features considered for the best split equals the number of all input features), and the minimum number of sample necessary for split: 2. The performance of the decision tree on the test set with the use of 4 features, namely the age of the person at the time of the accident, the day in the week, the point of time in the day and the information if the person is a worker or an employee was a Mean Squared Error (MSE) of 151222.99 and a Relative Root Mean Squared Error (RRMSE) of 1.15.

The learned rules of the decision tree trained with 4 features are depicted in Fig. 2. The decision tree rules can also be extracted and compared to previous research insights. According to the decision tree it can be shown that a rather small number of recovery hours is expected for younger employee and workers in dependence to the time of the accident; furthermore differences can be observed between the first days of the week and the rest of the week.

```
|--- Day <= 5.50
|   |--- Worker or employee <= 0.50
|   |   |--- Age at accident year <= 41.50
|   |   |   |--- Timepoint <= 1275.00
|   |   |   |   |--- Age at accident year <= 20.50
|   |   |   |   |   |--- value: [58.00]
|   |   |   |   |--- Age at accident year >  20.50
|   |   |   |   |   |--- value: [17.67]
|   |   |   |--- Timepoint >  1275.00
|   |   |   |   |--- Timepoint <= 1320.00
|   |   |   |   |   |--- value: [504.00]
|   |   |   |   |--- Timepoint >  1320.00
|   |   |   |   |   |--- value: [80.00]
|   |   |--- Age at accident year >  41.50
|   |   |   |--- Age at accident year <= 45.50
|   |   |   |   |--- Timepoint <= 525.00
|   |   |   |   |   |--- value: [484.80]
|   |   |   |   |--- Timepoint >  525.00
|   |   |   |   |   |--- value: [61.33]
|   |   |   |--- Age at accident year >  45.50
|   |   |   |   |--- Age at accident year <= 54.50
|   |   |   |   |   |--- value: [34.57]
|   |   |   |   |--- Age at accident year >  54.50
|   |   |   |   |   |--- value: [70.95]
...
|--- Day >  5.50
...
```

The input features importances are in accordance to previous research work [33,34] and are depicted in Fig. 3.

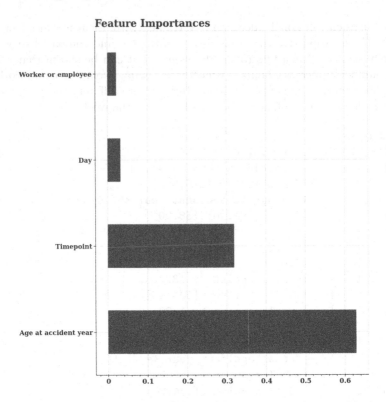

Fig. 3. The feature importances of the decision tree that is trained with 4 features.

Training with all 7 input features does have an impact on the MSE 164610.79 and RRMSE 1.83. The learned rules of the decision tree are more detailed, and the order of feature importances changes.

Random Forest. A random forest was also trained in a way similar to the decision tree as far as pre-processing, dataset split and grid search of the best parameters is concerned. 200 estimators were found, enough to achieve an acceptable prediction performance with MSE for 4 features: 121154.73 and RRMSE 1.84, whereas for 7 features: 114203.96 and 1.71 correspondingly.

The feature importances are more robust than the ones from the decision trees [9], but there is no tree to depict the rules. Random forests are not considered interpretable, but have generally better performance than decision trees.

Fully-Connected Neural Network. A fully-connected neural network [6] was implemented to also tackle the problem of regression. Four layers with 50 neurons each are enough to reach an MSE of 93215.32 and an RRMSE 35.39 with the 4 basic features, whereas 155141.41 and 37.97 with all 7 correspondingly.

The optimizer that was used was "RMSprop" and the activation unit was the Rectified Linear Unit (ReLU). The results of the prediction for 4 features can be seen in Fig. 4.

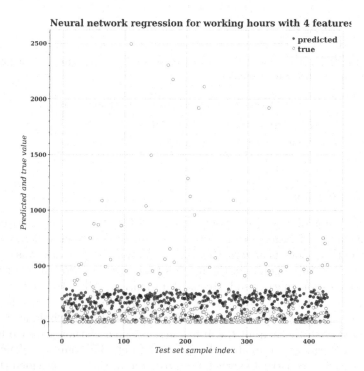

Fig. 4. The results of the neural network on the test set for 4 input features.

One way to compute the importance of features is the SHAP (Shapley) values Explainable AI method [30–32] as depicted in Fig. 5 for 4 features.

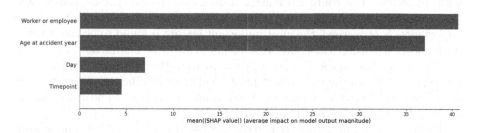

Fig. 5. The feature importances of the neural network that is trained with 4 features.

Table 1. Comparison of different regression models using 4 features for the prediction of lost working hours due to accidents

	Decision tree	Random forest	Neural network
MSE	151222.99	121154.73	93215.32
RRMSE	1.15	1.84	35.39

Table 2. Comparison of different regression models using all features for the prediction of lost working hours due to accidents.

	Decision tree	Random forest	Neural network
MSE	164610.79	114203.96	155141.41
RRMSE	1.83	1.71	37.97

3.2   Classification

The classification of fatal and non-fatal cases is a task that is characterized by its great imbalance - 9 fatal cases in 1965 data samples led to classifiers with high accuracy but very low, near zero Mutual Information (MI) - which was expected, as explained in [20]. Therefore, the metrics that were used for the classification were the confusion matrix and Mutual Information (MI) [9]. Another way to counteract the imbalance is to perform oversampling on the class with the smaller number of samples. The Synthetic Minority Oversampling TEchnique (SMOTE) [7] was used with the help of the `imbalanced-learn` Python package https:// imbalanced-learn.org to create a balanced dataset. The number of neighbors that were used to construct synthetic samples was 4, this has provided the most profitable performance results.

Decision Tree. The decision tree parameters were defined by a grid search similar to the one described in the regression Sect. 3.1. The Mutual Information (MI) is 0.60, and accuracy 0.98 for the balanced dataset after oversampling with 4 features. The confusion matrix is described by the true negatives: 463, true positives: 470, false positives: 17 and false negatives: 2. As far as the feature importances is concerned, the most important feature is found to be the "worker or employee" feature; since this is a binary feature and does not carry such discriminating information in comparison to the other categorical with more categories or the continuous ones, it must be of great significance. From the raw data 5 out of 9 fatal accidents occurred among employees. The rules can be extracted as text, equal to the regression task 3.1.

For 7 input features the MI is 0.67 and the accuracy is 0.99. The feature importances are depicted in the Fig. 6.

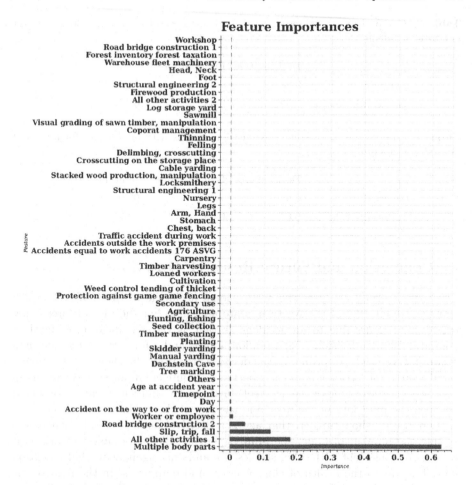

Fig. 6. The feature importances of the decision tree that is trained with all features.

Random Forest. MI with random forest, 100 estimators and 4 input features 0.63, whereas accuracy reaches 0.98. 7 input features reach MI 0.69, accuracy 1.0 with a "perfect" confusion matrix (no false positives or negatives), although this needs to be further examined with care since it might be an over-fitting indication.

Fully-Connected Neural Network. The architecture of the neural network is similar to the one used in the regression task 3.1. The fundamental difference lies in the use of a sigmoid non-linearity in the output layer. MI with 4 features: 0.64, with accuracy 0.98 and with 7 features: MI 0.67 and accuracy 0.99.

Table 3. Comparison of different classification models with 4 features for fatal and non-fatal accidents.

	Decision tree	Random forest	Neural network
Accuracy	0.98	0.98	0.98
Mutual Information	0.60	0.63	0.64

Table 4. Comparison of different classification models with all features for fatal and non-fatal accidents.

	Decision tree	Random forest	Neural network
Accuracy	0.99	1.0	0.99
Mutual Information	0.67	0.69	0.67

4 Conclusion and Future Research Questions

The current study analyzes the occupational accidents of the ÖBf AG, which manages about 10% of the Austrian state area [38]. For the task of forecasting the absence hours due to an accident as well as the classification of fatal or non-fatal cases, a decision tree, a random forest, and a fully-connected neuronal network were used. According to the decision tree it can be shown that a rather small number of recovery hours is expected for younger employee and workers in dependence to the time of the accident; furthermore differences can be observed between the first days of the week and the rest of the week.

Our results show further, that age is a decisive factor especially when predicting the number of hours the injured person will not be able to work. Our results are consistent with other studies [15] and once again show that attention must be paid to the health of elder workers and employees. In the light of the demographic change this statement is especially relevant. The found difference between workers and employees confirms the expectation, that manual work tends to cause more and more serious accidents. The time is also an important factor in our models. Peaks in occupational accidents occur between 10 and 12 am. A second smaller peak occurs between 2 and 4 pm. Our study is here in line with other studies of occupational accidents, which highlighted fatigue and dehydration as possible explanations [18,34] for different accident probabilities during the day.

In the future, by expanding the database, the methodology could be further refined and extended to include factors such as the size of the company, the level of training or the professional background of the people involved. Furthermore, it would be insightful to compare this data to data from other institutions throughout Europe and search for hidden patterns or correlations. A long-term goal of this research is a causal model [4,16,27]. As seen, the values of the categorical variables do have some relationship with each other; one of the most prominent examples is the "body part" feature, where one of the values encompasses accidents in several parts of the body whereas the rest concentrate on only one (especially before grouping). The existence, structure and parameters of relationships between working sector, cause of the accident and the affected body part are subject of future investigations and are expected to have higher data quality requirements in general.

Acknowledgements. The authors declare that there are no conflict of interests. This work does not raise any ethical issues. Parts of this work have been funded by the Austrian Science Fund (FWF), Project: P-32554, explainable AI.

The authors would like to thank ÖBf AG, and especially Mr. Stefan Trzesniowski, for the kind provision of accident data.

References

1. Akay, A.O.: Evaluation of occupational accidents in forestry in Europe and Turkey by k-means clustering analysis. Turkish J. Agric. Forest. **45**(4), 495–509 (2021). https://doi.org/10.3906/tar-2010-55
2. Allgemeine Unfallversicherungsanstalt (AUVA): Unfallstatistik 2021 forstwirtschaftliche arbeiten. AUVA Schwerpunktauswertungen (2022)
3. Allman, M., et al.: Work accidents during cable yarding operations in central Europe 2006–2014. Forest Syst. **26**(1), 13 (2017). https://doi.org/10.5424/fs/2017261-10365
4. Barber, D.: Bayesian Reasoning and Machine Learning. Cambridge University Press, Cambridge (2012)
5. Bentley, T.A., Parker, R.J., Ashby, L.: Understanding felling safety in the New Zealand forest industry. Appl. Egonom. **36**(2), 165–175 (2005). https://doi.org/10.1016/j.apergo.2004.10.009
6. Bishop, C.M., Nasrabadi, N.M.: Pattern recognition and machine learning. Springer (2006)
7. Chawla, N.V., Bowyer, K.W., Hall, L.O., Kegelmeyer, W.P.: Smote: synthetic minority over-sampling technique. J. Artif. Intell. Res. **16**, 321–357 (2002)
8. Galatioto, F., Catalano, M., Shaikh, N., McCormick, E., Johnston, R.: Advanced accident prediction models and impacts assessment. IET Intell. Transp. Syst. **12**(9), 1131–1141 (2018)
9. Géron, A.: Hands-on Machine Learning with Scikit-Learn, Keras, and TensorFlow: Concepts, Tools, and Techniques to Build Intelligent Systems. " O'Reilly Media, Inc." (2019)
10. Grzywiński, W., Skonieczna, J., Jelonek, T., Tomczak, A.: The influence of the privatization process on accident rates in the forestry sector in Poland. Int. J. Environ. Res. Public Health **17**(9), 3055 (2020). https://doi.org/10.3390/ijerph17093055
11. Holzinger, A.: The next frontier: AI we can really trust. In: Kamp, M. (ed.) Proceedings of the ECML PKDD 2021, CCIS 1524, pp. 1–14. Springer, Cham (2021). https://doi.org/10.1007/978-3-030-93736-2-33
12. Holzinger, A., et al.: Digital transformation in smart farm and forest operations needs human-centered AI: challenges and future directions. Sensors **22**(8), 3043 (2022). https://doi.org/10.3390/s22083043
13. Holzinger, A., Weippl, E., Tjoa, A.M., Kieseberg, P.: Digital transformation for sustainable development goals (SDGS) - a security, safety and privacy perspective on AI. In: Springer Lecture Notes in Computer Science, LNCS 12844, pp. 1–20. Springer (2021). https://doi.org/10.1007/978-3-030-84060-0

14. Ian, H.W., Eibe, F.: Data mining: Practical Machine Learning Tools And Techniques (2005)
15. Jankovský, M., Allman, M., Allmanová, Z.: What are the occupational risks in forestry? Results of a long-term study in Slovakia. Int. J. Environ. Res. Public Health **16**(24), 4931 (2019). https://doi.org/10.3390/ijerph16244931
16. Koller, D., Friedman, N.: Probabilistic Graphical Models: Principles And Techniques. MIT Press, London (2009)
17. Laschi, A., Marchi, E., Foderi, C., Neri, F.: Identifying causes, dynamics and consequences of work accidents in forest operations in an alpine context. Safety Sci. **89**, 28–35 (2016). https://doi.org/10.1016/j.ssci.2016.05.017
18. Lima, K.S., Meira Castro, A.C., Torres Costa, J., Baptista, J.S.: Occupational accidents in native and planted forests in Brazil: 2007–2018. Work **71**(3), 719–728 (2022). https://doi.org/10.3233/WOR-210543
19. López-Toro, A.A., Pardo-Ferreira, M.C., Martínez-Rojas, M., Carrillo-Castrillo, J.A., Rubio-Romero, J.C.: Analysis of occupational accidents during the chainsaws use in Andalucía. Saf. Sci. **143**, 105436 (2021). https://doi.org/10.1016/j.ssci.2021.105436
20. MacKay, D.J., Mac Kay, D.J., et al.: Information Theory, Inference and Learning Algorithms. Cambridge University Press , Cambridge (2003)
21. Martin, J.E., Rivas, T., Matías, J., Taboada, J., Argüelles, A.: A Bayesian network analysis of workplace accidents caused by falls from a height. Saf. Sci.Saf. Sci. **47**(2), 206–214 (2009)
22. Matías, J.M., Rivas, T., Martín, J., Taboada, J.: A machine learning methodology for the analysis of workplace accidents. Int. J. Comput. Math. **85**(3–4), 559–578 (2008)
23. Mohanta, B.K., Jena, D., Mohapatra, N., Ramasubbareddy, S., Rawal, B.S.: Machine learning based accident prediction in secure IoT enable transportation system. J. Intell. Fuzzy Syst. **42**(2), 713–725 (2022)
24. Montorselli, N.B., et al.: Relating safety, productivity and company type for motor-manual logging operations in the Italian Alps. Acc. Anal. Preven. **42**(6), 2013–2017 (2010). https://doi.org/10.1016/j.aap.2010.06.011
25. Oyedele, A., et al.: Deep learning and boosted trees for injuries prediction in power infrastructure projects. Appl. Soft Comput. **110**, 107587 (2021). https://doi.org/10.1016/j.asoc.2021.107587
26. Rickards, J.: The human factor in forest operations: engineering for health and safety. For. Chron. **84**(4), 539–542, (2008). https://doi.org/10.5558/tfc84539-4
27. Saranti, A., Taraghi, B., Ebner, M., Holzinger, A.: Insights into learning competence through probabilistic graphical models. In: Holzinger, A., Kieseberg, P., Tjoa, A.M., Weippl, E. (eds.) CD-MAKE 2019. LNCS, vol. 11713, pp. 250–271. Springer, Cham (2019). https://doi.org/10.1007/978-3-030-29726-8_16
28. Sarkar, S., Pramanik, A., Maiti, J., Reniers, G.: Predicting and analyzing injury severity: A machine learning-based approach using class-imbalanced proactive and reactive data. Safety science **125**, 104616 (2020). https://doi.org/10.1016/j.ssci.2020.104616
29. Sarkar, S., Vinay, S., Raj, R., Maiti, J., Mitra, P.: Application of optimized machine learning techniques for prediction of occupational accidents. Comput. Operat. Res. **106**, 210–224 (2019)
30. Shapley, L.S.: A value for n-person games. Classics Game Theory **69** (1997)
31. Staniak, M., Biecek, P.: Explanations of model predictions with live and breakdown packages. arXiv preprint arXiv:1804.01955 (2018)

32. Štrumbelj, E., Kononenko, I.: Explaining prediction models and individual predictions with feature contributions. Knowl. Inf. Syst. **41**(3), 647–665 (2013). https://doi.org/10.1007/s10115-013-0679-x

33. Tamascelli, N., Solini, R., Paltrinieri, N., Cozzani, V.: Learning from major accidents: A machine learning approach. Comput. Chem. Eng. **162**, 107786 (2022)

34. Tsioras, P.A., Rottensteiner, C., Stampfer, K.: Wood harvesting accidents in the Austrian state forest enterprise 2000–2009. Safety Sci. **62**, 400–408 (2014)

35. Venkat, A., KP, G.V., Thomas, I.S., et al.: Machine learning based analysis for road accident prediction. IJETIE **6**(2) (2020)

36. Zheng, A., Casari, A.: Feature Engineering for Machine Learning: Principles and Techniques for Data Scientists. " O'Reilly Media, Inc." (2018)

37. Zink, K.J.: Designing sustainable work systems: the need for a systems approach. Appl. Ergonom. **45**(1), 126–132 (2014). https://doi.org/10.1016/j.apergo.2013.03.023

38. Österreichische Bundesforste: Insights (2020),https://www.bundesforste.at/fileadmin/bundesforste/Zahlen__Fakten/2020/Oesterreichische_Bundesforste_Folder_Einblicke_EN_200630_Screen.pdf

Author Index

Printed in the United States
by Baker & Taylor Publisher Services

Printed in the United States
by Baker & Taylor Publisher Services